P9-DHG-639

CHEMISTRY

Fourth Edition

THE
EASY
WAY

Joseph A. Mascetta
Former Principal
Chemistry Teacher and Coordinator
Science Department
Mount Lebanon High School
Pittsburgh, Pennsylvania

Science Educational Consultant

BARRON'S

© Copyright 2003, 1996, 1989, 1983 by Barron's Educational Series, Inc.
© Copyright 1981, 1969 by Barron's Educational Series, Inc. under the
title *How to Prepare for College Board Achievement Tests: Chemistry.*

All rights reserved.
No part of this book may be reproduced
in any form, by photostat, microfilm, xerography,
or any other means, or incorporated into any
information retrieval system, electronic or
mechanical, without the written permission
of the copyright owner.

All inquiries should be addressed to:
Barron's Educational Series, Inc.
250 Wireless Boulevard
Hauppauge, New York 11788
http://www.barronseduc.com

Library of Congress Catalog Card No. 2003049614
International Standard Book No. 0-7641-1978-8

Library of Congress Cataloging-in-Publication Data
Mascetta, Joseph A.
   Chemistry the easy way / Joseph A. Mascetta.—4th ed.
      p.   cm.—(Barron's easy way series)
   Includes index.
   ISBN 0-7641-1978-8
   1. Chemistry—Outlines, syllabi, etc.   2. Chemistry—Examinations,
questions, etc.   I. Title.
II. Easy way.
QD41.M378 2003
540′.76—dc21

                                        2003049614

PRINTED IN THE UNITED STATES OF AMERICA
9 8 7 6 5 4

CONTENTS

Preface vi
Introduction vii

How to Use This Book vii
Diagnosing Your Needs vii
Taking a Test viii
Problem Solving: A Thinking Skill viii

PART I: AN INTRODUCTION / 1

1 Introduction to Chemistry 3

Matter 3
  Definition of Matter 3
  States of Matter 3
  Composition of Matter 4
  Chemical and Physical Properties 5
  Chemical and Physical Changes 5
  Conservation of Mass 6
Energy 6
  Definition of Energy 6
  Forms of Energy 7
  Types of Reactions (Exothermic versus Endothermic) 7
  Conservation of Energy 8
Conservation of Mass and Energy 8
Measurements and Calculations 8
  The Scientific Method 8
  Metric System 9
  Temperature Measurements 12
  Heat Measurements 13
  Scientific Notation 14
  Factor-Label Method of Conversion (Dimensional Analysis) 15
  Precision, Accuracy, and Uncertainty 16
  Significant Figures 17
  Calculations with Significant Figures 18

PART II: THE NATURE OF MATTER / 25

2 Atomic Structure and the Periodic Table 27

History 27
Electric Nature of Atoms 28
  Basic Electric Charges 28
  Bohr Model 30
  Components of Atomic Structure 31
  Calculating Average Atomic Mass 32
  Oxidation Number and Valence 33
  Metallic, Nonmetallic, and Noble Gas Structures 34
  Reactivity 34
Atomic Spectra 34
  Spectroscopy 35
  Mass Spectroscopy 37
The Wave-Mechanical Model 38
  Quantum Numbers 39
  Hund's Rule of Maximum Multiplicity 41
Sublevels and Electron Configuration 41
  Order of Filling and Notation 41
  Electron Dot Notation (Lewis Dot Structures) 44

Transition Elements and Variable Oxidation Numbers 44
Periodic Table of the Elements 45
  History 45
  Periodic Law 46
  The Table 47
Properties Related to the Periodic Table 47
  Radii of Atoms 48
  Atomic Radii in Periods 50
  Atomic Radii in Groups 50
  Ionic Radius Compared to Atomic Radius 50
  Electronegativity 50
  Ionization Energy 50

3 Bonding 56

Types of Bonds 57
  Ionic Bonds 57
  Covalent Bonds 58
  Metallic Bonds 60
Intermolecular Forces of Attraction 61
  Dipole-Dipole Attraction 61
  London Forces 61
  Hydrogen Bonds 61
Double and Triple Bonds 62
Resonance Structures 63
Electrostatic Repulsion (VSEPR) and Hybridization 63
  Electrostatic Repulsion—VSEPR 63
  VSEPR and Unshared Electron Pairs 64
  VSEPR and Molecular Geometry 65
  Hybridization 66
Sigma and Pi Bonds 70
Properties of Ionic Substances 71
Properties of Molecular Crystals and Liquids 71

PART III: USING ATOMS AND MOLECULES / 73

4 Chemical Formulas 75

Writing Formulas 75
  General Observations about Oxidation Numbers and Formula Writing 75
More About Oxidation Numbers 77
Naming Compounds 79
Chemical Formulas 81
Laws of Definite Composition and Multiple Proportions 84
Writing and Balancing Simple Equations 84
Showing Phases in Chemical Equations 85
Writing Ionic Equations 86

PART IV: THE STATES AND PHASES OF MATTER / 89

5 Gases and the Gas Laws 91

Introduction—Gases in the Environment 91
Some Representative Gases 92
  Oxygen 92
  Hydrogen 95
General Characteristics of Gases 98
  Measuring the Pressure of a Gas 98
  Kinetic Molecular Theory 100
  Some Particular Properties of Gases 100

Gas Laws and Related Problems 101
Graham's Law 101
Charles's Law 101
Boyle's Law 103
Combined Gas Law 104
Pressure versus Temperature 105
Dalton's Law of Partial Pressures 105
Corrections of Pressure 106
Ideal Gas Law 107
Ideal Gas Deviations 109

6 Chemical Calculations (Stoichiometry) and the Mole Concept 113

Solving Problems 113
The Mole Concept 113
Molar Mass and Moles 114
Mole Relationships 116
Gas Volumes and Molar Mass 116
Density and Molar Mass 117
Mass-Volume Relationships 120
Mass-Mass Problems 121
Volume-Volume Problems 123
Problems with an Excess of One Reactant 125

7 Liquids, Solids, and Phase Changes 130

Liquids 130
Importance of Intermolecular Interaction 130
Kinetics of Liquids 130
Viscosity 131
Surface Tension 131
Phase Equilibrium 132
Boiling Point 133
Critical Temperature and Pressure 133
Solids 133
Phase Diagrams 134
Water 135
History of Water 135
Purification of Water 135
Composition of Water 137
Properties and Uses of Water 139
Water Calorimetry Problems 139
Reactions of Water with Anhydrides 141
Polarity and Hydrogen Bonding 141
Solubility 142
General Rules of Solubility 143
Factors That Affect Rate of Solubility 144
Summary of Types of Solutes and Relationship of Type to Solubility 144
Water Solutions 144
Continuum of Water Mixtures 145
Expressions of Concentration 146
Using Specific Gravity in Solutions 147
Dilution 150
Colligative Properties of Solutions 151
Crystallization 154

PART V: CHEMICAL REACTIONS / 159

8 Chemical Reactions and Thermochemistry 161

Types of Reactions 161
Predicting Reactions 162
Combination (Synthesis) 162
Decomposition (Analysis) 163
Single Replacement 164
Double Replacement 164
Hydrolysis Reactions 165
Entropy 166
Thermochemistry 167
Changes in Enthalpy 167
Additivity of Reaction Heats and Hess's Law 169
Bond Dissociation Energy 171
Enthalpy from Bond Energies 172

9 Rates of Chemical Reactions 175

Measurements of Reaction Rates 175
Factors Affecting Reaction Rates 175
Collision Theory of Reaction Rates 177
Activation Energy 177
Reaction Rate Law 178
Reaction Mechanism and Rates of Reaction 178

10 Chemical Equilibrium 181

Reversible Reactions and Equilibrium 181
Le Châtelier's Principle 185
Effects of Changing Conditions 186
Effect of Changing Concentrations 186
Effect of Temperature on Equilibrium 186
Effect of Pressure on Equilibrium 186
Equilibria in Heterogeneous Systems 187
Equilibrium Constant for Systems Involving Solids 187
Acid Ionization Constants 187
Ionization Constant of Water 188
Solubility Products 189
Common Ion Effect 191
Factors Related to the Magnitude of K 191
Relation of Minimum Energy (Enthalpy) to Maximum Disorder (Entropy) 191
Change in Free Energy of a System—Gibbs Equation 192

11 Acids, Bases, and Salts 196

Definitions and Properties 196
Acids 196
Bases 197
Broader Acid-Base Theories 198
Conjugate Acids and Bases 199
Acid Concentration Expressed as pH 199
Indicators 201
Volumetric Analysis—Titration 201
Buffer Solutions 205
Salts 205
Amphoteric Substances 206
Acid Rain—An Environmental Concern 206

12 Oxidation-Reduction and Electrochemistry 210

Ionization 210
Oxidation-Reduction and Electrochemistry 211
Electrochemistry 211
Voltaic Cells 212
Electrode Potentials 213
Electrolytic Cells 216
Applications of Electrochemical Cells (Commercial Voltaic Cells) 218
Quantitative Aspects of Electrolysis 219
Relationship between Quantity of Electricity and Amount of Products 219
Balancing Redox Equations Using Oxidation Numbers 220
The Electron Shift Method 220
The Ion-Electron Method 222

PART VI: REPRESENTATIVE GROUPS AND FAMILIES / 229

13 Some Representative Groups and Families 231

Sulfur Family 231
Sulfuric Acid 233
Other Important Compounds of Sulfur 234
Halogen Family 236
Testing for Halides 237
Some Important Halides and Their Uses 237
Uses of Halogens 237
Nitrogen Family 238
Nitric Acid 238
Other Important Compounds of Nitrogen 239
Other Members of the Nitrogen Family 240
Metals 241
Properties of Metals 241
Some Important Reduction Methods 243
Alloys 244
Metalloids 245

PART VII: ORGANIC AND NUCLEAR CHEMISTRY / 251

14 Carbon and Organic Chemistry 253

Carbon 253
Forms of Carbon 253
Carbon Dioxide 255
Organic Chemistry 257
Hydrocarbons 257
Changing Hydrocarbons 268
Hydrocarbon Derivatives 270

15 Nucleonics 285

Radioactivity 285
The Nature of Radioactive Emissions 286
Methods of Detection of Alpha, Beta, and Gamma Rays 287
Decay Series and Transmutations 289
Radioactive Dating 290
Nuclear Energy 291
Conditions for Fission 291
Methods of Obtaining Fissionable Material 293
Fusion 293
Radiation Exposure 293
New Subatomic Particles 294

16 Representative Laboratory Setups 298

New Technology in the Laboratory 298
Some Basic Setups 299
Summary of Qualitative Tests 308

Practice Tests in Chemistry 311
General Information 311
Basic Topics and Abilities Tested 312
What Types of Questions Appear on the Test 313
Practice Test 1 323
Answers and Explanations for Test 1 338
Diagnosing Your Needs 343
Planning Your Study 344
Practice Test 2 349
Answers and Explanations for Test 2 364
Diagnosing Your Needs 369
Planning Your Study 370
Practice Test 3 375
Answers and Explanations for Test 3 389
Diagnosing Your Needs 394
Planning Your Study 395
Practice Test 4 399
Answers and Explanations for Test 4 413
Diagnosing Your Needs 417
Planning Your Study 418
Practice Test 5 423
Answers and Explanations for Test 5 438
Diagnosing Your Needs 445
Planning Your Study 446

Final Preparation—The Day Before the Test 448

Equations and Tables for Reference 449

Glossary of Common Terms 458

Index 469

PREFACE

This new edition has added four important elements to improve coverage of the material and to make it more useful for students in learning chemistry "the easy way."

1. THE MOST UP-TO-DATE TERMINOLOGY AND USAGE

All notations, word usage, units of measure, and terminology have been carefully checked to ensure that they reflect the most current practices employed in chemistry today. This new edition gives students an opportunity to learn chemistry using these latest innovations and to become confident that they are familiar with modern concepts and language should they decide to move on to more advanced material. The five practice tests also use the most current terminology and are formatted to resemble the current College Entrance Examination Board SAT II Chemistry Tests.

2. EXPANDED EXPLANATIONS, CHARTS, AND GRAPHS

Many of the explanations of major concepts and theories have been expanded to give students a more detailed, up-to-date understanding of the basic ideas of chemistry. These presentations are frequently reinforced with additional, concrete examples. Also included in this edition are more than 150 charts, diagrams, graphs, and illustrations designed to help students visualize and better understand the material.

3. MORE DETAILED SAMPLE PROBLEMS

The quantitative aspects of chemistry are explained in step-by-step fashion and then clarified with sample problems. The factor-label method of keeping track of units is used throughout this book. More sample problems have been added to this revision to aid in the development of problem-solving skills. "Problem Solving: A Thinking Skill" is a separate section in the **Introduction** intended to provide students with a methodology for attacking chemistry problems efficiently.

4. FIVE UP-TO-DATE PRACTICE TESTS WITH ANSWERS

This revised edition has five up-to-date tests covering the whole range of topics included in a high school chemistry course. These tests are modeled after the College Entrance Examination Board SAT II Chemistry Test, and each is accompanied by a detailed answer section. Also included is a self-evaluation section that enables students to identify their strengths and weaknesses. Finally, there is a study guide, referenced to individual sections of the text, that makes it easier to review the material.

INTRODUCTION

How to Use This Book

The purpose of this book is to introduce the student to the basic essentials of a good high school chemistry course. These are developed in a simplified manner with the visual reinforcement of charts, graphs, lists, and simplified drawings.

It is important that the student read and comprehend each section. To check this comprehension, each chapter has a review at the end of the text. This includes a variety of questions and usually a list of terms the student should know. The answers are given along with the chapter review, but all problems are explained in a separate section at the end of the book. A concise glossary of important chemical terms is also provided at the end of the book. The complete index assists the student in finding specific topics or information.

Since chemistry is a quantitative science, the mathematical solution of typical problems has been included. This process includes an introduction of the type of problem and a careful development of the solution.

Knowing laboratory setups and the procedures for conducting particular tests is another important aspect of chemistry. The last chapter of this book summarizes typical laboratory techniques and the basic procedures for specific laboratory tests. A reference section also gives the student access to the most important tables and charts of chemical properties.

The last portion of the book is devoted to five practice tests. These should be of use to the student in preparing for any chemistry achievement test—be it the Regents test, College Entrance Examination Board test, or one of the privately produced standardized tests. The general advice is the same.

To prepare for a test, the student should allow sufficient time so that last minute cramming will be unnecessary. Start early! Review each of the chapters in this book and then attempt to answer the questions in the review section at the end of the chapter. When you experience difficulty on a particular topic, go back over the information here or in any good high school text until you feel you have mastered that concept. When you have finished this, go on to the general tests in the last section of this book. Attempt to answer the questions as quickly and accurately as possible so that you get some practice in pacing yourself. Omit difficult questions you are not sure of and then go back to them after you have answered the easier ones. Do not attempt questions for which you do not have the background to make a judgment since, on an actual test, a percentage of wrong answers may be automatically subtracted from the number of right answers. When you have finished each test, check your answers to see how well you have done and review material you have missed or omitted.

Diagnosing Your Needs

You can use the practice tests as a means of assessing your achievement in each of the major areas tested and as a guide for studying particular topics. To do this, you should take the practice test simulating the conditions, particularly the length of time allowed, of the actual test. You should give yourself 1 hour to answer the questions in Practice Test 1. Check your answers against the correct answers given immediately after the test. In the section following the answers to this test, directions are given to arrive at subscores for each of the identified areas you need to emphasize by comparing your subscores. Plan to spend more time reviewing the areas in which your subscores were the lowest. However, do not omit any area in your review.

Taking a Test

Below are some suggestions for taking actual tests:

1. Get a good night's sleep. If you should become ill while taking the test, report this to the proctor so that he or she may record this information.
2. Read the instructions carefully and be sure you understand them.
3. Skip questions that seem too difficult for you. Go on to the other questions and come back to the omitted ones if time permits. An easy question counts as much as a hard question.
4. Avoid haphazard guessing since this probably will lower your score. If you can eliminate some of the choices to a question, however, it will be to your advantage to answer the question even if you must guess which of the remaining answers is correct. All questions that are answered correctly count the same toward your score.
5. It is important to pace yourself throughout the hour to answer as many questions as possible. Make the best use of your time! Don't be too upset then if you don't finish the test. At least you have done the best you could in the 1-hour time.

Problem Solving: A Thinking Skill*

Chemistry is a subject that deals with many problem situations that you, the student, must be able to solve. Solving problems may seem to be a natural process when the degree of difficulty is not very great. In these cases, you may not need to have a structured method to attack the problem. However, for complex problems you will need an orderly process to solve the problem. The following is such a problem-solving process. Each step is vital to the next step and to the final solution of the problem.

Step 1. Clarify the problem: to separate the problem into the facts, the conditions, and the questions that need to be answered, and to establish the goal.

Step 2. Explore: to examine the sufficiency of the data, to organize the data, and to apply previously acquired knowledge, skills, and understanding.

Step 3. Select a strategy: to choose an appropriate method to solve the problem.

Step 4. Solve: to apply the skills needed to carry out the strategy chosen.

Step 5. Review: to examine the reasonableness of the solution and to evaluate the effectiveness of the process.

The steps of the problem-solving process listed above should be followed in sequence. The subskills listed below for each step, however, are not in sequence. The order in which subskill patterns are used will differ with the nature of the problem and/or with the ways in

*Adapted with permission from *Thinking Skills Resource Guide*, a noncopyrighted publication of Mount Lebanon School District, Pittsburgh, Pa.

which the individual problem solver thinks. Also, not every subskill need be employed in solving every problem.

Clarify the Problem

Identify the facts.   What is known about the problem?

Identify the conditions.   What is the current situation?

Identify the questions.   What needs to be answered before the problem can be solved?

Visualize the problem.

- Make mental images of the problem.
- If desirable or necessary, draw a sketch or diagram, make an outline, or write down symbols or equations that correspond to the mental images.

Establish the goal.   The goal defines the specific result to be accomplished through the problem-solving process. It defines the purpose or function the solution is expected to achieve and serves as the basis for evaluating the solution.

Explore

Review previously acquired knowledge, skills, and understanding.   Determine whether the current problem is similar to a previously seen type of problem.

Estimate the sufficiency of the data.   Does there seem to be enough information to solve the problem?

Organize the data.   There are many ways in which data can be organized. Some examples are outline, written symbols and equations, chart, table, graph, map, diagram, and drawing. Determine whether the data organized in the way(s) you have chosen will enable you to partially or completely solve the problem. The organization of data may suggest what new data need to be collected.

Determine what new data, if any, need to be collected.   What new information may be needed to solve the problem? Can the existing data be reorganized to generate new information? Do other resources need to be consulted? This step may suggest possible strategies to be used to solve the problem.

Select a Strategy

A strategy is a goal-directed sequence of mental operations. Selecting a strategy is the most important and also the most difficult step in the problem-solving process. Although there may be several strategies that will lead to the solution of a problem, the skilled problem solver uses the most efficient strategy. The choice of the most efficient strategy is based on knowledge and experience as well as a careful application of the clarify-and-explore steps of the problem-solving method. Some problems may require the use of a combination of strategies.

The following search methods may help you to select a strategy. They do not represent all of the possible ways in which this can be done. Other methods of strategy selection are related to specific content areas.

Trial-and-error search.   Such a search either doesn't have or doesn't use information that indicates that one path is more likely to lead to the goal than any other path.

Trial-and-error search comes in two forms, blind and systematic. In *blind search*, the searchers pick paths to explore blindly, without considering whether they have already explored these paths. A preferable method is *systematic search*, in which the searchers keep track of the paths they have already explored and do not duplicate them. Because this method avoids multiple searches, systematic search is usually twice as efficient as blind search.

Reduction method.   This involves breaking the problem into a sequence of smaller parts by setting up subgoals. Subgoals make problem solving easier because they reduce the amount of search required to find the solution.

You can set up subgoals by working part way into a problem and then analyzing the partial goal to be achieved. In doing this, you can drop the problem restrictions that do not apply to the subgoal. By adding up all the subgoals, you can solve the "abstracted" problem.

Working backward.   When you have trouble solving a problem head-on, it is often useful to try to work backward. Working backward involves a simple change in representation or point of view. Your new starting point is the original goal. Working backward can be helpful because problems are often easier to solve in one direction than in another.

Knowledge-based method.   This strategy uses information stored in the problem solver's memory, or newly acquired information, to guide the search for the solution. The problem solver may have solved a similar problem and can use this knowledge in a new situation. In other cases, problem solvers may have to acquire needed knowledge. For example, they may solve an auxiliary problem to learn how to solve the one they are having difficulty with.

Searching for analogous (similar) problems is a very powerful problem-solving technique. When you are having difficulty with a problem, try to pose a related, easier one and hope to learn something that will help you solve the harder problem.

Solve

Use the strategy chosen to actually solve the problem. Executing the solution provides you with a very valuable check on the adequacy of your plan. Sometimes students will look at a problem and decide that, since they know how to solve it, they need not bother with the drudgery of actually executing the solution. Sometimes the students are right, but at other times they miss an excellent opportunity to discover that they were wrong.

Review

Evaluation.   The critical question in evaluation is Does the answer I propose meet all of the goals and conditions set by the problem? Thus, after the effort of finding a solution, you must turn back to the problem statement and check carefully to be sure your solution satisfies it.

With easy problems there is a strong temptation to skip evaluation because the probability of error seems small. In some cases, however, this can be costly. Evaluation may prove that errors were present.

Verifying the reasonableness of the answer.   It is easy to become so involved with the process and mathematics of a problem that an answer is recorded that is totally illogical. To avoid this mistake, you should simplify the numbers involved and solve for an answer. Having done this, compare your estimated result with your answer to ensure that your answer is feasible.

For example, a problem requires the following operations:

$$5.12 \times 10^5 \times 3.98 \times 10^6 \text{ divided by } 910$$

And doing all the math, you get an answer of

$$0.02239 \times 10^{11} \text{ or } 2.24 \times 10^9$$

To estimate the answer, first simplify the numbers to one significant figure (significant figures are discussed in Chapter 1). This gives

$$5 \times 10^5 \times 4 \times 10^6 \text{ divided by } 9 \times 10^2$$

which is

$$20 \times 10^{11} \text{ divided by } 9 \times 10^2 = 2.2 \times 10^9$$

This is the estimated answer, which validates the answer above.

When you are dealing with test items that provide multiple-choice answers, you can often use estimation to arrive at the answer without doing the more complicated mathematics.

Consolidation.   Here the basic question to be answered is What can I learn from the experience of solving this problem? The following more specific questions may help you to answer this general one:

- Why was this problem difficult?
- Was it difficult to follow a plan?
- Was it difficult to decide on a plan? If so, why?
- Did I take the long way to the answer?
- Can I use this plan again in similar problems?

The important thing is to reflect on the process that you used in order to make future problem solving easier.

PART I:
AN INTRODUCTION

Chapter 1

INTRODUCTION TO CHEMISTRY

Matter

Definition of Matter

Matter is defined as anything that occupies space and has mass. *Mass* is the quantity of matter that a substance possesses and, depending on the gravitational force acting on it, has a unit of *weight* assigned to it. Although the weight can then vary, the mass of the body is a constant and can be measured by its resistance to a change of position or motion. This property of mass to resist a change of position or motion is called *inertia*. Since matter does occupy space, we can compare the masses of various substances that occupy a particular unit volume. This relationship of mass to a unit volume is called the *density* of the substance. It can be shown in a mathematical formula as $D = m/V$. The basic unit of mass (m) in chemistry is the gram (g), and of volume (V) is the cubic centimeter (cm^3) or milliliter (mL).

An example of how density varies can be shown by the difference in the volume occupied by 1 g of a metal, such as gold, and 1 g of styrofoam. Both have the same mass, that is, 1 g, but the volume occupied by the styrofoam is much larger. Therefore the density of the metal will be much larger than that of the styrofoam. When dealing with gases in chemistry, the standard units for the density of gases are grams per liter at a standard temperature and pressure. This aspect of the density of gases is dealt with in Chapter 6. Basically then, density can be defined as the mass per unit volume.

States of Matter

Matter occurs in three states: solid, liquid, and gas. A *solid* has both a definite size and shape. A *liquid* has a definite volume but takes the shape of the container, and a *gas* has neither a definite shape nor a definite volume. These states of matter can often be changed by the addition of heat energy. An example of this is ice changing to liquid water and finally to steam.

Composition of Matter

Matter can be subdivided into two general categories: distinct substances and mixtures. Distinct substances are substances that can be subdivided into the smallest particle that still has the properties of the substance. At that point, if the substance is made up of only one kind of atom, it is called an *element*. Atoms are considered to be the basic building blocks of matter that cannot be easily created or destroyed. The word *atom* comes from the Greek and means the smallest possible piece of something. Today there are approximately 109 different kinds of atoms, each with its own unique composition. These atoms then are the building blocks of elements when only one kind of atom makes up the substance. If, however, there are two or more kinds of atoms joined together in definite grouping, this distinct substance is called a *compound*. Compounds are made by combining elements in a definite proportion (or ratio) by mass and are made up of two or more kinds of atoms. This is called the *Law of Definite Composition (or Proportions)*. The smallest natural occurring unit of a compound is called a *molecule* of that compound. A molecule of a compound has a definite shape that is determined by how the atoms are bonded to or combine with each other. This bonding is described in Chapter 3. An example is the compound water: it always occurs in a two hydrogen atoms to one oxygen atom relationship. *Mixtures*, however, can vary in their composition.

In general, then:

Mixtures	Distinct Substances
1. Composition is indefinite (generally heterogeneous).* (Example: marble) 2. Properties of the constituents are retained. 3. Parts of the mixture react differently to changed conditions.	**Elements** 1. Composition is made up of one kind of atom. (Examples: nitrogen, gold, neon) 2. All parts are the same throughout (homogeneous). **Compounds** 1. Composition is definite (homogeneous). (Examples: water, carbon dioxide) 2. All parts react the same. 3. Properties of the compound are distinct and different from the properties of the individual elements that are combined in its makeup.

*Solutions are mixtures, such as sugar in water, but since the substance, like sugar, is distributed evenly throughout the water, it can be said to be a homogeneous mixture.

The following chart shows a classification scheme for matter.

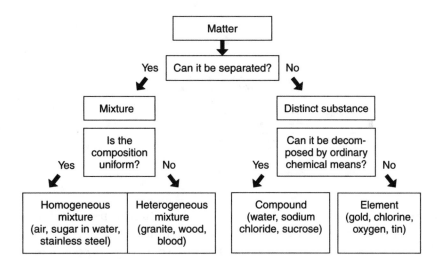

Chemical and Physical Properties

Physical properties of matter are those properties that can usually be observed with our senses. They include everything about a substance that can be noted when no change is occurring in the type of structure that makes up its smallest component. Some common examples are physical state, color, odor, solubility in water, density, melting point, taste, boiling point, and hardness.

Chemical properties are those properties that can be observed in regard to whether or not a substance reacts with other substances. For example, iron rusts in moist air, nitrogen does not burn, gold does not rust, sodium reacts with water, silver does not react with water, and water can be decomposed by an electric current.

Chemical and Physical Changes

The changes matter undergoes are classified as either physical or chemical. In general, a *physical change* alters the physical properties of matter, but the composition remains constant. The most often altered properties are form and state. Some examples are breaking glass, cutting wood, melting ice, and magnetizing a piece of metal. In some cases, the process that caused the change can be easily reversed and the substance regains its original form.

Chemical changes are changes in the composition and structure of a substance. They are always accompanied by energy changes. If the energy released in the formation of a new structure exceeds the chemical energy in the original substances, energy will be given off, usually in the form of heat or light or both. This is called an *exothermic reaction*. If, however, the new structure needs to absorb more energy than is available from the reactants, the result is an *endothermic reaction*. This can be shown graphically.

Notice that in Figures 1 and 2 the term *activation energy* is used. The activation energy is the energy necessary to get the reaction going by increasing the energy of the reactants

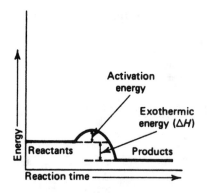

Figure 1. An Exothermic Reaction.

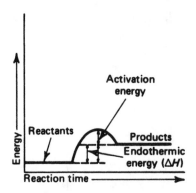

Figure 2. An Endothermic Reaction.

so that they can combine. You know you have to heat paper before it burns. This heat raises the energy of the reactants so that the burning can begin; then enough energy is given off from the burning so that an external source of energy is no longer necessary.

Conservation of Mass

When ordinary chemical changes occur, the mass of the reactants equals the mass of the products. This can be stated another way: In a chemical change, matter can neither be created nor destroyed, but only changed from one form to another. This is referred to as the *Law of Conservation of Matter* (Lavoisier—1785). This law is extended by the Einstein mass-energy relationship, which states that matter and energy are interchangeable (see page 8).

Energy
Definition of Energy

The concept of energy plays an important role in all of the sciences. In chemistry all physical and chemical changes have energy considerations associated with them. To understand how and why these changes happen, an understanding of energy is required.

Energy is defined as the capacity to do work. Work is done whenever a force is applied over a distance. Thus anything that can force matter to move, to change speed, or to change direction has energy. The following example will help you understand this definition of energy. When you charge a battery with electricity, you are storing energy in the form of chemical energy. The charged battery has a capacity to do work. If you use the battery to operate a toy car, the energy stored is transformed into mechanical energy which exerts a force on the mechanism that turns the wheels and makes the car move. This continues until the "charge" or stored energy is used up. In its uncharged condition, the battery no longer has the capacity to do work.

Just as work itself is measured in *joules* (J), so is energy. In some problems, it may be expressed in *kilocalories* (kcal). The relationship between these two units is that 4.18×10^3 J is equal to 1 kcal.

Forms of Energy

Energy may appear in a variety of forms. Most commonly, energy in reactions is evolved as *heat*. Some other forms of energy are *light*, *sound*, *mechanical energy*, *electrical energy*, and *chemical energy*. Energy can be converted from one form to another, as when the heat from burning fuel is used to vaporize water to steam. The energy of the steam is used to turn the wheels of a turbine to produce mechanical energy. The turbine turns the generator armature to produce electricity, which is then available in homes for use as light or heat or for the operation of many modern appliances.

Two general classifications of energy are *potential energy* and *kinetic energy*. Potential energy is due to position; kinetic energy is energy of motion. The difference can be illustrated by a boulder sitting on the side of a mountain. It has a high potential energy because of its position above the valley floor. If it falls, however, its potential energy is converted to kinetic energy. This illustration is very similar to the situation of electrons cascading to lower energy levels in the atomic model.

This concept can also be applied at the molecular level. For example, the molecule of N_2 in the diagram below shows three variations of kinetic energy. The N_2 molecule also possesses potential energy because of the chemical bond within the molecule. To break this bond would require energy. The potential energy in each isolated atom of nitrogen would then be greater. This would be similar to raising the boulder to a higher position on the mountainside. Both systems have a higher potential energy and have a tendency to fall back to the lower state, which is more stable.

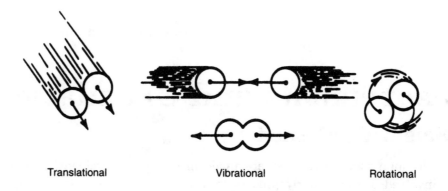

Translational	Vibrational	Rotational

Types of Reactions (Exothermic versus Endothermic)

When physical or chemical changes occur, energy is involved. If the heat content of the product(s) is higher than that of the reactants, the reaction is *endothermic*. If, on the other hand, the heat content of the product(s) is less than that of the reactants, the reaction is *exothermic*. This change of heat content can be designated as ΔH. The heat content (H) is sometimes referred to as the *enthalpy*. Every system has a certain amount of heat that changes during the course of a physical or chemical change. The change in heat content (ΔH) is the difference between the heat content of the products and that of the reactants. The equation is

$$\Delta H = H_{products} - H_{reactants}$$

If the heat content of the products is greater than the heat content of the reactants, ΔH is a positive quantity ($\Delta H > 0$) and the reaction is endothermic. If, however, the heat content of the products is less than the heat content of the reactants, ΔH is a negative quantity ($\Delta H < 0$) and the reaction is exothermic. This relationship is shown graphically in Figures 1 and 2 on page 6. This topic is developed in detail in Chapter 8.

Conservation of Energy

Experiments have shown that energy is neither gained nor lost during physical or chemical changes. This principle is known as the *Law of Conservation of Energy* and is often stated as follows: Energy is neither created nor destroyed in ordinary physical and chemical changes.

Conservation of Mass and Energy

With the introduction of atomic theory and a more complete understanding of the nature of both mass and energy, it was found that a relationship exists between these two concepts. Einstein formulated the *Law of Conservation of Mass and Energy*, which states that mass and energy are interchangeable under special conditions. The conditions have been created in nuclear reactors and accelerators, and the law has been verified. This relationship can be expressed by Einstein's famous equation:

$$E = mc^2$$

Energy = Mass $\times$ (velocity of light)^2

Measurements and Calculations
The Scientific Method

Although some discoveries are made in science by accident, in most cases the scientists involved use an orderly process to work on their projects and discoveries. The process researchers use to carry out their investigations is often called the *scientific method*. It is a logical approach to solving problems by observing and collecting data, formulating a hypothesis, and constructing theories supported by the data. The formulating of a hypothesis consists of carefully studying the data collected and organized to see if a testable statement can be made with regard to the data. The hypothesis takes the form of an "if . . . then" statement. If certain data is true, then a prediction can be made concerning the outcome. The next step is to test the prediction to see if it withstands the experimentation. The hypothesis can go through several revisions as the process continues. If the data from experimentation shows that the predictions of the hypothesis are successful, scientists usually try to explain the phenomena by constructing a *model*. The model can be a visual, verbal, or mathematical means of explaining how the data is related to the phenomena.

The stages of this process can be illustrated by the diagram shown below:

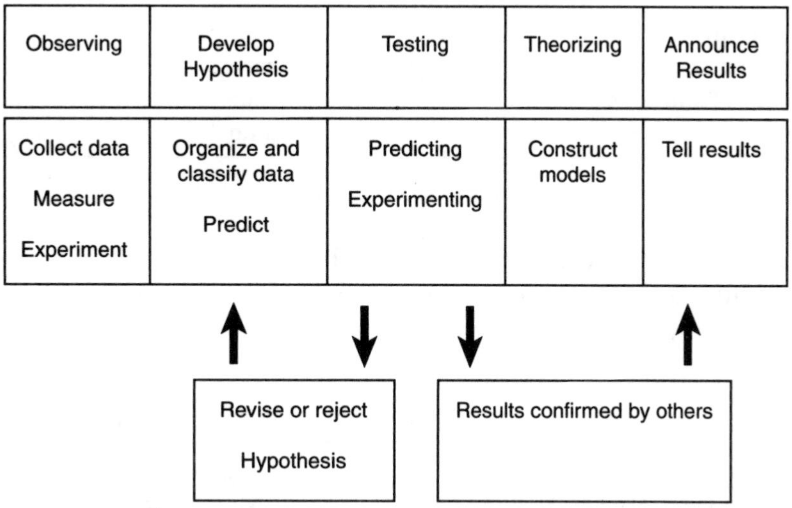

Observing	Develop Hypothesis	Testing	Theorizing	Announce Results
Collect data Measure Experiment	Organize and classify data Predict	Predicting Experimenting	Construct models	Tell results

Revise or reject Hypothesis	Results confirmed by others

The student of chemistry must be able to use the correct measurement terms accurately and to solve problems that require mathematical skill, as well as proper terminology, for their correct solution. The following sections review these topics.

Metric System

It is important that scientists around the world use the same units when communicating information. For this reason, scientists use the modernized metric system, designated in 1960 by the General Conference on Weights and Measures as the International System of Units. This is commonly known as the *SI* system, an abbreviation for the French *Système International d'Unités*. It is now the most common system of measurement in the world.

The reason it is so widely accepted is twofold. First, SI uses the decimal system as its base. Second, in many cases units for various quantities are defined in terms of units for simpler quantities.

There are seven basic units that can be used to express the fundamental properties of measurement. These are called the SI base units and are shown in the following table.

SI BASE UNITS

Property	Unit	Abbreviation
mass	kilogram	kg
length	meter	m
time	second	s
electric current	ampere	A
temperature	kelvin	K
amount of substance	mole	mol
luminous intensity	candela*	cd

*The candela is rarely used in chemistry.

Other SI units are derived by combining prefixes with a root unit. The prefixes represent multiples or fractions of 10. The following table gives some basic prefixes used in the metric system.

PREFIXES USED WITH SI UNITS

Prefix	Symbol	Meaning	Scientific Notation
exa-	E	1,000,000,000,000,000,000	10^{18}
peta-	P	1,000,000,000,000,000	10^{15}
tera-	T	1,000,000,000,000	10^{12}
giga-	G	1,000,000,000	10^{9}
mega-	M	1,000,000	10^{6}
kilo-	k	1,000	10^{3}
hecto-	h	100	10^{2}
deka-	da	10	10^{1}
—	—	1	10^{0}
deci-	d	0.1	10^{-1}
centi-	c	0.01	10^{-2}
milli-	m	0.001	10^{-3}
micro-	μ	0.000 001	10^{-6}
nano-	n	0.000 000 001	10^{-9}
pico-	p	0.000 000 000 001	10^{-12}
femto-	f	0.000 000 000 000 001	10^{-15}
atto-	a	0.000 000 000 000 000 001	10^{-18}

For an example of how the prefix works in conjunction with the root word, consider the term *kilometer*. The prefix *kilo-* means "multiply the root word by 1000," and so a kilometer is 1000 m. By the same reasoning, a millimeter is 1/1000 m.

Because of the prefix system, all units and quantities can be easily related by some factor of 10. Here is a brief table of some metric unit equivalents.

LENGTH

10 millimeters (mm) = 1 centimeter (cm) $10^{-2}/10^{-3}$
100 cm = 1 meter (m) $10^{0}/10^{-2}$
1000 m = 1 kilometer (km) $10^{3}/10^{0}$

VOLUME

1000 milliliters (mL) = 1 liter (L)
1000 cubic centimeters (cm^3) = 1 liter
1 mL = 1 cm^3

MASS

1000 milligrams (mg) = 1 gram (g) $10^{0}/10^{-3}$
1000 g = 1 kilogram (kg) $10^{3}/10^{0}$

A unit of length, used especially in expressing the length of light waves, is the nanometer, abbreviated as nm and equal to 10^{-9} meter.

Because measurements are occasionally reported in units of the English system, it is important to be aware of some metric-to-English system equivalents. Some common conversion factors are shown in the following table.

2.54 centimeters = 1 inch (in)	1 kilogram = 2.2 pounds
1 meter = 39.37 inches (10% longer than a yard [yd])	0.946 liter = 1 quart (qt)
28.35 grams = 1 ounce (oz)	1 liter (5% larger than a quart) = 1.06 quarts
454 grams = 1 pound (1 b)	

The metric system standards were chosen as natural standards. The meter was first described as the distance marked off on a platinum-iridium bar but is now defined as the length of the path traveled by light in a vacuum during a time interval of $1/2.99792458 \times 10^8$ second.

There are some interesting relationships between volume and mass units in the metric system. Since water is most dense at 4°C, the gram was intended to be 1 cm^3 of water at this temperature. This means, then, that:

1000 cm^3 = 1 L of water at 4°C.
1000 cm^3 of water weighs
1000 g at 4°C.

Therefore

1 L of water at 4°C weighs 1 kg

and

1 mL of water at 4°C weighs 1 g.

When 1 L is filled with water at 4°C, it has a mass of 1 kg.

Temperature Measurements

The most commonly used temperature scale in scientific work is the Celsius scale. It gets its name from the Swedish astronomer Anders Celsius and dates back to 1742. For a long time it was called the centigrade scale because it is based on the concept of dividing the distance on a thermometer between the freezing point of water and its boiling point into 100 equal markings or degrees.

There is another scale that is based on the lowest theoretical temperature (called absolute zero). This temperature has never actually been reached, but scientists in laboratories have recorded temperatures within about a thousandth of a degree above absolute zero. Lord William Kelvin proposed this scale, on which a degree is the same size as a Celsius degree and which is referred to as the Kelvin scale. Through experiments and calculations, it has been determined that absolute zero is 273.15° below zero on the Celsius scale. This figure is usually rounded off to –273°C.

The diagram and formulas that follow give the graphic and algebraic relationships between the temperature scales encountered in chemistry.

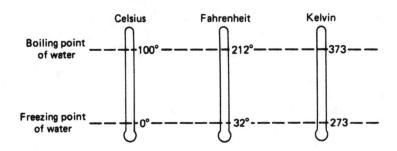

CONVERSION FORMULAS

$$°F = \tfrac{9}{5}°C + 32°$$

$$°C = \tfrac{5}{9}(°F - 32°)$$

$$K = °C + 273°$$

$$°C = K - 273$$

Example 1

$30°C = $ _____ $°F$

Solution:

$$°F = \tfrac{9}{5}(30°) + 32° = 86°$$

Example 2

$68°F = $ _____ $°C$

Solution:

$°C = \frac{5}{9}(68° - 32°) = 20°$

Example 3

$10°C = $ _____ K

Solution:
$K = 10 + 273 = 283$

Example 4

$200\ K = $ _____ $°C$

Solution:
$°C = 200 - 273 = -73°$

Note: In Kelvin notation, the degree sign is omitted: $283\ K$. The unit is the kelvin, abbreviated K.

Heat Measurements

The scales discussed above are used to measure the degree of heat. A pail of water and a thimble full of water can both be filled with water at 100°C. They both have the same measurement of the degree of heat of the water. However, the pail of water has a greater quantity of heat. This can be easily demonstrated by the amount of ice that can be melted by the water in these two containers. Obviously, the pail of water at 100°C will melt more ice than a thimble full of water at the same temperature. We say that the pail of water contains a greater number of *calories* (cal) of heat. The calorie unit is used to measure the quantity of heat. It is defined as the amount of heat needed to raise the temperature of 1 g of water by 1° on the Celsius scale. This is a rather small unit for the quantities of heat that are involved in most chemical reactions. Therefore, the *kilocalorie* (kcal) is more often used. The kilocalorie is equal to 1000 cal. It is the quantity of heat that will increase the temperature of 1 kg of water by 1° on the Celsius scale. As a unit of heat energy, 1 cal is approximately 4.18 J.

Problems involving heat transfers in water are called water calorimetry problems and are explained on page 139.

Scientific Notation

When students must do mathematical operations with numerical figures, the *scientific notation system* is very useful. Basically this system uses the exponential means of expressing figures. With large numbers, such as 3,630,000, move the decimal point to the left until only one digit remains to the left (3.630000) and then indicate the number of moves of the decimal point as the exponent of 10 giving you 3.63×10^6. With a very small number such as 0.000000123, move the decimal point to the right until only one digit is to the left 0000001.23 and then express the number of moves as the negative exponent of 10 giving you 1.23×10^{-7}.

With numbers expressed in this exponential form, you can now use your knowledge of exponents in mathematical operations. An important fact to remember is that in multiplication you add the exponents of 10, and in division you subtract the exponents. Addition and subtraction can be performed only if the values have the same exponent.

Multiplication:

Example 1

$(2.3 \times 10^5)(5.0 \times 10^{-12}) =$

Solution:
Multiplying the first number in each, you get 11.5, and the addition of the exponents gives 10^{-7}. Now changing to a number with only one digit to the left of the decimal point gives you 1.15×10^{-6} for the answer.

Example 2

$(5.1 \times 10^{-6})(2 \times 10^{-3}) =$

Solution:
$(5.1 \times 10^{-6})(2 \times 10^{-3}) = 10.2 \times 10^{-9} = 1.02 \times 10^{-8}$

Example 3

$(3 \times 10^5)(6 \times 10^3) =$

Solution:
$(3 \times 10^5)(6 \times 10^3) = 18 \times 10^8 = 1.8 \times 10^9$

Division:

(Notice that in division the exponents of 10 are subtracted.)

Example 1

$(1.5 \times 10^3) \div (5.0 \times 10^{-2}) =$

Solution:
$(1.5 \times 10^3) \div (5.0 \times 10^{-2}) = 0.3 \times 10^5 = 3 \times 10^4$

Example 2

$(2.1 \times 10^{-2}) \div (7.0 \times 10^{-3}) =$

Solution:
$(2.1 \times 10^{-2}) \div (7.0 \times 10^{-3}) = 0.3 \times 10^1 = 3$

Addition and Subtraction:

Example 1

$4.2 \times 10^4 \text{ kg} + 7.9 \times 10^3 \text{ kg} =$

Solution:
$4.2 \times 10^4 \text{ kg} + (0.79 \times 10^4 \text{ kg So that the exponents of 10 are the same}) =$
$4.2 \times 10^4 \text{ kg} + 0.79 \times 10^4 \text{ kg} = 4.99 \times 10^4 \text{ kg} =$
$5.0 \times 10^4 \text{ kg}$

Apply the same process to subtraction problems.

Factor-Label Method of Conversion (Dimensional Analysis)

When you are working problems that involve numbers with units of measurement, it is convenient to use this method so that you do not become confused in the operation of multiplication or division. For example, if you are changing 0.001 kg to milligrams, you set up each conversion as a fraction so that all the units will factor out except the one you want in the answer.

Example 1

$$1 \times 10^{-3} \text{ kg} \times \frac{1 \times 10^3 \text{ g}}{1 \text{ kg}} \times \frac{1 \times 10^3 \text{ mg}}{1 \text{ g}} = 1 \times 10^3 \text{ mg}$$

Solution:
Notice that the kilogram has the denominator in the first fraction to be factored with the original kilogram unit. The numerator is equal to the denominator except that it is expressed in smaller units. The second fraction has the gram unit in the denominator to be factored with the gram unit in the preceding fraction. The answer is in milligrams since this is the only unit remaining and it assures you that the correct operations have been performed in the conversion.

Example 2

1 foot (ft) = ? centimeters

Solution:

$$1 \text{ ft} \times \frac{12 \text{ in.}}{1 \text{ ft}} \times \frac{2.54 \text{ cm}}{1 \text{ in.}} = 30.48 \text{ cm}$$

This method is used in examples throughout this book.

Precision, Accuracy, and Uncertainty

Two other factors to consider in measurement are *precision* and *accuracy*. *Precision* indicates the reliability or reproducibility of a measurement. *Accuracy* indicates how close a measurement is to its known or accepted value.

For example, suppose you are taking a reading of the boiling point of pure water at sea level. Using the same thermometer in three trials, you record 96.8, 96.9, and 97.0°C. Since these figures show a high reproducibility, you can say that they are precise. However, the values are considerably off from the accepted value of 100°C, and so we say they are not accurate. In this example we probably would suspect that the inaccuracy was the fault of the thermometer.

Regardless of precision and accuracy, all measurements have a degree of *uncertainty*. This is usually dependent on two factors—the limitation of the measuring instrument and the skill of the person making the measurement. This can best be shown by example.

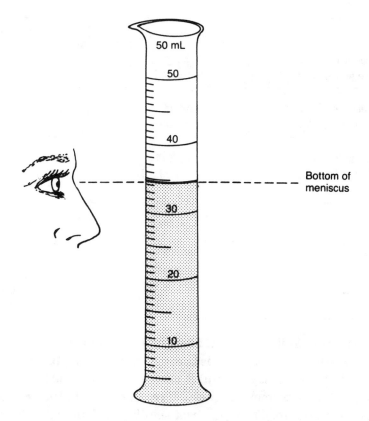

Graduated Cylinder Reading 34.3 mL

The graduated cylinder in the illustration contains a quantity of water to be measured. It is obvious that the quantity is between 30 and 40 mL because the *meniscus* lies between these two marked quantities. Now, checking to see where the bottom of the meniscus lies with reference to the ten intervening subdivisions, we see that it is between the fourth and fifth. This means that the volume lies between 34 and 35 mL. The next step introduces the uncertainty. We have to guess how far the reading is between these two markings. We can make an approximate guess, or estimate, that the level is more than 0.2 but less than 0.4 of the distance. We therefore report the volume as 34.3 mL. The last digit in any measurement is an estimate of this kind and is uncertain.

Significant Figures

Any time a measurement is recorded, it includes all the digits that are certain plus one uncertain digit. These certain digits plus the one uncertain digit are referred to as *significant figures*. The more digits you are able to record in a measurement, the less relative uncertainty there is in the measurement. The following table summarizes the rules of significant figures.

Rule	Example	Number of Significant Figures	
All digits other than zeros are significant.	25 g 5.471 g	2 4	
Zeros between nonzero digits are significant.	309 g 40.06 g	3 4	
Final zeros to the right of the decimal point are significant.	6.00 mL 2.350 mL	3 4	
In numbers smaller than 1, zeros to the left or directly to the right of the decimal point are not significant.	0.05 cm 0.060 cm	(1) The zeros merely mark the position of the decimal point. (2) The first two zeros mark the position of the decimal point. The final zero is significant.	

One last rule deals with final zeros in a whole number. These zeros may or may not be significant, depending on the measuring instrument. For instance, if an instrument that measures to the nearest mile (mi) is used, the number 3000 mi has four significant figures. If, however, the instrument in question records miles to the nearest thousands, there is only one significant figure. The number of significant figures in 3000 can be one, two, three, or four, depending on the limitation of the measuring device.

This problem can be avoided by using the system of scientific notation. For this example, the following notations indicate the numbers of significant figures:

$$3 \times 10^3 \quad \text{one significant figure}$$
$$3.0 \times 10^3 \quad \text{two significant figures}$$
$$3.00 \times 10^3 \quad \text{three significant figures}$$
$$3.000 \times 10^3 \quad \text{four significant figures}$$

Calculations with Significant Figures

When you do calculations involving numbers that do not have the same number of significant figures in each, it is important to keep these two rules in mind.

In multiplication and division, a simple rule that usually holds is that the number of significant figures in a product or a quotient obtained from manipulating figures of measured quantities is the same as the number of significant figures in the quantity having the smaller number of significant figures.

Example 1

4.29 cm × 3.24 cm =

Solution:

Unrounded answer = 13.8996 cm^2.

Answer rounded to the correct number of significant figures = 13.9 cm^2.

Both measured quantities have three significant figures. Therefore, the answer should be rounded to three significant figures.

Example 2

4.29 cm × 3.2 cm =

Solution:

Unrounded answer = 13.728 cm^2.

Answer rounded to the correct number of significant figures = 14 cm^2.

One of the measured quantities has only two significant figures. Therefore, the answer should be rounded to two significant figures.

Example 3

8.47 cm^2/4.26 cm =

Solution:

Unrounded answer = 1.9882629 cm.

Answer rounded to the correct number of significant figures 1.99 cm.

Both measured quantities have three significant figures. Therefore, the answer should be rounded to three significant figures.

In addition and subtraction the simple rule is that when adding or subtracting measured quantities, the sum or difference should be rounded to the same number of decimal places as in the quantity having the least number of decimal places.

Example 1

      3.56 cm
      2.6 cm
   + 6.12 cm
   Total =

Solution:

Unrounded answer = 12.28 cm.

Answer rounded to the correct number of significant figures = 12.3 cm.

One of the quantities added has only one decimal place. Therefore, the answer should be rounded to only one decimal place.

Example 2

3.514 cm
$\underline{-\,2.13 \quad \text{cm}}$
Difference =

Solution:
Unrounded answer = 1.384 cm.
Answer rounded to the correct number of significant figures = 1.38 cm.
One of the quantities has only two decimal places so the answer should be rounded to two decimal places.

Chapter 1 Review

1. 1.2 mg   = _____ g

2. 6.3 cm   = _____ mm

3. 32°C   = _____ K

4. 12 L   = _____ qt

5. 6.111 mL   = _____ L

6. 1 km   = _____ mm

7. 1.03 kg   = _____ g

8. 0.003 g   = _____ kg

9. 22.4 L   = _____ mL

10. 10,013 cm = _____ km

11. The density of CCl_4 (carbon tetrachloride) is 1.58 g/mL. What will 100 mL of CCl_4 weigh?

12. A piece of sulfur weighs 227 g. When it was submerged in a graduated cylinder containing 50 mL of H_2O, the level rose to 150 mL. What is the density (g/mL) of the sulfur?

13. (a) A box 20 cm × 20 cm × 5.08 in. has what volume in cubic centimeters?

    (b) What weight of H_2O at 4°C will the box hold?

14. Set up the following using the *factor-label method*:
$$\frac{5\ \text{cm}}{\text{second}} = \frac{\text{kilometers}}{\text{hour}}$$

15. How many significant figures are in each of the following?
    (A) 1.01
    (B) 200.0
    (C) 0.0021
    (D) 0.0230

16. If the graphic representation of the energy levels of the reactants and products in a chemical reaction looks like this

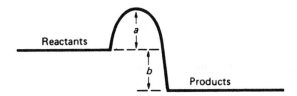

    (A) there is an exothermic reaction
    (B) there is an endothermic reaction
    (C) the a portion is the energy given off
    (D) the b portion is called the activation energy

17. The amount of mass per unit volume refers to the
    (A) density
    (B) specific weight
    (C) volume
    (D) weight

18. A baking powder can carry the statement, "Ingredients: corn starch, sodium bicarbonate, calcium acid phosphate, and sodium aluminum sulfate." Therefore, this baking powder is
    (A) a compound
    (B) a mixture
    (C) a molecule
    (D) a mixture of elements

19. Which of the following is a physical property of sugar?
    (A) It decomposes readily.
    (B) Its composition is carbon, hydrogen, and oxygen.
    (C) It turns black with concentrated H_2SO_4.
    (D) It can be decomposed with heat.
    (E) It is a white, crystalline solid.

20. A substance that can be further simplified may be either
    (A) an element or a compound
    (B) an element or a mixture
    (C) a mixture or a compound
    (D) a mixture or an atom

21. A substance composed of two or more elements chemically united is called
    (A) an isotope
    (B) a compound
    (C) an element
    (D) a mixture

22. An example of a chemical change is the
    (A) breaking of a glass bottle
    (B) sawing of a piece of wood
    (C) rusting of iron
    (D) melting of an ice cube

23. A substance that cannot be further decomposed by ordinary chemical means is
    (A) water
    (B) air
    (C) sugar
    (D) silver

24. An example of a physical change is
    (A) the fermenting of sugar to alcohol
    (B) the rusting of iron
    (C) the burning of paper
    (D) a solution of sugar in water

25. Chemical action may involve all of the following except
    (A) combining of atoms of elements to form a molecule
    (B) separation of the molecules in a mixture
    (C) breaking down compounds into elements
    (D) reacting a compound and an element to form a new compound and a new element

26. The energy of a system can be
    (A) easily changed to mass
    (B) transformed into a different form
    (C) measured only as potential energy
    (D) measured only as kinetic energy

27. If the ΔH of a reaction is a negative quantity, the reaction is definitely
    (A) endothermic
    (B) unstable
    (C) exothermic
    (D) reversible

In the following list, write E for the elements, C for the compounds, and M for the mixtures.

28. Water	32. Aluminum oxide	36. Hydrochloric acid
29. Wine	33. Hydrogen	37. Nitrogen
30. Soil	34. Carbon dioxide	38. Tin
31. Silver	35. Air	39. Potassium chloride

Answers

1. 0.0012 g

2. 63 mm

3. 512 cm

4. 305 K

5. 0.006111 L

6. 1,000,000 or 1×10^6 mm

7. 1030 or 1.03×10^3 g

8. 0.000003 or 3×10^{-6} kg

9. 22400 or 2.24×10^4 mL

10. 0.10013 km

11. 158 g

12. 2.27 g/mL

    To find the volume of 227 g sulfur, subtract the volume of water before from the volume after.

    150 mL − 50 mL = 100 mL

    So, $\dfrac{227\ g}{100\ mL} = 2.27$ g/mL

13. (a) 5160 cm³   (b) 5160 g

    Converting 5.08 in. to cubic centimeters,

    $$5.08\ \cancel{in.} \times \frac{2.54\ cm}{1\ \cancel{in.}} = 12.9\ cm$$

    (a) Then 20 cm × 20 cm × 12.9 cm = 5160 cm³
    Since 1 cm³ of water at 4°C weighs 1 g:
    (b) 5160 cm³ = 5160 g

14. $\dfrac{5\ \cancel{cm}}{\cancel{s}} \times \dfrac{1\ \cancel{m}}{100\ \cancel{cm}} \times \dfrac{1\ km}{1000\ \cancel{m}} \times \dfrac{60\ s}{1\ \cancel{min}} \times \dfrac{60\ \cancel{min}}{1\ hr} =$

15. (a) 3   (b) 4   (c) 2   (d) 3

16. **A**	21. **B**	26. **B**	31. **E**	36. **C**
17. **A**	22. **C**	27. **C**	32. **C**	37. **E**
18. **B**	23. **D**	28. **C**	33. **E**	38. **E**
19. **E**	24. **D**	29. **M**	34. **C**	39. **C**
20. **C**	25. **B**	30. **M**	35. **M**	

Terms You Should Know

accuracy	mass
activation energy	matter
Celsius	meniscus
chemical change	mixture
chemical property	physical change
compound	physical property
density	potential energy
element	precision
endothermic	scientific notation
exothermic	significant figures
Fahrenheit	SI units
gas	solid
heterogeneous	uncertainty
homogeneous	weight
inertia	Law of Conservation of Energy
Kelvin	Law of Conservation of Matter
kilocalorie	Law of Conservation of Mass and Energy
kinetic energy	Law of Definite Composition or Proportion
liquid	

ENTHALPY — Heat Content

CALORIE — quantity of heat

T — degree of heat

CALORIMETRY

PART II:
THE NATURE OF MATTER

Chapter 2

ATOMIC STRUCTURE AND THE PERIODIC TABLE

History

The idea of small, invisible particles being the building blocks of matter can be traced back more than 2000 years ago to the Greek philosophers Democritus and Leucippus. These particles were supposed to be so small and indestructible that they could not be divided into smaller particles. They were called *atoms*, the Greek word for "indivisible." The English word *atom* comes from this Greek word. This early concept of atoms was not based upon experimental evidence but was simply a result of thinking and reasoning on the part of the philosophers. It was not until the eighteenth century that experimental evidence in favor of the atomic hypothesis began to accumulate. Finally, about 1805, John Dalton proposed some basic assumptions about atoms based on what was known through scientific experimentation and observation at that time. These assumptions were very closely related to what we presently know about atoms. Because of this, Dalton is often referred to as the father of modern atomic theory. Some of these basic ideas were

1. All matter is made up of very small, discrete particles called atoms.
2. All atoms of an element are alike in weight and different from the weight of any other kind of atom.
3. Atoms cannot be subdivided, created, or destroyed.
4. Atoms of different elements combine in simple whole-number ratios to form chemical compounds.
5. In chemical reactions, atoms are combined, separated, or rearranged.

By the second half of the 1800s, many scientists believed that all the major discoveries related to the elements had been made. The only thing left for young scientists to do was to refine what was already known. This came to a suprising halt when J. J. Thomson discovered the electron beam in a cathode ray tube in 1897. Soon afterward, Henri Becquerel announced his work with radioactivity, and Marie Curie and her husband, Pierre, set about trying to isolate the source of radioactivity in their laboratory in France.

During the late nineteenth and early twentieth centuries, more and more physicists turned their attention to the structure of the atom. In 1913 the Danish physicist Niels Bohr published a theory explaining the line spectrum of hydrogen. He proposed a planetary model that quantized the energy of electrons to specific orbits. The work of Louis de Broglia and others in the 1920s and 1930s showed that quantum theory had a more probabilistic model of where the electrons could be found that resulted in the theory of orbitals. Enough was learned about nuclear structure to make practical use of atomic nuclei, as in nuclear power generators that use fission reactions and in fission and fusion bombs in the 1940s and 1950s. Today the search still goes on to study the particle physics of the atom and to attempt to control nuclear fusion reactions as a source of energy.

Electric Nature of Atoms

From the beginning of the twentieth century, scientists have been gathering evidence about the structure of atoms and fitting the information into a model of the atomic structure.

Basic Electric Charges

The discovery of the electron as the first subatomic particle is credited to J. J. Thomson (England, 1897). He used an evacuated tube connected to a spark coil as shown in Figure 3. As the voltage across the tube was increased, a beam of light became visible. This was referred to as a cathode ray. Thomson found that the beam was deflected by both electrical and magnetic fields. From this, he concluded that the cathode rays were made up of very small negatively charged particles, which he named *electrons*.

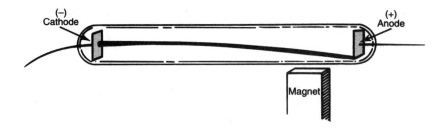

Figure 3. Electron Experiment.

Further experimentation led Thomson to find the ratio of the electrical charge of the electron to its mass. This was a major step toward understanding the nature of the particle. He was awarded a Nobel Prize in 1906 for his accomplishment.

It was an American scientist, Robert Millikan, who in 1909 was able to measure the charge on an electron using the apparatus pictured below.

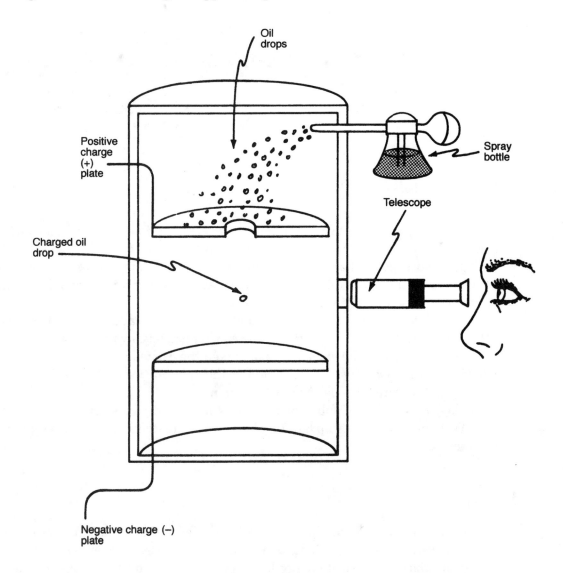

Oil droplets were sprayed into the chamber and in the process became randomly charged by gaining or losing electrons. The electric field was adjusted so that a negatively charged drop would move slowly upward in front of the grid in the telescope. Knowing the rate at which the drop was rising, the strength of the field, and the weight of the drop, Millikan was able to calculate the charge on the drop. Combining the information with Thomson's results, he was able to calculate a value for the mass of a single electron. Eventually, this number was found to be 9.11×10^{-28} g.

Ernest Rutherford (England, 1911) performed a gold foil experiment (Figure 4) which had tremendous implications for atomic structure.

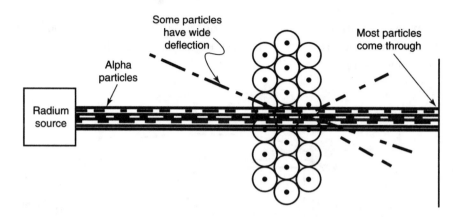

Figure 4. Rutherford Experiment.

Alpha particles (helium nuclei) passed through the foil with few deflections. However, some deflections (1 per 8000) were almost directly back toward the source. This was unexpected and suggested an atomic model with mostly empty space between a *nucleus*, in which most of the mass of the atom was located and which was positively charged, and the electrons that defined the volume of the atom.

Further experiments showed that the nucleus was made up of still smaller particles called protons. Rutherford realized, however, that protons, by themselves, could not account for the entire mass of the nucleus. He predicted the existence of a new nuclear particle that would be neutral and would account for the missing mass. In 1932, James Chadwick (England) discovered this particle, the *neutron*.

Today the number of subatomic particles identified as discrete units has risen to well over 90 named units. An update on the newest subatomic particles research is given in Chapter 15.

Bohr Model

In 1913 Neils Bohr (Denmark) proposed his model of the atom. This pictured the atom as having a dense, positively charged nucleus and negatively charged electrons in specific spherical orbits, also called energy levels or shells, around this nucleus. These shells are arranged concentrically around the nucleus, and each shell is designated by a letter: K, L, M, N, O. . . . The closer to the nucleus, the less energy an electron needs in one of these shells, but it has to gain energy to go from one shell to another that is further away from the nucleus.

Because of its simplicity and general ability to explain chemical change, the Bohr model still has some usefulness today.

Principal Energy Level	Maximum Number of Electrons ($2n^2$)
1	2
2	8
3	18
4	32
5	32

Components of Atomic Structure

The following table lists the basic particles of the atom. For the latest information on sub-atomic particles, see "New Subatomic Particles," page 294.

Particle	Charge	Symbol	Actual Mass	Relative Mass Compared to Proton	Discovery
Electron	– (e⁻)	$_{-1}^{0}e$	9.109×10^{-28} g	1/1,837	J. J. Thomson–1897
Proton	+ (p⁺)	$_{1}^{1}H$	1.673×10^{-24} g	1	———— early 1900s
Neutron	0 (n⁰)	$_{0}^{1}n$	1.675×10^{-24} g	1	J. C. Chadwick–1932

(There are now some 30 or more named particles or units of atomic structure, but the above are the most commonly used.)

When these components are used in the Bohr model, we show the protons and neutrons in the nucleus. These particles are known as *nucleons*. The electrons are shown in the shells.

The number of protons in the nucleus of an atom determines the *atomic number*. All atoms of the same element have the same number of protons and therefore the same atomic number; atoms of different elements have different atomic numbers. Thus, the atomic number identifies the element. An English scientist, Henry Moseley, first determined the atomic numbers of the elements through the use of X-rays.

Since the actual masses of subatomic particles and atoms themselves are very small numbers when expressed in grams, scientists use atomic mass units (amu) instead. An *atomic mass unit* is defined as 1/12 the mass of a carbon-12 atom. Thus, the mass of any atom is expressed relative to the mass of one atom of carbon-12, which is sometimes called the *unified mass unit* (u) or the *dalton*.

The sum of the number of protons and the number of neutrons in the nucleus is called the *mass number*.

Table 1 summarizes the relationships just discussed. Notice that the outermost energy level can contain no more than eight electrons. The explanation of this is given in the next section.

There are cases where different types of atoms of the same element have different masses. Three types of hydrogen atoms are known. The most common type is sometimes called protium and accounts for 99.985% of the hydrogen atoms found on earth. The nucleus of a protium atom consists of only one proton, and has one electron moving about it. There are two other known forms of hydrogen. One is called deuterium and accounts for 0.015% of the earth's hydrogen atoms. Each deuterium atom has a nucleus containing one proton and one neutron. The third form of hydrogen is known as tritium and is radioactive. It exists in very small amounts in nature, but it can be prepared artificially. Each tritium atom contains one proton, two neutrons, and one electron.

Protium, deuterium, and tritium are isotopes of hydrogen. *Isotopes* are atoms of the same element that have different masses. The isotopes of a particular element all have the same number of protons and electrons but different numbers of neutrons. In all three isotopes of hydrogen, the positive charge of the single proton is balanced by the negative

TABLE 1

TABLE OF THE FIRST 21 ELEMENTS*

Element	Atomic No.	Mass No.	Number of Protons	Number of Neutrons	Number of Electrons	Electrons in Energy Levels			
						1	2	3	4
Hydrogen	1	1	1	0	1	1			
Helium	2	4	2	2	2	2			
Lithium	3	7	3	4	3	2	1		
Beryllium	4	9	4	5	4	2	2		
Boron	5	11	5	6	5	2	3		
Carbon	6	12	6	6	6	2	4		
Nitrogen	7	14	7	7	7	2	5		
Oxygen	8	16	8	8	8	2	6		
Fluorine	9	19	9	10	9	2	7		
Neon	10	20	10	10	10	2	8		
Sodium	11	23	11	12	11	2	8	1	
Magnesium	12	24	12	12	12	2	8	2	
Aluminum	13	27	13	14	13	2	8	3	
Silicon	14	28	14	14	14	2	8	4	
Phosphorus	15	31	15	16	15	2	8	5	
Sulfur	16	32	16	16	16	2	8	6	
Chlorine	17	35	17	18	17	2	8	7	
Argon	18	40	18	22	18	2	8	8	
Potassium	19	39	19	20	19	2	8	8	1
Calcium	20	40	20	20	20	2	8	8	2
Scandium	21	45	21	24	21	2	8	9	2

*A complete list of the names and symbols of the known elements can be found in the Tables for Reference section.

charge of the electron. Most elements consist of mixtures of isotopes. Tin has ten stable isotopes, for example, the most of any element.

The percentage of each isotope in the naturally occurring element on earth is nearly always the same no matter where the element is found. The percentage at which each of an element's isotopes occurs in nature is taken into account when calculating the element's average atomic mass. *Average atomic mass* is the weighted average of the atomic masses of the naturally occurring isotopes of an element.

Calculating Average Atomic Mass

The average atomic mass of an element depends on both the mass and the relative abundance of each of the element's isotopes. For example, naturally occurring copper consists of 69.17% copper-63, which has an atomic mass of 62.919598 amu, and 30.83% copper-65, which has an atomic mass of 64.927793 amu. The average atomic mass of copper can be calculated by multiplying the atomic mass of each isotope by its relative abundance (expressed in decimal form) and adding the results:

$$0.6917 \times 62.929599 \text{ amu} + 0.3083 \times 64.927793 \text{ amu} = 63.55 \text{ amu}$$

Therefore, the calculated average atomic mass of naturally occurring copper is 63.55 amu. The average atomic mass is included for the elements listed in the periodic table rounded to one decimal place for use in calculations and to four decimal places in the table of elements in the reference section.

Oxidation Number and Valence

Each atom attempts to have its outer energy level complete and accomplishes this by borrowing, lending, or sharing its electrons. The electrons found in the outermost energy level are called *valence electrons*. The absolute number of electrons gained, lost, or borrowed is referred to as the valence of the atom. When valence electrons are lost or partially lost by sharing, the valence number is assigned a + sign for that element and is called its *oxidation number*. If valence electrons are gained or partially gained by an atom, its valence is assigned a – sign for its oxidation number.

These + and – signs go to the right of the number.

Example 1

$$_{17}\text{Cl} = \overset{\text{nucleus}}{\bullet} \Big) 2 \Big) 8 \Big) 7 \leftarrow \text{valence electrons}$$

Solution:
This picture can be simplified to $\cdot\ddot{\underset{..}{\text{Cl}}}\!:$, showing only the valence electrons as dots. This is called the *Lewis dot structure* of the atom and was devised in 1916 by G. N. Lewis. To complete its outer orbit to eight electrons, Cl must borrow one from another atom. Its valence number then is 1. When electrons are gained, we assign a – sign to this number so the oxidation number of Cl is –1.

Example 2

$$_{11}\text{Na} = \overset{\text{nucleus}}{\bullet} \Big) 2 \Big) 8 \Big) 1 \leftarrow \text{valence electrons}$$

Na • (Lewis dot structure)

Solution:
Since Na tends to lose this electron, its oxidation number is +1.

Some other general rules for oxidation numbers are

1. Atoms of free elements have an oxidation number of zero.
2. Hydrogen has an oxidation number of +1 except in metallic hydrides, where it is –1.
3. Oxygen has an oxidation number of –2 except in peroxides, where it is –1. In combination with fluorine, it is +2.

A table of oxidation numbers is given on page 76. Information on oxidation numbers in chemical combinations is given on page 77.

Metallic, Nonmetallic, and Noble Gas Structures

On the basis of atomic structure, atoms are classified as *metals* if they tend to lend electrons, as *nonmetals* if they tend to borrow electrons, and as *noble gases* if they tend neither to borrow nor to lend electrons and have a complete outer energy level.

Reactivity

The fewer electrons an atom tends to borrow, lend, or share to fill its outer energy level, the more reactive it tends to be in chemical reactions.

Atomic Spectra

The Bohr model was based on a simple postulate. Bohr applied to the hydrogen atom the concept that the electron can exist only in certain energy levels without an energy change, but that when the electron changes its state, it must absorb or emit the exact amount of energy that will bring it from the initial state to the final state. The *ground state* is the lowest energy state available to the electron. The *excited state* is any level higher than the ground state. The formula for changes in energy (ΔE) is

$$\Delta E_{electron} = E_{final} - E_{initial}$$

When the electron moves from the ground state to an excited state, it must absorb energy. When it moves from an excited state to the ground state, it emits energy. This release of energy is the basis for *atomic spectra*. (See Figure 5.)

The energy values were calculated from Bohr's equation

$$E_{mole\ of\ electrons} = \frac{-313.6\ \text{kcal}}{n^2}$$

n (indicating each energy level) = 1, 2, 3, 4, ...

These values are also expressed in electron volts (eV) by using the relationship

$$\frac{\text{kilocalories}}{\text{mole of atoms}} = 23.1 \times \frac{\text{electron volts}}{\text{atom}}$$

Both values are given in Figure 5 along with the SI unit, joules. The equation for arriving at these values in joules is given below. Although the derivation of the equation is not shown, it, too, comes from the Bohr model that describes the energy levels available to the electron in the hydrogen atom

$$E = -2.178 \times 10^{-18}\ \text{J}\left(\frac{Z^2}{n^2}\right)$$

Where n again represents the energy level and Z is the nuclear charge. The values in joules for the principal energy levels were calculated for Figure 6 by using this equation.

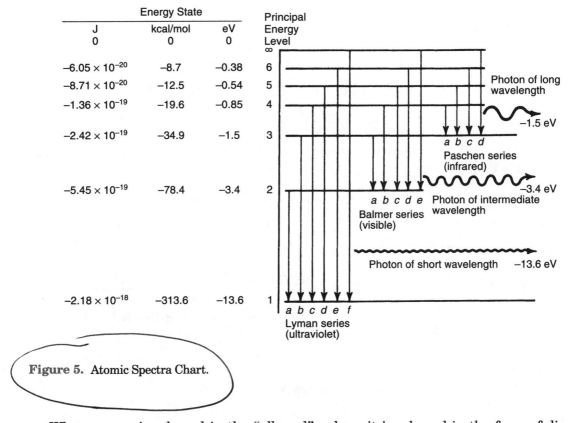

Figure 5. Atomic Spectra Chart.

When energy is released in the "allowed" values, it is released in the form of discrete radiant energy called *photons*. Each of the first three levels has a particular name associated with the emissions that occur when an electron reaches its ground state on that level. The emissions, consisting of ultraviolet radiation, that occur when an electron cascades from a level higher than the first level down to $n = 1$ are known as the Lyman series. Note in Figure 5 that the next two higher levels have the names Balmer (for $n = 2$) and Paschen ($n = 3$) series, respectively.

Spectroscopy

When the light emitted by energized atoms is examined with an instrument called a *spectroscope*, the prism or diffraction grating in the spectroscope disperses the light to allow an examination of the *spectra* or distinct colored lines. Since only particular energy jumps are available in each type of atom, each element has its own unique emission spectra made up of only the lines of specific wavelength that correspond to its atomic structure.

Example

Hydrogen can have an electron drop from the $n = 4$ to the $n = 2$ level. What visible spectral line in the Balmer series will result from this emission of energy?

Solution:

$$\Delta E_{\text{evolved}} = E_{n=2} - E_{n=4}$$

From Figure 6, $E_2 = -5.45 \times 10^{-19}$ J and $E_4 = -1.36 \times 10^{-19}$ J. Then,

$$E_{\text{evolved}} = -5.45 \times 10^{-19} \text{ J} - (-1.36 \times 10^{-19} \text{ J}) = -4.09 \times 10^{-19} \text{ J}$$

Since ΔE is negative, energy is released. The formula for the relationship of ΔE to the emission frequency is

$$\Delta E = \frac{h \, (\text{Planck's constant}) \; c \, (\text{velocity of light})}{\lambda \, (\text{wavelength})}$$

Substituting, we have

$$-4.09 \times 10^{-19} \text{ J} = \frac{(6.626 \times 10^{-34} \text{ J} \cdot \text{s})(2.9979 \times 10^8 \text{ m/s})}{\lambda}$$

$$\lambda = \frac{(6.626 \times 10^{-34} \text{ J} \cdot \text{s})(2.9979 \times 10^8 \text{ m/s})}{4.09 \times 10^{-19} \text{ J}}$$

$$= 4.87 \times 10^{-7} \text{ m}$$

This line is in the blue-green sector of the spectrum.

Visible Light Spectra Wavelengths

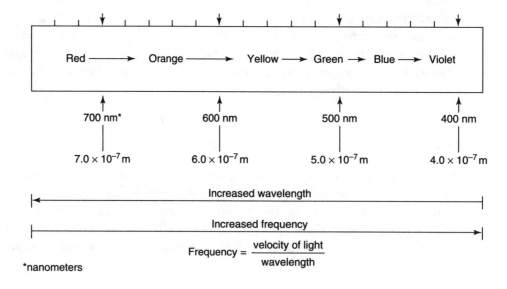

Frequency = $\dfrac{\text{velocity of light}}{\text{wavelength}}$

*nanometers

A partial atomic spectrum for hydrogen would look like this

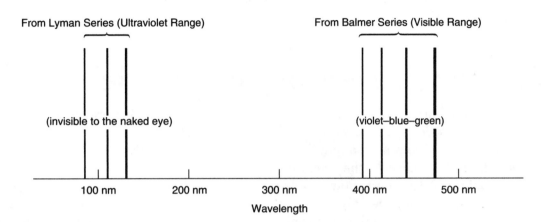

The right-hand group is in the visible range and is part of the Balmer series. The left-hand group is in the ultraviolet region and belongs to the Lyman series.

Investigating spectral lines like these can be used in the identification of unknown specimens.

Mass Spectroscopy

Another tool used to identify specific atomic structures is mass spectroscopy, which is based on the concept that differences in mass cause differences in the degree of bending that occurs in a beam of ions passing through a magnetic field. This is shown in Figure 6.

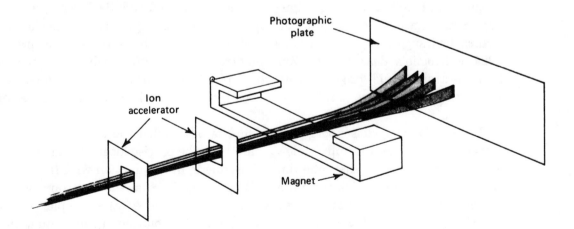

Figure 6. Mass Spectroscope.

The intensity on the photographic plate indicates the amount of each particular *isotope*. Other collectors may be used in place of the photographic plate to collect and interpret these data.

The Wave-Mechanical Model

In the early 1920s, it was becoming apparent that there were some difficulties with the Bohr model of the atom. One difficulty was that although Bohr used classical mechanics (which is the branch of physics that deals with the motion of bodies under the influence of forces) to calculate the orbits of the hydrogen atom, it could not be used to explain the ability of electrons to stay in only certain energy levels without the loss of energy. Nor could it explain that the only change of energy occurred when an electron "jumped" from one energy level to another and could not exist in the atom at any energy level between these levels. According to Newton's laws, the kinetic energy of a body always changed smoothly and continuously, not in sudden jumps. The idea of only certain "quantized" energy levels being available in the Bohr atom was a very important one. The energy levels explained the existence of atomic spectra in the previous sections.

Another difficulty with the Bohr model was that it only worked well for the hydrogen atom with its single electron. It did not work with atoms that had more electrons. A new approach to the laws governing the behavior of electrons inside the atom was needed and such an approach was developed in the 1920s by the combined work of many scientists. Their work dealt with a more mathematical model usually referred to as *quantum mechanics* or *wave mechanics*. By this time Albert Einstein had already proposed a relativity mechanics to deal with the relative nature of mass as its speed approached the speed of light. In the same manner a quantum/wave mechanics was now needed to fit the data of the atomic model. Max Planck suggested in his quantum theory of light that it had particle-like properties as well as wavelike characteristics. In 1924 Louis de Broglie, a young French physicist, suggested that if light can have both wavelike and particle-like characteristics as Planck had suggested, then perhaps particles can also have wavelike characteristics. In 1927 de Broglie's ideas were proven to be true experimentally when investigators showed that electrons could produce diffraction patterns, a property associated with waves. Diffraction patterns are patterns produced by waves as they pass through small holes or narrow slits.

In 1927 Werner Heisenberg stated what is now called the *uncertainty principle*. This principle states that it is impossible to know both the precise location and precise velocity of a subatomic particle at the same time. Heisenberg, in conjunction with the Austrian physicist Erwin Schrödinger, joined in the de Broglie concept that the electron was bound to the nucleus similarly to a standing wave, and they developed the complex equations that describe the *wave mechanical model* of the atom. The solution of these equations gave specific wave functions called *orbitals*. These were not related at all to the Bohr orbits. The electron was not moving in a circular orbit in this model. Rather, the orbital is a three-dimensional region around the nucleus that indicates the probable location of an electron. They give no information on the pathway. Notice that the drawings in Figures 7 and 8 are only probability distribution representations of where the electrons in these orbitals might be found.

Quantum Numbers

Each electron orbital of the atom may be described by a set of four quantum numbers in this model. They give the position with respect to the nucleus, the shape of the orbital, its spatial orientation, and the spin of the electron in the orbital.

Principal Quantum Number (n)

1, 2, 3, 4, 5, etc.

Refers to average distance of the orbital from the nucleus. 1 is closest to the nucleus and has the least energy. These numbers correspond to the orbits in the previous model. They are called energy levels.

Angular Momentum (ℓ) Quantum Number

s, p, d, f
(in order of increasing energy)

Refers to the shape of the orbital. The number of possible shapes is limited by the principal quantum number. The first energy level has only one possible shape, the s orbital. The second has two possible shapes, s and p.

Shapes (see Figures 7 and 8)
Magnetic Quantum Number (m_ℓ)

$s = 1$ space-oriented orbital
$p = 3$ space-oriented orbitals
$d = 5$ space-oriented orbitals
$f = 7$ space-oriented orbitals

The drawings in Figure 7 show the s-orbital shape, which is a sphere, and the p orbitals, which have dumbbell shapes with three possible orientations on the axis shown. The number of spatial orientations of orbitals is referred to as the magnetic quantum number. The possible orientations are listed. Figure 8 represents the d orbitals.

s-orbital shape:

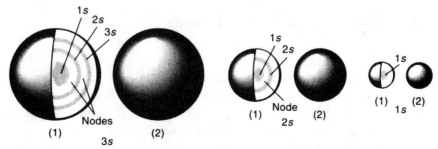

The preceding drawings show two representations of the hydrogen 1s, 2s, and 3s orbitals. (1) The electron probability distribution; the nodes indicated regions of zero probability. (2) The surface that contains 90% of the total electron probability (the size of the orbital, by definition).

p-orbital shapes:

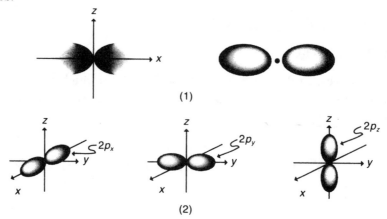

Representation of the 2p orbitals. (1) The electron probability distribution and (2) the boundary surface representations of all three orbitals.

Figure 7. Representations of s and p Orbitals.

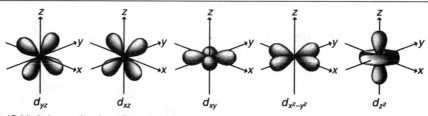

d_{yz} d_{xz} d_{xy} $d_{x^2-y^2}$ d_{z^2}

(Orbitals in y and z planes)
Representations of the 3d orbitals in terms of their boundary surfaces.
The subscripts of the first four orbitals indicate the plane in which the four lobes are centered.

Figure 8. Representation of d Orbitals.

Spin Quantum Number (ms)

+ spin – spin

Electrons are assigned one more quantum number, called the spin quantum number. This describes the spin in either of two possible directions. Each orbital can be filled by only two electrons, each with an opposite spin. The main significance of electron spin is explained by the postulate of Wolfgang Pauli. It states that in a given atom no two electrons can have the same set of four quantum numbers (n, ℓ, m_ℓ, and m_s). This is referred to as the *Pauli Exclusion Principle*. Therefore each orbital in Figures 7 and 8 can hold only two electrons.

Quantum numbers are summarized in the table below.

SUMMARY OF QUANTUM NUMBERS FOR THE FIRST FOUR LEVELS
OF ORBITALS IN THE HYDROGEN ATOM

Principal Quantum No., n	Angular Momentum Quantum No., ℓ	Orbital Shape Designation	Magnetic Quantum No., m_ℓ	Number of Orbitals	Total Electrons
1	0	1s	0	1	2
2	0	2s	0	1	2
	1	2p	−1, 0, +1	3	6
3	0	3s	0	1	2
	1	3p	−1, 0, +1	3	6
	2	3d	−2, −1, 0, +1, +2	5	10
4	0	4s	0	1	2
	1	4p	−1, 0, +1	3	6
	2	4d	−2, −1, 0, +1, +2	5	10
	3	4f	−3, −2, −1, 0, +1, +2, +3	7	14

Limits of Quantum Numbers		
$n = 1, 2, 3, \ldots$	$\ell = 0, 1, \ldots (n-1)$	$m_\ell = -\ell, \ldots, 0, \ldots, +\ell$

Hund's Rule of Maximum Multiplicity

It is important to remember that, when there are more than one orbital at a particular energy level, such as three p orbitals or five d orbitals, only one electron will fill each orbital until each has one electron. This principle, that an electron occupies the lowest energy orbital that can receive it, is called the *Aufbau Principle*. After this, pairing will occur with the addition of one more electron to each orbital. This is called *Hund's Rule of Maximum Multiplicity* and is shown in Table 2, where each slant line indicates an electron (∅)

If each orbital is indicated in an energy diagram as a circle (○), we can show the relative energies in a diagram like Figure 9. If this figure represents a ravine with the energy levels as ledges onto which stones can come to rest only in numbers equal to the circles for orbitals, then pushing stones into the ravine would cause them to lose their kinetic energy as they drop to the lowest level available to them. Much the same is true for electrons.

Sublevels and Electron Configuration

Order of Filling and Notation

The sublevels do not fill up in numerical order, and the pattern of filling is shown on the right side of the approximate relative energy levels diagram (Figure 9). In the first instance of failure to follow numerical order, the 4s fills before the 3d. (Study Figure 9 carefully before going on.)

TABLE 2

ORBITAL NOTATIONS

Chemical Symbol	Atomic No.	Orbital Notation			Electron Configuration Notation
		1s	2s	2p	
H	1	⊡	☐	☐☐☐	$1s^1$
He	2	⊡	☐	☐☐☐	$1s^2$
Li	3	⊡	⊡	☐☐☐	$1s^2, 2s^1$
Be	4	⊡	⊡	☐☐☐	$1s^2, 2s^2$
B	5	⊡	⊡	⊡☐☐	$1s^2, 2s^2, 2p^1$
C	6	⊡	⊡	⊡⊡☐	$1s^2, 2s^2, 2p^2$
N	7	⊡	⊡	⊡⊡⊡	$1s^2, 2s^2, 2p^3$
O	8	⊡	⊡	⊡⊡⊡	$1s^2, 2s^2, 2p^4$
F	9	⊡	⊡	⊡⊡⊡	$1s^2, 2s^2, 2p^5$
Ne	10	⊡	⊡	⊡⊡⊡	$1s^2, 2s^2, 2p^6$

Maximum electrons in orbitals at a particular sublevel:

$s = 2$ (one orbital)
$p = 6$ (three orbitals)
$d = 10$ (five orbitals)
$f = 14$ (seven orbitals)

If each orbital is indicated in an energy diagram as a square (☐), we can show relative energies in a chart such as Figure 9. If this drawing represented a ravine with the energy levels as ledges onto which stones could come to rest only in numbers equal to the squares for orbitals, then pushing stones into the ravine would cause the stones to lose their kinetic energy as they dropped to the lowest level available to them. Much the same is true for electrons.

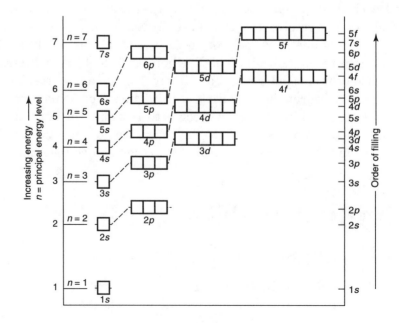

Figure 9. Approximate Relative Energy Levels of Subshells.

$_{19}$K $1s^2, 2s^2, 2p^6, 3s^2, 3p^6, 4s^1$

$_{20}$Ca $1s^2, 2s^2, 2p^6, 3s^2, 3p^6, 4s^2$

$_{21}$Sc $1s^2, 2s^2, 2p^6, 3s^2, 3p^6, 4s^2, 3d^1$ (note that $4s$ filled before $3d$)

There is a more stable configuration at a half-filled or filled sublevel, and so at atomic number 24 the $3d$ sublevel becomes half-filled by taking a $4s$ electron

$_{24}$Cr $1s^2, 2s^2, 2p^6, 3s^2, 3p^6, 3d^5, 4s^1$

and at atomic number 29 the $3d$ becomes filled by taking a $4s$ electron

$_{29}$Cu $1s^2, 2s^2, 2p^6, 3s^2, 3p^6, 3d^{10}, 4s^1$

Table 3 shows the electron configurations of the elements. A triangular mark ▶ indicates the phenomenon of an outer-level electron dropping back to a lower unfilled orbital. These are exceptions to the Aufbau Principle. By following the atomic numbers through this chart, you can establish the same order of filling as shown in Figure 9.

TABLE 3

ELECTRON CONFIGURATIONS OF THE ELEMENTS

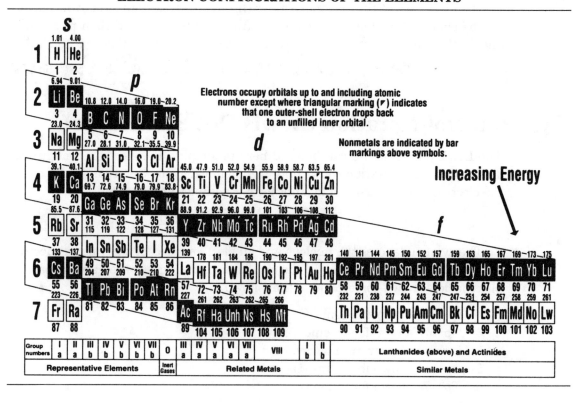

By following the atomic numbers in numerical order in Table 3 you can plot the order of filling of the orbitals for every element shown. A simplified method of showing the order of filling of the orbitals is to use the following diagram. It works for all naturally occurring elements through lanthanium, atomic number 88.

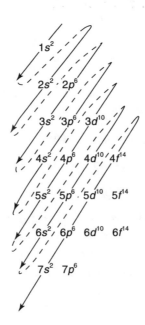

Start by drawing the diagonal arrows through the diagram as shown. The order of filling can be charted by following each arrow from tail to head and then to the tail of the next one. In this way you get the same order of filling as shown in Figure 9 and Table 3. It is

$$1s^2, 2s^2, 2p^6, 3s^2, 3p^6, 4s^2, 3d^{10}, 4p^6, 5s^2, 4d^{10}, 5p^6, 6s^2, 4f^{14}, 5d^{10}, 6p^6, 7s^2$$

Electron Dot Notation (Lewis Dot Structures)

The Lewis dot structure can be used here to simplify the electron configuration notation. The electron dot notation shows only the chemical symbol surrounded by dots to represent the electrons in the incomplete outer level. Examples are

$$\dot{K}, \ \cdot \dot{As} \cdot, \ \ddot{Sr}, \ \vdots \ddot{I} \vdots, \ \vdots \ddot{Rn} \vdots$$

The symbol denotes the nucleus and all electrons except the valence electrons. The dots are arranged at the four sides of the symbol and are paired when appropriate. In the examples the dots shown correspond to the following:

$4s^1$ is shown for potassium (K).
$4s^2, 4p^3$ are shown for arsenic (As).
$5s^2$ is shown for strontium (Sr).
$5s^2, 5p^5$ are shown for iodine (I).
$6s^2, 6p^6$ are shown for radon (Rn).

Transition Elements and Variable Oxidation Numbers

The elements involved with the filling of a d sublevel with electrons after two electrons are in the s sublevel of the next principal energy level are often referred to as the *transition ele-*

ments. The first examples of these are the elements between calcium, atomic number 20, and gallium, atomic number 31. Their electron configurations are the same in the 1*s*, 2*s*, 2*p*, 3*s*, and 3*p* sublevels. It is the filling of the 3*d* and changes in the 4*s* sublevels that are of interest. This is shown in the table below.

| Element | Atomic No. | Electron Configuration | |
		$1s^2$, $2s^2$, $2p^6$, $3s^2$, $3p^6$, 3*d*	4*s*
Scandium	21	1	2
Titanium	22	2	2
Vanadium	23	All 3	2
Chromium	24	the 5	1*
Manganese	25	same 5	2
Iron	26	6	2
Cobalt	27	7	2
Nickel	28	8	2
Copper	29	10	1*
Zinc	30	10	2

The asterisk (*) shows where a 4*s* electron is promoted into the 3*d* sublevel. This is due to the fact that the 3*d* and 4*s* sublevels are very close in energy and that there is a state of greater stability in half-filled and filled sublevels. There, chromium gains stability by the movement of an electron from the 4*s* sublevel into the 3*d* sublevel to give a half-filled 3*d* sublevel. It then has one electron in each of the five orbitals of the 3*d* sublevel. In copper, the movement of one 4*s* electron into the 3*d* sublevel gives the 3*d* sublevel a completely filled configuration.

The fact that the electrons in the 3*d* and 4*s* sublevels are so close in energy levels leads to the possibility of some or all of the 3*d* electrons being involved in chemical bonding. With the variable number of electrons available for bonding, it is not surprising that transition elements can exhibit variable oxidation numbers. An example is manganese with possible oxidation numbers of +2, +3, +4, +6, and +7, which correspond, respectively, to the use of none, one, two, four, and five electrons from the 3*d* sublevel.

The transition elements in the other periods of the table show this same type of anomaly, as they have *d* sublevels filling in the same manner.

Periodic Table of the Elements

History

The history of the development of a systematic pattern for the elements includes the work of a number of scientists such as John Newlands who in 1863 proposed the idea of repeating octaves of properties.

Dimitry I. Mendeleev in 1869 proposed a table containing seventeen columns and is usually given credit for the first periodic table since he arranged elements in groups according to their atomic weights and properties. It is interesting to note that Lothar Meyer proposed a similar arrangement at about the same time. In 1871 Mendeleev rearranged

TABLE 4

PERIODIC TABLE—PROPERTIES

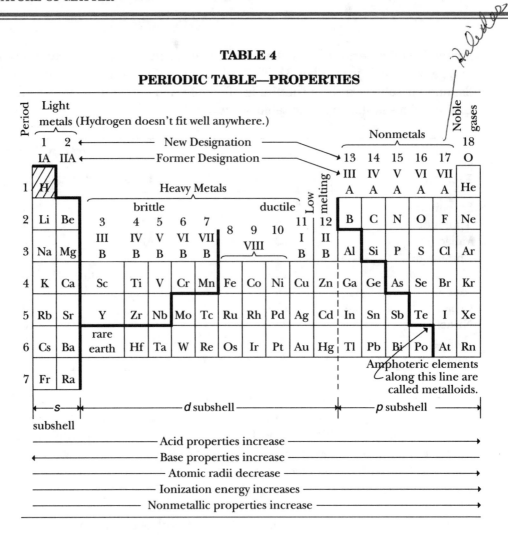

some elements and proposed a table of eight columns, obtained by splitting each of the long periods across into a period of seven elements, an eighth group containing the three central elements (such as Fe, Co, Ni), and a second period of seven elements. The first and second periods of seven across were later distinguished by use of the letters a and b attached to the group symbols, which are Roman numerals. This nomenclature of periods (Ia, IIa, etc.) appears slightly revised in the present periodic table, even in the extended form of assigning Arabic numbers from 1 through 18 as shown in Table 4.

Mendeleyev's table had the elements arranged by atomic weights with properties recurring in a periodic manner. Where atomic weight placement disagreed with the properties that should occur in a particular spot in the table, he gave preference to the element with the correct properties. He even predicted elements for places that were not yet occupied in the table. These predictions proved to be amazingly accurate.

Periodic Law

Henry Moseley stated, after his work with X-ray spectra in the early 1900s, that the properties of elements are a periodic function of their atomic numbers, thus changing the basis of the periodic law from atomic weight to atomic number. This is the present statement of the periodic law.

The Table

The horizontal rows of the periodic table are called *periods* or *rows*. There are seven periods, each of which begins with an atom having only one valence electron and ends with a complete outer shell structure of an inert gas. The first three periods are short, consisting of 2, 8, and 8 elements, respectively. Periods 4 and 5 are long, with 18 each, while period 6 has 32 elements, and period 7 is incomplete with 22 elements, most of which are radioactive and do not occur in nature.

In Table 4, you should note the relationship of the length of periods to the orbital structure of the elements. In the first period, the $1s^2$ orbital is filled with the noble gas helium. The second period begins with the filling of the $2s^2$ orbital and then fills the $2p^6$ and again ends with a noble gas neon, Ne. The same pattern is repeated in period 3 going from $3s^1$ to $3p^6$. These eight elements from sodium, Na, to argon, Ar, complete the filling of the $n = 3$ energy level with $3s^2$ and $3p^6$. In the fourth period, the first two elements indicate the filling of the $4s^2$ orbital. Beyond calcium, Ca, the pattern becomes more complicated. As we discussed in the section "Order of Filling and Notation," the next orbitals to be filled are the five $3d$ orbitals whose elements represent transition elements. Then the three $4p$ orbitals are filled and end with the noble gas krypton, Kr. The fifth period is similar to the fourth period. The $5s^2$ orbital filling is represented by Rb and Sr, both of which resemble the elements directly above them on the table. Next come the transition elements to fill the five $4d$ orbitals before the next group of elements from In to Xe complete the three $5p$ orbitals. (Table 3 should be consulted for the irregularities that occur as the d orbitals fill.) The sixth period follows much the same pattern and has the order of filling as $6s^2$, $4f^{14}$, $5d^{10}$, $6p^6$. Here again irregularities occur, and these can best be followed by using Table 3.

The vertical columns of the periodic table are called *groups* or *families*. The elements in a group exhibit similar or related properties. The Roman numeral group number gives an indication of the number of electrons probably found in the outer shell of the atom and thus an indication of one of its possible valence numbers. In 1984 the International Union of Pure and Applied Chemistry (IUPAC) agreed that the Roman numerals of groups would be replaced by Arabic numbers 1 through 18. Most periodic tables today show this method of identifying groups, but some still show both.

Properties Related to the Periodic Table

Metals are found on the left of the chart (see Table 4) with the most active metal in the lower left corner. Nonmetals are found on the right side with the most active nonmetal in the upper right-hand corner. The noble gases are on the far right. Since the most active metals react with water to form bases, the Group 1 metals are called alkali metals. As you proceed to the right, the base-forming property decreases and the acid-forming properties increase. The metals in the first two groups are the light metals, and those toward the center are heavy metals.

The elements found along the dark line in the periodic chart (Table 4) are called metalloids. These elements have certain characteristics of metals and other characteristics of nonmetals. Some examples of metalloids are boron, silicon, arsenic, and tellurium.

Here are some important general summary statements:

- Acid-forming properties increase moving to the right side of the table.
- Base-forming properties are high on the left side and decrease moving right.
- The atomic radii of elements decrease from left to right cross a period.
- First ionization energies increase from left to right across a period.
- Metallic properties are greatest on the left side of the table and decrease to the right.
- Nonmetallic properties are greatest on the right side of the table and decrease to the left.

Study Table 4 carefully because it summarizes many of these properties. For a more detailed description of metals, alloys, and metalloids, see page 245 in Chapter 13.

Radii of Atoms

The size of an atom is difficult to describe because atoms have no definite outer boundary. Unlike a volleyball, the atom does not have a definite circumference.

To overcome this problem, the size of an atom is estimated by describing its radius. In metals, this is done by measuring the distance between two nuclei in the solid state and dividing this distance by 2. Such measurements can be made with X-ray diffraction. For nonmetallic elements that exist in pure form as molecules, such as chlorine, measurements can be made of the distance between nuclei for two atoms covalently bonded together. Half of this distance is referred to as the *covalent radius*. The method for finding the covalent radius of the chlorine atom is illustrated below.

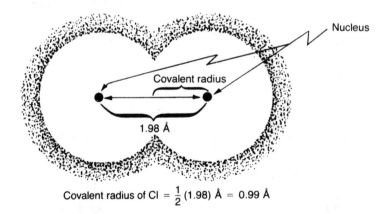

Covalent radius of Cl $= \frac{1}{2}$ (1.98) Å $= 0.99$ Å

Figure 10 shows the relative atomic and ionic radii for some elements. As you review this chart, you should note two trends:

1. Atomic radii decrease from left to right across a period in the periodic chart (until the noble gases).
2. Atomic radii increase from top to bottom in a group or family.

The reason for these trends will become clear in the following discussions.

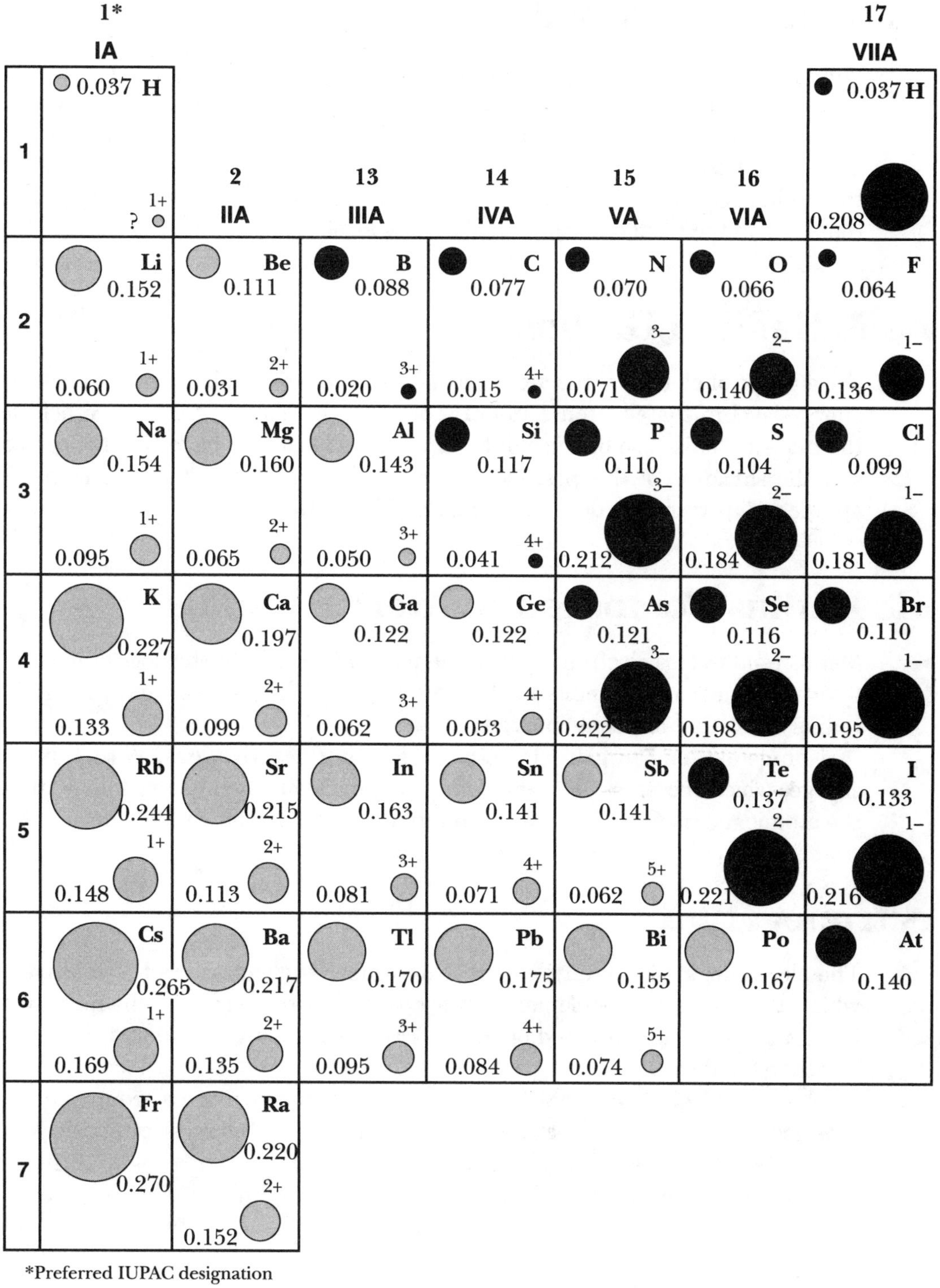

1* IA	2 IIA	13 IIIA	14 IVA	15 VA	16 VIA	17 VIIA
H 0.037 ? 1+						H 0.037 0.208
Li 0.152 0.060 1+	Be 0.111 0.031 2+	B 0.088 0.020 3+	C 0.077 0.015 4+	N 0.070 0.071 3–	O 0.066 0.140 2–	F 0.064 0.136 1–
Na 0.154 0.095 1+	Mg 0.160 0.065 2+	Al 0.143 0.050 3+	Si 0.117 0.041 4+	P 0.110 0.212 3–	S 0.104 0.184 2–	Cl 0.099 0.181 1–
K 0.227 0.133 1+	Ca 0.197 0.099 2+	Ga 0.122 0.062 3+	Ge 0.122 0.053 4+	As 0.121 0.222 3–	Se 0.116 0.198 2–	Br 0.110 0.195 1–
Rb 0.244 0.148 1+	Sr 0.215 0.113 2+	In 0.163 0.081 3+	Sn 0.141 0.071 4+	Sb 0.141 0.062 5+	Te 0.137 0.221 2–	I 0.133 0.216 1–
Cs 0.265 0.169 1+	Ba 0.217 0.135 2+	Tl 0.170 0.095 3+	Pb 0.175 0.084 4+	Bi 0.155 0.074 5+	Po 0.167	At 0.140
Fr 0.270	Ra 0.220 0.152 2+					

*Preferred IUPAC designation

Figure 10. Radii of Some Atoms and Ions (in nanometers).

Note: The atomic radius is usually given for metal atoms, which are shown in gray, and the covalent radius is usually given for atoms of nonmetals, which are shown in black.

Atomic Radii in Periods

Since the number of electrons in the outer principal energy level increases as you go from left to right in each period, the corresponding increase in the nuclear charge due to the additional protons pulls the electrons more tightly around the nucleus. This attraction more than balances the repulsion between the added electrons and the other electrons, and the radius is generally reduced. The inert gas has a slight increase because of the electron repulsion in the filled outer principal energy level.

Atomic Radii in Groups

For a group of elements, the atoms of each successive member have another outer principal energy level in the electron configuration, and the electrons there are held less tightly by the nucleus. This is so because of their increased distance from the nuclear positive charge and the shielding of this positive charge by all the intermediate electrons. Therefore the atomic radius increases down a group. See Figure 10.

Ionic Radius Compared to Atomic Radius

Metals tend to lose electrons in forming positive ions. With this loss of negative charge, the positive nuclear charge pulls in the remaining electrons closer and thus reduces the ionic radius below that of the atomic radius.

Nonmetals tend to gain electrons in forming negative ions. With this added negative charge, which increases the inner electron repulsion, the ionic radius is increased beyond the atomic radius. See Figure 10 for relative atomic and ionic radii values.

Electronegativity

The electronegativity of an element is a number that measures the relative strength with which the atoms of the element attract valence electrons in a chemical bond. This electronegativity number is based on an arbitrary scale going from 0 to 4. In general, a value of less than 2 indicates a metal.

Notice in Table 5 that the electronegativity decreases down a group and increases across a period. The lower the electronegativity number, the more electropositive an element is said to be. The most electronegative element is in the upper right corner—F. The most electropositive is in the lower left corner of the chart—Fr.

Ionization Energy

The atoms hold their valence electrons, then, with different amounts of energy. If enough energy is supplied to one outer electron to remove it from its atom, this amount of energy is called the *first ionization energy*. With the first electron gone, the removal of succeeding electrons becomes more difficult because of the imbalance between the positive nuclear charge and the remaining electrons. The lowest ionization energy is found with the least electronegative atom.

TABLE 5

**ELECTRONEGATIVITIES OF THE ELEMENTS
AND IONIZATION ENERGIES**

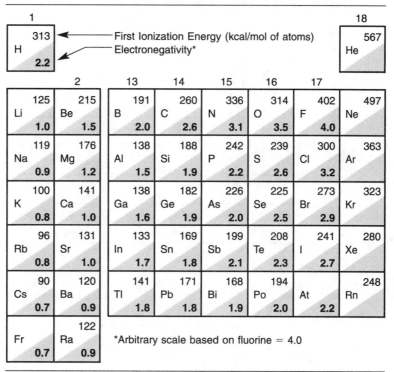

It is not surprising then that the highest peaks on the chart shown on the next page occur with the ionization energy needed to remove the first electron from the outer energy level of the noble gases, that is, He, Ne, Ar, Kr, Xe, and Rn, because of the stability of the filled p orbitals in the outer energy level. Notice that even among these elements there is a gradual decline in the energy needed. This can be explained by considering the distance the energy level involved is from the positively charged nucleus. With each succeeding noble gas, a more distant p orbital is involved, therefore making it easier to remove an electron from the positive attraction of the nucleus. Besides this consideration, as more energy levels are added to the atomic structure as the atomic number increases, the additional negative fields associated with the additional electrons screen out some of the positive attraction of the nucleus. Within a period such as from Li to Ne, the ionization energy generally increases. The lowest occurs when there is a lone electron in the outer s orbital as in Li. As the s orbital fills with two electrons at atomic number 3, Be, there is the added stability of a filled $2s$ orbital which explains the small peak at 3. At atomic number 4, B, there is one lone electron in a $2p$ orbital which can be removed with less energy, and therefore a dip occurs in the chart. With the $2p$ orbitals filling according to Hund's Rule (refer to Table 2), with only one electron in each orbital before pairing occurs, there is again a slightly more stable situation and therefore another small peak at atomic number 7. After this peak, a dip and a continual increase occur until the $2p$ orbitals are completely filled with paired electrons at the noble gas Ne. As you continue to associate the atomic number with the line in the chart, you find peaks occurring in the same general pattern and always related to the state of filling of the orbitals involved and the distance of these orbitals from the nucleus.

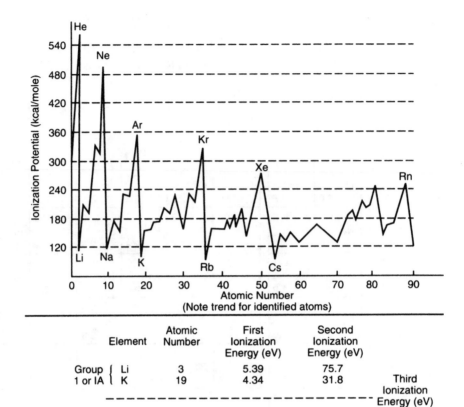

Element	Atomic Number	First Ionization Energy (eV)	Second Ionization Energy (eV)	Third Ionization Energy (eV)
Group 1 or IA { Li	3	5.39	75.7	
K	19	4.34	31.8	
Group 2 or IIA { Be	4	9.32	18.2	154
Mg	12	7.64	15.1	80.3

*Sample Ionization Energies for
Second and Third Electron Removal*

This relationship of bonding to the valence electrons of atoms can be further explained by studying the electron structure of the atoms involved. As already mentioned, the noble gases are monoatomic molecules. The reason for this can be seen in the electron distribution of these noble gases as shown below:

Noble Gas	Electron Distribution						Electrons in Valence Energy Level
Helium	$1s^2$						2
Neon	$1s^2$	$2s^2, 2p^6$					8
Argon	$1s^2$	$2s^2, 2p^6$	$3s^2, 3p^6$				8
Krypton	$1s^2$	$2s^2, 2p^6$	$3s^2, 3p^6, 3d^{10}$	$4s^2, 4p^6$			8
Xenon	$1s^2$	$2s^2, 2p^6$	$3s^2, 3p^6, 3d^{10}$	$4s^2, 4p^6, 4d^{10}$	$5s^2, 5p^6$		8
Radon	$1s^2$	$2s^2, 2p^6$	$3s^2, 3p^6, 3d^{10}$	$4s^2, 4p^6, 4d^{10}, 4f^{14}$	$5s^2, 5p^6, 5d^{10}$	$6s^2, 6p^6$	8

The distinguishing factor in these very stable configurations is the arrangement of two s electrons and six p electrons in the valence energy level of the atoms. This arrangement is called a *stable octet*. All other elements, other than the noble gases, have one to seven electrons in their outer energy level, and these elements are reactive to varying degrees. When they do react to form chemical bonds, the electrons usually shift in such a way that stable octets are formed. In other words, in bond formation, atoms usually attain the stable electron structure of one of the noble gases. The type of bond formed is directly related to whether this structure is achieved through gaining, losing, or sharing of electrons.

Chapter 2 Review

1. The two main parts of an atom are the
   (A) principal energy levels and energy sublevels
   (B) nucleus and kernel
   (C) nucleus and energy levels
   (D) planetary electrons and energy levels

2. The lowest principal quantum number that an electron can have is
   (A) 0
   (B) 1
   (C) 2
   (D) 3

3. The sublevel that has only one orbital is identified by the letter
   (A) s
   (B) p
   (C) d
   (D) f

4. The sublevel that can be occupied by a maximum of ten electrons is identified by the letter
   (A) d
   (B) f
   (C) p
   (D) s

5. An orbital may never be occupied by
   (A) 1 electron
   (B) 2 electrons
   (C) 3 electrons
   (D) 0 electron

6. An atom of beryllium consists of 4 protons, 5 neutrons, 4 electrons. The mass number of this atom is
   (A) 13
   (B) 9
   (C) 8
   (D) 5

7. The number of orbitals in the second principal energy level, $n = 2$, of an atom is
   (A) 1
   (B) 9
   (C) 16
   (D) 4

8. An electron-dot symbol consists of the symbol representing the element and an arrangement of dots that usually shows
   (A) the atomic number
   (B) the atomic mass
   (C) the number of neutrons
   (D) the electrons in the outermost energy level

9. Chlorine is represented by the electron-dot symbol $\cdot\overset{\cdot\cdot}{\underset{\cdot\cdot}{Cl}}\colon$. The atom that would be represented by an identical electron-dot arrangement has the atomic number
   (A) 7
   (B) 9
   (C) 15
   (D) 19

Using the periodic chart, answer the following questions:

10. Name the element of the first 20 whose atom gives up an electron the most readily.
    (A) Li
    (B) F
    (C) Cl
    (D) K

11. Name the element whose atom shows the greatest affinity for an additional electron.
    (A) Li
    (B) F
    (C) Cl
    (D) O

12. Name the most active nonmetal in period 3.
    (A) Na
    (B) Cl
    (C) Ar
    (D) S

13. Which of the following elements is most likely to form covalent compounds?
    (A) Na
    (B) Mg
    (C) Cs
    (D) C

14. What is the most probable oxidation number for silicon in a compound?
    (A) +1
    (B) +2
    (C) +3
    (D) +4

15. Where in a periodic series do you find strong base formers?
    (A) left
    (B) right
    (C) middle
    (D) inert gases

Answers

1. **C**	4. **A**	7. **D**	10. **D**	13. **D**
2. **C**	5. **C**	8. **D**	11. **B**	14. **D**
3. **A**	6. **B**	9. **B**	12. **B**	15. **A**

Terms You Should Know

atomic mass	metallic atoms
atomic number	Moseley
atomic radii	neutron
Aufbau Principle	nonmetallic atoms
Bohr model	nucleus
covalent radius	oxidation number
Dalton's atomic theory	Pauli Exclusion Principle
electron	periodic law
electronegativity	period or row
group or family	proton
Hund's Rule	quantum numbers
inert atoms	s, p, d, f, orbitals
ionic radii	stable octet
ionization energy	transition elements
isotopes	valence electrons
Lewis (electron-dot) structure	wave mechanical model
Mendeleev	

Hydrogen — Protium   99.985%
Deuterium
Tritium

MONOATOMIC MOLECULES — NOBLE GASES

Chapter 3

BONDING

Some elements show no tendency to combine with either like atoms or other kinds of elements. These elements are said to be monoatomic molecules; three examples are helium, neon, and argon. A *molecule* is defined as the smallest particle of an element or a compound that retains the characteristics of the original substance. Water is a triatomic molecule since two hydrogen atoms and one oxygen atom must combine to form the substance water with its characteristic properties. When atoms do combine to form molecules, there is a shifting of valence electrons, that is, the electrons in the outer energy level of each atom. Usually, this results in completion of the outer energy level of each atom. This more stable form may be achieved by the gain or loss of electrons or by the sharing of pairs of electrons. The resulting attraction of the atoms involved is called a *chemical bond*. When a chemical bond forms, energy is released; when this bond is broken, energy is absorbed.

This relationship of bonding to the valence electrons of atoms can be further explained by studying the electron structure of the atoms involved. As already mentioned, the noble gases are monoatomic molecules. The reason for this can be seen in the electron distribution of these noble gases as shown below:

Noble Gas	Electron Distribution	Electrons in Valence Energy Level
Helium	$1s^2$	2
Neon	$1s^2, 2s^2, 2p^6$	8
Argon	$1s^2, 2s^2, 2p^6, 3s^2, 3p^6$	8
Krypton	$1s^2, 2s^2, 2p^6, 3s^2, 3p^6, 3d^{10}, 4s^2, 4p^6$	8
Xenon	$1s^2, 2s^2, 2p^6, 3s^2, 3p^6, 3d^{10}, 4s^2, 4p^6, 4d^{10}, \quad 5s^2, 5p^6$	8
Radon	$1s^2, 2s^2, 2p^6, 3s^2, 3p^6, 3d^{10}, 4s^2, 4p^6, 4d^{10}, 4f^{14}, 5s^2, 5p^6, 5d^{10}, 6s^2, 6p^6$	8

The distinguishing factor in these very stable configurations is the arrangement of two *s* electrons and six *p* electrons in the valence energy level in five of the six atoms. (Note that helium, He, has only a single *s* valence energy level that is filled with two electrons, making He a very stable atom.) This arrangement is called a *stable octet*. All other elements, other than the noble gases, have one to seven electrons in their outer energy level. These elements are reactive to varying degrees. When they do react to form chemical bonds, usually the electrons shift in such a way that stable octets form. In other words, in bond formation, atoms usually attain the stable electron structure of one of the noble gases. The type of bond formed is directly related to whether this structure is achieved through gaining, losing, or sharing of electrons.

Types of Bonds

Ionic Bonds

When the electronegativity values of two kinds of atoms differ by 1.7 or more (especially greater differences than 1.7), the more electronegative atom will borrow the electrons it needs to fill its energy level, and the other kind of atom will lend electrons until it, too, has a complete energy level. Because of this exchange, the borrower becomes negatively charged and the lender becomes positively charged. They are now referred to as *ions*, and the bond or attraction between them is called an *ionic* or *electrovalence bond*. These ions do not retain the properties of the original atoms. An example can be seen in Figure 11.

The reaction:

$$\text{Li (atom)} + \text{F (atom)} \longrightarrow \text{Li}^+ + \text{F}^- \text{ (Ionic compound formed)}$$

Electron notation representations:

$$\text{Li } 1s^2, 2s^1 + \text{F } 1s^2, 2s^2, 2p^5 \longrightarrow \text{Li}^+ 1s^2 + \text{F}^- 1s^2, 2p^2, 2p^6$$

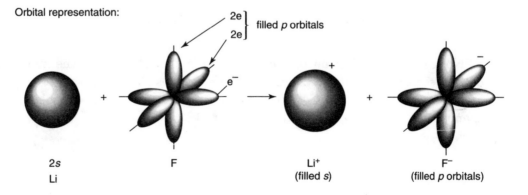

Orbital representation:

Figure 11. Three Representations of the Ionic Bonding of LiF.

These ions do not form an individual molecule in the liquid or solid phase but are arranged into a crystal lattice or giant ion-molecule containing many of these ions. Ionic solids like this tend to have high melting points and do not conduct a current of electricity until they are in the molten state.

Covalent Bonds

When the electronegativity difference between two or more atoms is zero or very small (not greater than about 0.5), they tend to share the valence electrons in the respective outer energy levels. This attraction is called a *nonpolar covalent bond*. Here is an example using electron-dot notation and orbital notation.

fluorine atoms ⟶ fluorine molecule

These covalent bonded molecules do not have an electrostatic charge like the ionic bonded substances. In general, covalent compounds are gases, or liquids having fairly low boiling points, or solids that melt at relatively low temperatures. Unlike ionic compounds, they do not conduct electric currents.

When the electronegativity difference is between 0.4 and 1.7, there is no equal sharing of electrons between the atoms involved. The shared electrons will be more strongly attracted to the atom of greater electronegativity. As the difference in the electronegativities of the two elements increases above 0.4, the polarity or degree of ionic character increases. At a difference of 1.7 or more the bond has more than 50% ionic character. In fact, such a bond is then considered to be ionic. However, when the difference is between 0.5 and 1.7, the bond is called a *polar covalent bond*. An example:

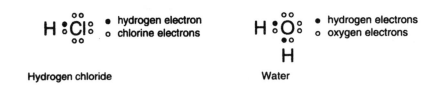

Hydrogen chloride   Water

Notice that the electron pair in the bond is shown closer to the more electronegative atom. Because of this unequal sharing, the molecules shown are said to be polar molecules, or *dipoles*. There are cases of polar covalent bonds existing in nonpolar molecules. Some examples are CO_2, CH_4, and CCl_4. (See Figure 12.)

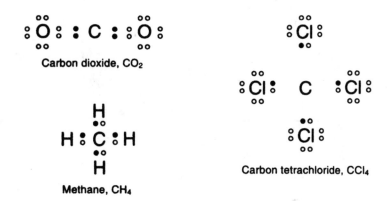

Carbon dioxide, CO₂

Methane, CH₄

Carbon tetrachloride, CCl₄

Figure 12. Polar Covalent Bonds in Nonpolar Molecules.

In all these examples the bonds are polar covalent bonds, but the important thing is that they are symmetrically arranged in the molecule. This results in a nonpolar molecule.

In the *covalent* bonds described so far, the shared electrons in the pair are contributed one each from the atoms bonded. In some cases, however, both electrons for the shared pair are supplied by only one of the atoms. These bonds are called *coordinate covalent bonds*. Two examples are NH_4^+ and H_2SO_4. (See Figure 13.)

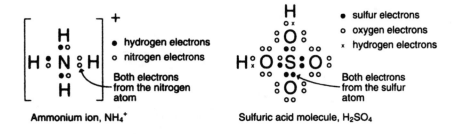

Ammonium ion, NH₄⁺

Sulfuric acid molecule, H₂SO₄

Figure 13. Coordinate Covalent Bonds.

The formation of a covalent bond can be described in a graphic form and related to the potential energy of the atoms involved. Using the formation of the hydrogen molecule as an example, we can show how the potential energy changes as the two atoms approach and form a covalent bond. In the illustration that follows, (1), (2), and (3) show the effect on potential energy as the atoms move closer to each other. In (3), the atoms have reached the condition of lowest potential energy, but their inertia pulls them even closer, as shown in (4). The repulsion between them then forces the two nucleii into a stable position, as shown in (5).

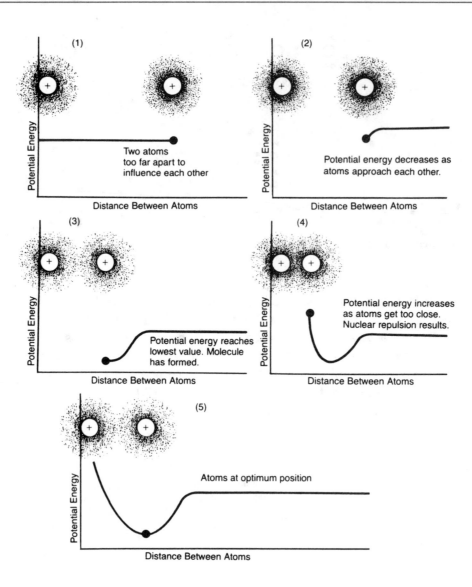

Metallic Bonds

In the case of most metals, one or more of the valence electrons become detached from the atom and migrate into a "sea" of free electrons among the positive metal ions. The metallic bond strength varies with the nuclear charge of the metal atoms and the number of electrons in this electron sea. Both of these factors are reflected in the amount of heat required to vaporize a metal. The strong attraction between these differently charged particles forms a *metallic bond*. Because of this firm bonding, metals usually have high melting points, show great strength, and are good conductors of electricity.

Intermolecular Forces of Attraction

The term *intermolecular forces* refers to attractions *between* molecules. Although it is proper to refer to all intermolecular forces as *van der Waals forces*, this concept should be expanded for clarity.

Dipole-Dipole Attraction

One component of van der Waals forces is *dipole-dipole attraction*. It was shown in the discussion of polar covalent bonding that the unsymmetrical distribution of electronic charges leads to positive and negative charges in the molecules, which are referred to as *dipoles*. In polar molecular substances, the dipoles line up so that the positive pole of one molecule attracts the negative pole of another. This is much like the lineup of small bar magnets. The force of attraction between polar molecules is called *dipole-dipole attraction*. These attractive forces are less than the full charges carried by ions in ionic crystals.

London Forces

Another component of van der Waals forces is called *London force*. Found in both polar and nonpolar molecules, it can be attributed to the fact that an atom that usually is nonpolar sometimes becomes polar because the constant motion of its electrons may cause uneven charge distribution at any one instant. When this occurs, the atom has a temporary dipole. This dipole can then cause a second, adjacent atom to be distorted and to have its nucleus attracted to the negative end of the first atom. London forces are about one tenth the force of most dipole interactions and are the weakest of all the electrical forces that act between atoms or molecules. These forces help to explain why nonpolar substances such as noble gases and the halogens condense into liquids and then freeze into solids when the temperature is lowered sufficiently. In general, they also explain why liquids composed of discrete molecules with no permanent dipole attraction have low boiling points relative to their molecular masses. It is also true that compounds in the solid state that are bound mainly by this type of attraction have rather soft crystals, are easily deformed, and vaporize easily. Because of the low intermolecular forces, the melting points are low and evaporation takes place so easily that it may occur at room temperature. Examples of such solids are iodine crystals and mothballs (paradichlorobenzene and naphthalene).

Hydrogen Bonds

A proton or hydrogen nucleus has a high concentration of positive charge. When a hydrogen atom is bonded to a highly electronegative atom, its positive charge will have an attraction for neighboring electron pairs. This special kind of dipole-dipole attraction is called a *hydrogen bond*. The more strongly polar the molecule is, the more effective the hydrogen bonding is in binding the molecules into a larger unit. As a result, the boiling points of such molecules are higher than those of similar nonpolar molecules. Good examples are water and hydrogen fluoride. Studying Figure 14 shows that in the formation of H_2O, H_2S, H_2Se, and H_2Te there is an unusual rise in the boiling point of H_2O that is not in keeping with the slow increase in boiling point with the increase in molecular mass. Instead of an expected slope between H_2O and H_2S, shown here as a dashed line, the actual boiling point is quite a bit higher at 100°C. The explanation is that hydrogen bonding occurs in H_2O but not to any significant degree in the other compounds.

This same phenomenon occurs with the hydrogen halides (HF, HCl, HBr, and HI). Note in Figure 14 that hydrogen fluoride, which has strong hydrogen bonding, shows an unexpectedly high boiling point.

Hydrogen bonding also explains why some substances have unexpectedly low vapor pressure, high heats of vaporization, and high melting points. In order for vaporization or melting to take place, molecules must be separated. Energy must be expended to break hydrogen bonds and thus break down the larger clusters of molecules into separate molecules. Like the boiling point, the melting point of H_2O is abnormally high when compared to the melting point of the hydrogen compounds of the other elements having six valence electrons, which are chemically similar but have no apparent hydrogen bonding. The hydrogen bonding effect in water is covered on page 141.

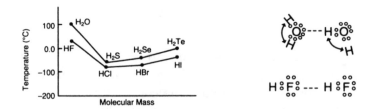

Figure 14. Boiling Points of Hydrogen Compounds with Similar Electron-Dot Structures.

Double and Triple Bonds

To achieve the *octet* structure, which is an outer energy level resembling the noble gas configuration of eight electrons, it is necessary for some atoms to share two or even three pairs of electrons. Sharing two pairs of electrons produces a *double bond*. An example is

$$\overset{\times}{\underset{\times\times}{\times}}\!O\!\overset{\times}{\underset{\times}{\times}}\; \overset{\circ}{\underset{\circ}{\circ}}\!C\!\overset{\circ}{\underset{\circ}{\circ}}\; \overset{\times}{\underset{\times\times}{\times}}\!O\!\overset{\times}{\underset{\times}{\times}}$$

carbon dioxide   , and by a line formula $O\!=\!C\!=\!O$

In the line formula, only the shared pair of electrons is indicated by a bond (—). The sharing of three electron pairs results in a *triple* bond. An example is

$$H\overset{\times}{\underset{\circ}{\,}}C\overset{\circ\circ}{\underset{\circ\circ}{\,}}C\overset{\times}{\underset{\circ}{\,}}H$$

acetylene   , and by a line formula $H\!-\!C\!\equiv\!C\!-\!H$

It can be assumed from these structures that there is a greater electron density between the nuclei involved and hence a greater attractive force between the nuclei and the shared electrons. Experimental data verify that greater energy is required to break double bonds than single bonds, and triple bonds than double bonds. Also, since these stronger bonds tend to pull atoms closer together, the atoms joined by double and triple bonds have smaller interatomic distances and greater bond strengths, respectively.

Resonance Structures

It is not always possible to represent the bonding structure by either the Lewis dot structure or the line drawing because data about the bonding distance and bond strength are between possible drawing configurations and really indicate a hybrid condition. To represent this, the possible alternatives are drawn with arrows in between. Classic examples are sulfur trioxide and benzene. These are shown in Chapters 13 and 14, respectively, but are repeated here as examples.

Sulfur trioxide resonance structures:

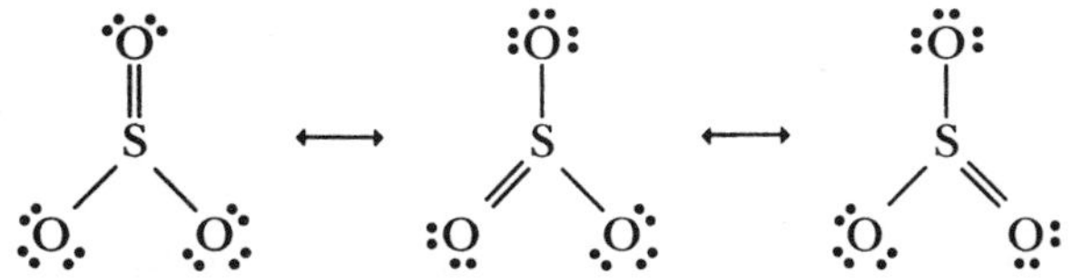

Benzene resonance structures:

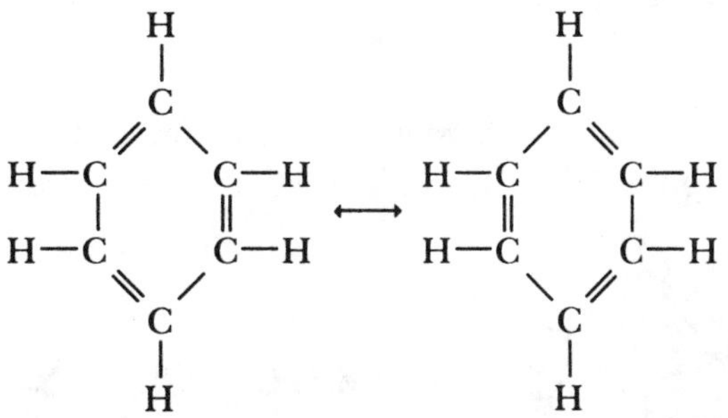

Electrostatic Repulsion (VSEPR) and Hybridization

Electrostatic Repulsion—VSEPR

Data have shown that bond angles for atoms in molecules with p orbitals in the outer energy level do not conform to the expected 90° separation of an x, y, z axis orientation. This variation can be expressed by *electrostatic repulsion between valence electron charge clouds* or by the concept of *hybridization*. The valence energy level electron pair repulsion model is sometimes called the VSEPR model. It is an abbreviation of *valence shell electron pair repulsion*.

VSEPR uses as its basis the fact that like charges will orient themselves in such a way as to diminish the repulsion between them.

1. Mutual repulsion of two electron clouds forces them to the opposite sides of a sphere. This is called a *linear* arrangement.

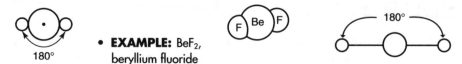

•   **EXAMPLE:** BeF_2, beryllium fluoride

2. Minimum repulsion between three electron pairs occurs when the pairs are at the vertices of an equilateral triangle inscribed in a sphere. This is called *trigonal planar*.

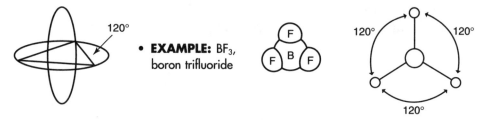

•   **EXAMPLE:** BF_3, boron trifluoride

3. Four electron pairs are farthest apart at the vertices of a tetrahedron inscribed in a sphere. This is called a *tetrahedral* shape.

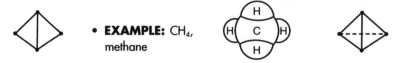

•   **EXAMPLE:** CH_4, methane

4. Mutual repulsion of six identical electron clouds directs them to the corners of an inscribed regular octahedron. This structure is said to have *octahedral* geometry.

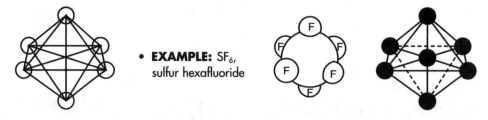

•   **EXAMPLE:** SF_6, sulfur hexafluoride

VSEPR and Unshared Electron Pairs

Ammonia, NH_3, and water, H_2O, are examples of molecules in which the central atom has both shared and unshared electron pairs. Here is how the VSEPR theory accounts for the geometries of these molecules.

The Lewis structure of ammonia shows that in addition to the three electron pairs it shares with three hydrogen atoms, the central nitrogen atom has one unshared pair of electrons.

$$H \!:\! \overset{\cdot\cdot}{\underset{\cdot\cdot}{N}} \!:\! H$$
$$H$$

VSEPR theory postulates that the lone pair occupies space around the nitrogen atom just as the bonding pairs do. Thus, as in the methane molecule shown above, the electron pairs maximize their separation by assuming the four corners of a tetrahedron. Lone pairs occupy space, but our description of the observed shape of a molecule refers to the positions of only the atoms. Consequently, as shown in the drawing below, the molecular geometry of an ammonia molecule is that of a pyramid with a triangular base. The general VSEPR formula for molecules such as ammonia (NH_3) is AB_3E, where A replaces N, and B replaces H, and E represents the unshared electron pair.

A water molecule has two unshared electron pairs. It can be represented as an AB_2E_2 molecule. Here, the oxygen atom is at the center of a tetrahedron, with two corners occupied by hydrogen atoms and two by the unshared pairs as shown below. Again, VSEPR theory states that the lone pairs occupy space around the central atom but that the actual shape of the molecule is determined by the positions of only the atoms. In the case of water, this results in a "bent," or angular, molecule.

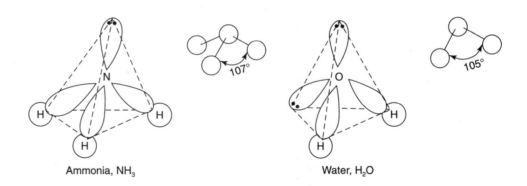

Ammonia, NH_3 Water, H_2O

VSEPR and Molecular Geometry

The following table summarizes the molecular shapes associated with particular types of molecules. Notice that in VSEPR theory, double and triple bonds are treated in the same way as single bonds. It is helpful to use the Lewis structures and this table together to predict the shapes of molecules with double and triple bonds as well as the shapes of polyatomic ions.

SUMMARY OF MOLECULAR SHAPES

Type of Molecule	Molecular Shape	Atoms Bonded to Central Atom	Lone Pairs of Electrons	Formula Example	Lewis Structure
Linear		2	0	BeF_2	$:\ddot{F}—Be—\ddot{F}:$
Bent		2	1	$SnCl_2$ *TIN*	Sn, :Cl, Cl:
Trigonal-planar		3	0	BF_3	F, F, B, F
Tetrahedral		4	0	CH_4	H–C–H
Trigonal-pyramidal		3	1	NH_3	N, H H H
Bent		2	2	H_2O	O, H H
Trigonal-bipyramidal	90° 120°	5	0	PCl_5	Cl–P–Cl
Octahedral	90° 90°	6	0	SF_6	F S F

Hybridization

These same configurations can also be arrived at through the concept of *hybridization*. Briefly stated, this means that two or more pure atomic orbitals (usually s, p, and d) can be mixed to form two or more hybrid atomic orbitals that are identical. This can be illustrated as follows:

sp hybrid orbitals

Beryllium fluoride spectroscopic measurements reveal a bond angle of 180° and equal bond lengths.

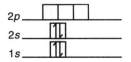

The ground state of beryllium is

2p ☐☐☐
2s ⇅
1s ⇅

To accommodate the experimental data we theorize that a 2s electron is excited to a 2p orbital; then the two orbitals hybridize to yield two identical orbitals called sp orbitals. Each contains one electron but is capable of holding two electrons.

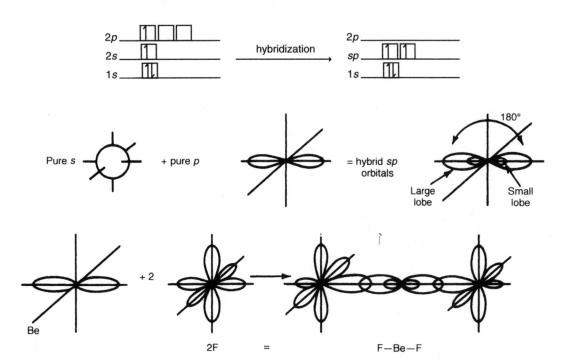

sp² hybrid orbitals

Boron trifluoride has bond angles of 120° of equal strength. To accommodate these data the boron atom hybridizes from its ground state of $1s^2 2s^2 2p^1$ to

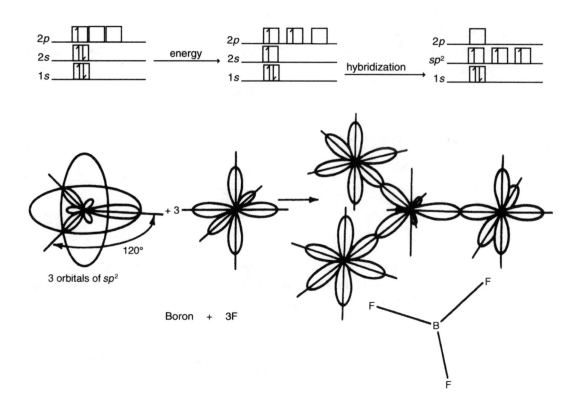

sp³ hybrid orbitals

Methane, CH_4, can be used to illustrate this hybridization. Carbon has a ground state of $1s^2 2s^2 2p^2$. One $2s$ electron is excited to a $2p$ orbital, and the four involved orbitals then form four new identical sp^3 orbitals.

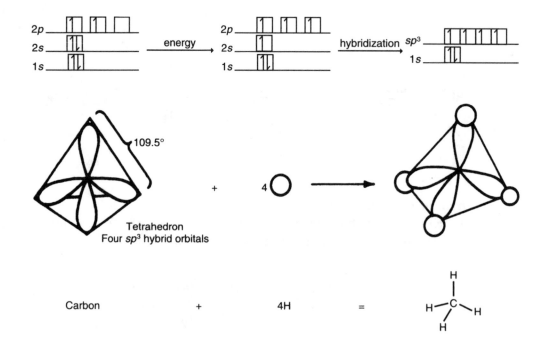

Tetrahedron
Four sp^3 hybrid orbitals

Carbon   +   4H   =

In some compounds where only certain sp^3 orbitals are involved in bonding, distortion in the bond angle occurs because of unbonded electron repulsion. Examples are

a. Water, H_2O.

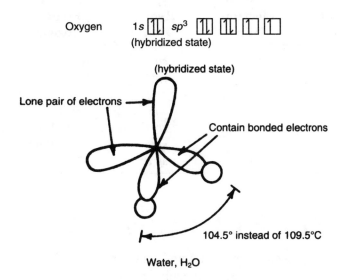

Oxygen $1s$ ⊓⊔ sp^3 ⊓⊔ ⊓⊔ ⊓ ⊓
(hybridized state)

(hybridized state)

Lone pair of electrons

Contain bonded electrons

104.5° instead of 109.5°C

Water, H_2O

b. Ammonia, NH_3.

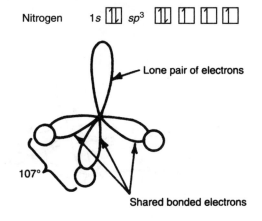

Nitrogen $1s$ ⊓⊔ sp^3 ⊓⊔ ⊓ ⊓ ⊓

Lone pair of electrons

107°

Shared bonded electrons

sp^3d^2 *hybrid orbitals*

These orbitals are formed from the hybridization of an s and a p electron promoted to d orbitals and transformed into six equal sp^3d^2 orbitals. The spatial form is

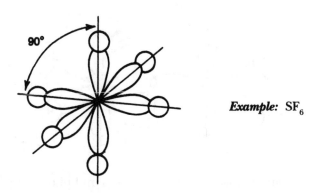

90°

Example: SF_6

SUMMARY OF HYBRIDIZATION

Number of Bonds	Number of Unused Electron Pairs	Type of Hybrid Orbital	Angle Between Bonded Atoms	Geometry	Example
2	0	sp	180°	Linear	BeF_2
3	0	sp^2	120°	Trigonal planar	BF_3
4	0	sp^3	109.5°	Tetrahedral	CH_4
3	1	sp^3	90° to 109.5°	Pyramidal	NH_3
2	2	sp^3	90° to 109.5°	Angular	H_2O
6	0	sp^3d^2	90°	Octahedral	SF_6

Sigma and Pi Bonds

When bonding occurs between s and p orbitals, each bond is identified by a special term. A *sigma bond* is a bond between s orbitals or between an s orbital and another orbital such as a p orbital. It includes bonding between hybrids of s orbitals such as sp, sp^2, and sp^3.

In the methane molecule, the sp^3 orbitals are each bonded to hydrogen atoms. These are sigma bonds.

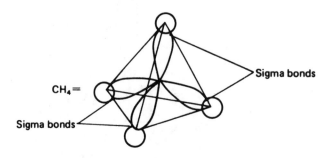

When two p orbitals share electrons in a covalent bond, this is called a *pi bond*. An example

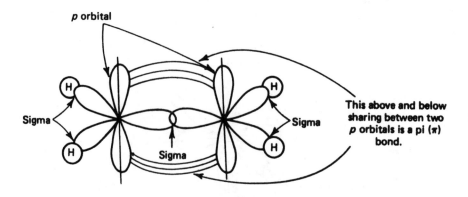

Chapter 14 gives more examples of sigma and pi bonding.

Properties of Ionic Substances

Laboratory experiments reveal that, in general, ionic substances are characterized by the following properties:

1. In the solid phase at room temperature they do not conduct appreciable electric current.
2. In the liquid phase they are relatively good conductors of electric current. The conductivity of ionic substances is much smaller than that of metallic substances.
3. They have relatively high melting and boiling points. There is a wide variation in the properties of different ionic compounds. For example, potassium iodide (KI) melts at 686°C and boils at 1330°C, while magnesium oxide (MgO) melts at 2800°C and boils at 3600°C. Both KI and MgO are ionic compounds.
4. They have relatively low volatilities and low vapor pressures. In other words, they do not vaporize readily at room temperatures.
5. They are brittle and easily broken when stress is exerted on them.
6. Those that are soluble in water form electrolytic solutions which are good conductors of electricity. This is, however, a wide range in the solubilities of ionic compounds. For example, at 25°C, 92 g of sodium nitrate ($NaNO_3$) dissolves in 100 g of water, while only 0.0002 g of $BaSO_4$ dissolves in the same mass of water.

Properties of Molecular Crystals and Liquids

Experiments have shown that the following are general properties of these substances:

1. Neither the liquids nor the solids conduct electric current appreciably.
2. Many exist as gases at room temperature and atmospheric pressure, and many solids and liquids are relatively volatile.
3. The melting points of solid crystals are relatively low.
4. The boiling points of the liquids are relatively low.
5. The solids are generally soft and have a waxy consistency.
6. A large amount of energy is often required to decompose the substance chemically into simpler substances.

Chapter 3 Review

Use the following choices for questions 1 through 7: (1) ionic (2) covalent (3) polar covalent (4) coordinate covalent (5) metallic (6) van der Waals forces (7) hydrogen bonding.

1. When the electronegativity difference between two atoms is 2, what type of bond can be predicted?

2. If two atoms are bonded in such a way that one member of the pair is supplying both electrons that are shared, what is this type of bond called?

3. If the seven choices above were ordered from the strongest bond to the weakest, which would be the last one?

4. Which of the above bonds explains water's abnormally high boiling point?

5. When electron pairs are shared equally between two atoms, what is the bond called?

6. If the sharing of an electron pair is unequal, what is this sharing called?

7. The force of attraction of the electrons of one atom for the protons of another atom in close proximity is which choice above?

8. The VSEPR model of the BF_3 molecule results in the same trigonol planar structure as that of the hybridized form represented by
   (A) sp
   (B) sp^2
   (C) sp^3
   (D) sp^3d^2

Answers

1. **1**		3. **6**		5. **2**		7. **6**	
2. **4**		4. **7**		6. **3**		8. **B**	

Terms You Should Know

covalent bond

dipole-dipole attraction

electrostatic repulsion

hybridization

hydrogen bond

ionic bond

London forces

metallic bond

pi bond

resonance structure

stable octet

sigma bond

van der Waals forces

VSEPR

Electronegativity   *1.7 +*   *ION Bond*
                        0.4 −   *Nonpolar Covalent Bond*
                        0.4−1.7   *Polar Covalent Bond*

PART III:
USING ATOMS AND
MOLECULES

Chapter 4

CHEMICAL FORMULAS

Writing Formulas

From the knowledge of oxidation numbers and valence and the understanding of atomic structure, it is now possible to write chemical formulas. Table 6 is a list of oxidation numbers often encountered in a first-year chemistry course.

General Observations about Oxidation Numbers and Formula Writing

An oxidation number is a small whole-number superscript preceded by a + or − sign. An ionic charge, on the other hand, is a small whole-number superscript followed by a + or − sign to indicate the charge of the ion. Table 6 is a list of oxidation numbers often encountered in a first-year chemistry course.

1. The symbols of the metals have + signs, while those of the nonmetals and all the polyatomic ions *except* ammonium have − signs.
2. When an element exhibits two possible oxidation states, the lower state can be indicated with the suffix *-ous*, and the higher one with *-ic*. Another method of indicating this difference is to use the Roman numeral of the oxidation state in parentheses after the name of the element. For example, ferrous iron or iron(II) for the +2 oxidation state of iron.
3. A *polyatomic ion* is a group of elements that act like a single atom in the formation of a compound. The bonds within these polyatomic ions are predominantly covalent, but the groups of atoms as a whole have an excess of electrons when combined and thus are usually negative ions. These groups were referred to as *radicals*, but using this term has fallen into disfavor.

When you attempt to write a formula, it is important to know whether the substance actually exists. For example, one can easily write the formula of carbon nitrate, but no chemist has ever prepared this compound. Here are the basic rules for writing formulas with three examples carried through each step.

USING ATOMS AND MOLECULES

TABLE 6

TABLE OF OXIDATION NUMBERS

	Monovalent I	Bivalent II	Trivalent III	Tetravalent IV	V
Metals Cations (+)	Hydrogen H Potassium K Sodium Na Silver Ag Mercury (ous) Hg Copper (ous) Cu gold (aurous) Au †Ammonium (NH$_4$)	Barium Ba Calcium Ca Cobalt Co Magnesium Mg Lead Pb Zinc Zn Mercury (ic)* Hg Copper (cupric)* Cu Iron (ferrous)* Fe Manganese (ous)* Mn Tin (stannous)* Sn	Aluminum Al Gold (auric)* Au Arsenic (ous)* As Chromium Cr Iron (ferric)* Fe Phosphorus (ous)* P Antimony (ous)* Sb Bismuth (ous)* Bi	Carbon C Silicon Si Manganese (ic)* Mn Tin (stannic)* Sn Platinum Pt Sulfur S	Arsenic (ic)* As Phosphorus (ic)* P Antimony (ic)* Sb Bismuth (ic)* Bi
				*Use of the Roman numeral, instead of the suffix, is now preferred, for example, iron(II) oxide instead of ferrous oxide.	
	†polyatomic ion				
Nonmetals‡ Anions (−)	Fluorine F Chlorine Cl Bromine Br Iodine I	Oxygen O Sulfur S	Nitrogen N Phosphorus P	Carbon C	
	‡Last syllable of nonmetal is changed to -*ide* in binary compound.				
Polyatomic Ions (−)	Hydroxide (OH) Bicarbonate (HCO$_3$) Nitrite (NO$_2$) Nitrate (NO$_3$) Hypochlorite (ClO) Chlorate (ClO$_3$) Chlorite (ClO$_2$) Perchlorate (ClO$_4$) Acetate (C$_2$H$_3$O$_2$) Permanganate (MnO$_4$) Bisulfate (HSO$_4$)	Carbonate (CO$_3$) Sulfite (SO$_3$) Sulfate (SO$_4$) Tetraborate (B$_4$O$_7$) Silicate (SiO$_3$) Chromate (CrO$_4$) Oxalate (C$_2$O$_4$)	Borate (BO$_3$) Phosphate (PO$_4$) Phosphite (PO$_3$) Ferricyanide [Fe(CN)$_6$]	Ferrocyanide [Fe(CN)$_6$]	

1. Represent the symbols of the components, writing the positive part first and then the negative part.

Sodium chloride   Calcium oxide   Ammonium sulfate

$NaCl$ CaO NH_4SO_4

2. Indicate the oxidation number above and to the right of each symbol. (Enclose polyatomic ions in parentheses for the time being.)

$Na^{+1}Cl^{-1}$ $Ca^{+2}O^{-2}$ $(NH_4)^{+1}(SO_4)^{-2}$

3. Write a subscript number equal to the oxidation number of the other element or polyatomic ion. This is the same as the mechanical crisscross method.

$Na_1^{1+}Cl_1^{1-}$ $Ca_2^{2+}O_2^{2-}$ $(NH_4)_2^{1+}(SO_4)_1^{2-}$

Since the positive oxidation number shows the number of electrons that may be lost or shared and the negative oxidation number shows the number of electrons that may be gained or shared, you must have just as many electrons lost (or partially lost in sharing) as are gained (or partially gained in sharing).

4. Now rewrite the formulas, omitting the subscript 1, the parentheses of the polyatomic ions that have the subscript 1, and the plus and minus numbers.

$NaCl$ Ca_2O_2 $(NH_4)_2SO_4$

5. As a general rule, reduce the subscript numbers in the final formula to their lowest terms. There are, however, certain exceptions, such as hydrogen peroxide (H_2O_2) and acetylene (C_2H_2). For these exceptions, you must know more specific information about the compound.

The only way to become proficient at writing formulas is to memorize the oxidation numbers of common elements (or learn to use the period chart group numbers) and practice writing formulas.

More about Oxidation Numbers

Because it is important to know how to assign oxidation numbers in order to write formulas correctly, the basic rules concerning the assignment of oxidation numbers need to be listed. The oxidation number concept was introduced in Chapter 2 (Atomic Structure). Keep in mind that oxidation numbers are written with the sign preceding the number, whereas ionic charges have the sign following the number.

The basic rules are:

1. The oxidation number of a monoatomic ion is the same as the charge on the ion. Therefore, in NaCl, the oxidation number assigned to sodium is +1, and the oxidation number of chlorine is −1.

2. The oxidation number of any element is zero when the atoms of that element are in the elemental state. An atom is in its elemental state when its atoms are not chemically combined with atoms of other elements.

3. In a covalent compound of two elements, the electronegative element has a negative oxidation number and the other element has a positive oxidation number. For example, in

ammonia, NH_3, the nitrogen is more electronegative. Each of the valence electrons from the three hydrogen atoms is drawn nearer to the nitrogen atom. The oxidation number of the nitrogen in ammonia is therefore –3. Since each hydrogen atom is partially losing one electron, the oxidation number of hydrogen is +1.

4. The oxidation number of oxygen in most of its compounds is –2. In peroxides there is an exception, and it is assigned –1.

5. The oxidation number of hydrogen in most compounds is +1. In hydrides there is an exception, and it is assigned –1.

6. In neutral compounds, for all the atoms represented by the formula, the quantitative sum of the positive oxidation numbers must equal the quantitative sum of the negative oxidation numbers and their sum must equal 0. This can be best illustrated in an example.

Example 1

In Na_2SO_4, what is the oxidation number of sulfur?

Solution:
We know the oxidation number of one atom of Na is +1. There are 2 atoms.

$$Na_2SO_4, \; 2(+1) = +2$$

We know one atom of oxygen is assigned –2. There are 4 atoms.

$$\text{So, } 4(-2) = -8$$

Since the positive sum plus the negative sum must equal 0,

$$(+2) + x + (-8) = 0$$
$$x = +6$$

The sulfur must be using a +6 oxidation state.

Example 2

What is the oxidation number of chromium in $K_2Cr_2O_7$?

Solution:

$$K = 2(+1) = +2$$
$$Cr = 2(x) = 2x$$
$$O = 7(-2) = -14$$

$$(+2) + (2x) + (-14) = 0$$
$$2x = +12$$
$$x = +6$$

So, Cr is using a +6 oxidation number.

7. In a polyatomic ion, the algebraic sum of the positive and negative oxidation numbers of all the atoms in the formula must equal the charge on the ion.

Example 1

SO_4^{2-}

Solution:

$$S = 1(x) = 1x$$
$$O = 4(-2) = -8$$

$$1x + (-8) = -2 \text{ (the charge of the ion)}$$
$$x = +6$$

Example 2

In the NO_3^- ion, find the oxidation number of N.

Solution:

$$x + 3(-2) = -1 \text{ (the charge of the ion)}$$
$$x = +5$$

N is using an oxidation number of +5.

Naming Compounds

A binary compound consists of two elements. The name of the compound also consists of the two elements, the second name having its ending changed to *-ide*, such as NaCl = sodium chloride; AgCl = silver chloride. If the metal has two different oxidation numbers, this can be indicated by the use of the suffix *-ous* for the lower one and *-ic* for the higher one. As previously stated, the more modern way is to use a Roman numeral after the name to indicate the oxidation state. This is called the Stock system. For example

$FeCl_2$ = ferrous chloride or iron(II) chloride
$FeCl_3$ = ferric chloride or iron(III) chloride

If elements combine in varying proportions, thus forming two or more compounds of varying compositions, the name of the second element may be preceded by a prefix, such as *mon(o)-* (one), *di-* (two), *tri-* (three), *pent(a)-* (five). (These prefixes do not use the letter in parentheses when preceding a vowel in the word to which it is attached.) Some examples are carbon dioxide, CO_2; carbon monoxide, CO; phosphorous trioxide, P_2O_3; phosphorous pentoxide, P_2O_5. Notice that when these prefixes are used, it is not necessary to indicate the oxidation state of the first element in the name since it is given indirectly by the prefixed second element.

A ternary compound consists of three components, usually an element and a polyatomic ion. To name the compound, you merely name each component in the order of positive first and negative second.

Binary acids use the prefix *hydro-* in front of the stem or full name of the nonmetallic element and add the ending *-ic*. Examples are *hydro*chlor*ic* acid (HCl) and *hydro*sulfur*ic* acid (H$_2$S).

Ternary acids usually contain hydrogen, a nonmetal, and oxygen. Since the amount of oxygen often varies, the name of the most common form of the acid in the series consists of merely the stem of the nonmetal with the ending *-ic*. The acid containing one less atom of oxygen than the most common acid is designated by the ending *-ous*. The name of the acid containing one more atom of oxygen than the most common acid has the prefix *per-* and the ending *-ic*; that of the acid containing one less atom of oxygen than the *-ous* acid has the prefix *hypo-* and the ending *-ous*.

You can remember the names of the common acids and their salts by learning the following simple rules:

Rule	Example
-ic acids form *-ate* salts.	Sulfuric acid forms sulfate salts.
-ous acids form *-ite* salts.	Sulfurous acid forms sulfite salts.
hydro-(stem)-*ic* acids form *-ide* salts.	Hydrochloric acid forms chloride salts.

When the name of the ternary acid has the prefix *hypo-* or *per-*, that prefix is retained in the name of the salt (hypochlorous acid = sodium hypochlorite).

The names and formulas of some common acids and bases are listed in Table 7.

TABLE 7

FORMULAS OF COMMON ACIDS AND BASES

Acids, Binary		Acids, Ternary	
Name	**Formula**	**Name**	**Formula**
Hydrofluoric	HF	Nitric	HNO$_3$
Hydrochloric	HCl	Nitrous	HNO$_2$
Hydrobromic	HBr	Hypochlorous	HClO
Hydriodic	HI	Chlorous	HClO$_2$
Hydrosulfuric	H$_2$S	Chloric	HClO$_3$
Bases		Perchloric	HClO$_4$
		Sulfuric	H$_2$SO$_4$
Sodium hydroxide	NaOH	Sulfurous	H$_2$SO$_3$
Potassium hydroxide	KOH	Phosphoric	H$_3$PO$_4$
Ammonium hydroxide	NH$_4$OH	Phosphorous	H$_3$PO$_3$
Calcium hydroxide	Ca(OH)$_2$	Carbonic	H$_2$CO$_3$
Magnesium hydroxide	Mg(OH)$_2$	Acetic	HC$_2$H$_3$O$_2$
Barium hydroxide	Ba(OH)$_2$	Oxalic	H$_2$C$_2$O$_4$
Aluminum hydroxide	Al(OH)$_3$	Boric	H$_3$BO$_3$
Ferrous hydroxide	Fe(OH)$_2$	Silicic	H$_2$SiO$_3$
Ferric hydroxide	Fe(OH)$_3$		
Zinc hydroxide	Zn(OH)$_2$		
Lithium hydroxide	LiOH		

Chemical Formulas

The chemical formula is an indication of the makeup of a compound in terms of the kinds of atoms and their relative numbers. It also has some quantitative applications. By using the atomic masses assigned to the elements, we can find the *formula mass* of a compound. If we are sure that the formula represents the actual makeup of one molecule of the substance, the term *molecular mass* may be used as well. In some cases the formula represents an ionic lattice and no discrete molecule exists, as in the case of table salt, NaCl, or the formula merely represents the simplest ratio of the combined substances and not a molecule of the substance. For example, CH_2 is the simplest ratio of carbon and hydrogen united to form the actual compound ethylene, C_2H_4. This simplest ratio formula is called the *empirical formula*, and the actual formula is the *true formula*. The formula mass is determined by multiplying the atomic mass (in whole numbers) by the subscript for that element in the formula.

Example

$Ca(OH)_2$ (one calcium atomic mass + two hydrogen and two oxygen atomic masses = formula mass).

Solution:

$$1 \text{ Ca (at. mass} = 40) = 40$$
$$2 \text{ O (at. mass} = 16) = 32$$
$$\underline{2 \text{ H (at. mass} = 1) = 2}$$
$$\text{Formula mass } Ca(OH)_2 = 74$$

Fe_2O_3

$$2 \text{ Fe (at. mass} = 56) = 112$$
$$\underline{3 \text{ O (at. mass} = 16) = 48}$$
$$\text{Formula mass } Fe_2O_3 = 160$$

It is sometimes useful to know what percentage of the total mass of a compound is made up of a particular element. This is called finding the *percentage composition*. The simple formula for this is

$$\frac{\text{Total mass of the element in the compound}}{\text{Total formula mass}} \times 100\% = \begin{array}{c}\text{Percentage composition}\\ \text{of that element}\end{array}$$

Example

To find the percentage composition of calcium in calcium hydroxide in the example above, we set the formula up as follows:

$$\frac{Ca = 40}{Formula\ mass = 74} \times 100\% = 54\%\ calcium$$

To find the percentage composition of oxygen in calcium hydroxide:

$$\frac{O = 32}{Formula\ mass = 74} = 100\% = 43\%\ oxygen$$

To find the percentage composition of hydrogen in calcium hydroxide:

$$\frac{H = 2}{Formula\ mass = 74} \times 100\% = 2.7\%\ hydrogen$$

Example

Find the percentage compositions of Cu and H_2O in the compound $CuSO_4 \bullet 5H_2O$ (the dot is read "with").

Solution:
First, we calculate the formula mass

$$1\ Cu = 64$$
$$1\ S = 32$$
$$4\ O = 64\ (4 \times 16)$$
$$\underline{5\ H_2O = 90\ (5 \times 18)}$$
$$250$$

and then find the percentages
    Percentage Cu:

$$\frac{Cu = 64}{Formula\ mass = 250} \times 100\% = 26\%$$

Percentage H_2O:

$$\frac{5\ H_2O = 90}{Formula\ mass = 250} \times 100\% = 36\%$$

When you are given the percentage of each element in a compound, you can find the empirical formula as shown with the following example:

Example 1

Given that a compound is composed of 60.0% Mg and 40.0% O, find the empirical formula of the compound.

Solution:
It is easiest to think of 100 mass units of this compound. In this case, the 100 mass units are composed of 60 mass units of Mg and 40 mass units of O. Since you know that 1 unit of

Mg is 24 mass units (from its atomic mass) and, likewise, 1 unit of O is 16, you can divide 60 by 24 to find the number of units of Mg in the compound and divide 40 by 16 to find the number of units of O in the compound.

$$
\begin{array}{cc}
\text{Mg} & \text{O} \\
24\overline{)60} & 16\overline{)40} \\
2.5 \text{ units Mg} & 2.5 \text{ units O}
\end{array}
$$

Now, since we know formulas are made up of whole-number units of the elements which are expressed as subscripts, we must manipulate these numbers to get whole numbers. This is usually accomplished by dividing these numbers by the smallest quotient. In this case they are equal, and so we divide by 2.5.

$$
\begin{array}{cc}
\text{Mg} & \text{O} \\
2.5\overline{)2.5} & 2.5\overline{)2.5} \\
1 & 1
\end{array}
$$

Our empirical formula, then, is one Mg and one O. Therefore, MgO is the formula.

Example 2

Given: Ba = 58.81%, S = 13.73%, and O = 27.46%.
Find the empirical formula.

Solution:
Divide each percentage by the atomic weight of the element.

$$
\begin{array}{ccc}
\text{Ba} & \text{S} & \text{O} \\
137\overline{)58.8} & 32\overline{)13.7} & 16\overline{)27.5} \\
0.43 & 0.43 & 1.72
\end{array}
$$

Manipulate the numbers to get small whole numbers. Try dividing them all by the smallest first.

$$
\begin{array}{ccc}
\text{Ba} & \text{S} & \text{O} \\
0.43\overline{)0.43} & 0.43\overline{)0.43} & 0.43\overline{)1.72} \\
1 & 1 & 4
\end{array}
$$

The formula is $BaSO_4$.

In some cases you may be given the true formula mass of the compound. To check if your empirical formula is correct, add up the formula mass of the empirical formula and compare it to the given formula mass. If it is *not* the same, multiply the empirical formula by the small whole number that gives you the correct formula mass. For example, if your empirical formula is CH_2 (which has a formula mass of 14) and the true formula mass is given as 28, you can see that you must double the empirical formula by doubling all the subscripts. The true formula is C_2H_4.

Laws of Definite Composition and Multiple Proportions

In the problems involving percent composition, we have depended on two things: each unit of an element has the same atomic mass, and every time the particular compound forms, it forms in the same percent composition. That this latter statement is true no matter the source of the compound is the *Law of Definite Composition*. There are some compounds formed by the same two elements in which the mass of one element is constant but the mass of the other varies. In every case, however, the mass of the other element is present in a small whole-number ratio to the mass of the first element. This is called the *Law of Multiple Proportions*. An example is H_2O and H_2O_2.

In H_2O the proportion of $H : O = 2 : 16$ or $1 : 8$.
In H_2O_2 the proportion of $H : O = 2 : 32$ or $1 : 16$.

The ratio of the mass of oxygen in each is $8 : 16$ or $1 : 2$ (a small whole-number ratio).

Writing and Balancing Simple Equations

An equation is a simplified way of recording a chemical change. Instead of words, chemical symbols and formulas are used to represent the reactants and the products. Here is an example of how this can be done. The following is the word equation for the reaction of burning hydrogen with oxygen:

Hydrogen + oxygen yields water.

Replacing the words with the chemical formulas, we have

$$H_2 + O_2 \rightarrow H_2O$$

We replaced hydrogen and oxygen with the formulas for their diatomic molecular states and wrote the appropriate formula for water based on the respective oxidation (valence) numbers for hydrogen and oxygen. Note that the word *yields* was replaced with an arrow.

Although the chemical statement tells what happened, it is not an equation because the two sides are not equal. While the left side has two atoms of oxygen, the right side has only one. Knowing that the Law of Conservation of Matter dictates that matter cannot easily be created or destroyed, we must get the number of atoms of each element represented on the left side to equal the number on the right. To do this, we can only use numbers, called *coefficients*, in front of the formulas. It is important to note that in attempting to balance equations THE SUBSCRIPTS IN THE FORMULAS MAY NOT BE CHANGED.

Looking again at the skeleton equation, we notice that if a 2 is placed in front of H_2O, the numbers of oxygen atoms represented on the two sides of the equation are equal. However, there are now 4 hydrogens on the right side with only 2 on the left. This can be corrected by using a coefficient of 2 in front of H_2. Now we have a balanced equation

$$2H_2 + O_2 \rightarrow 2H_2O$$

This equation tells us more than merely that hydrogen reacts with oxygen to form water. It has quantitative meaning as well. It tells us that two molecular masses of hydrogen react with one molecular mass of oxygen to form two molecular masses of water. Because molecular masses are indirectly related to grams, we may also relate the masses of reactants and products in grams.

$$2H_2 \quad + \quad O_2 \quad \rightarrow \quad 2H_2O$$
$$2(2) \qquad\qquad 32 \qquad\qquad 2(18)$$
$$4 \text{ units} \quad + \quad 32 \text{ units} \quad = \quad 36 \text{ units}$$
$$4 \text{ grams of } H_2 + 32 \text{ grams of } O_2 = 36 \text{ grams of water}$$

This aspect will be important in solving problems related to the weights of substances in a chemical equation.

Here is another, more difficult example: Write the balanced equation for the burning of butane (C_4H_{10}) in oxygen. First, we write the skeleton equation

$$C_4H_{10} + O_2 \text{ yields } CO_2 + H_2O.$$

Looking at the oxygens, we see that there are an even number on the left but an odd number on the right. This is a good place to start. If we use a coefficient of 2 for H_2O, that will even out the oxygens but introduce 4 hydrogens on the right while there are 10 on the left. A coefficient of 5 will give us the right number of hydrogens but introduces an odd number of oxygens. Therefore we have to go to the next even multiple of 5, which is 10. Ten gives us 20 hydrogen atoms on the right. By placing another coefficient of 2 in front of C_4H_{10}, we also have 20 hydrogen atoms on the left. Now the carbons need to be balanced. By placing an 8 in front of CO_2, we have 8 carbons on both sides. The remaining step is to balance the oxygens. We have 26 on the right side, and so we need a coefficient of 13 in front of the O_2 on the left to give us 26 oxygens on both sides. Our balanced equation is

$$2C_4H_{10} + 13O_2 \rightarrow 8CO_2 + 10H_2O$$

For more practice in balancing equations, see page 129. (The balancing of equations using the electron exchange method is introduced at the end of Chapter 12.)

Showing Phases in Chemical Equations

Once the equation is balanced, you may choose to give additional information in an equation. This can be done by indicating the phases of substances, stating whether the substance is in the liquid phase (ℓ), the gaseous phase (g), or the solid phase (s). Since many solids will not react to any appreciable extent unless they are dissolved in water, the notation (aq) is used to indicate that the substance exists in a water (aqueous) solution. Information concerning phase is given in parentheses following the formula for each substance. Several illustrations of this notation are given on the following page:

Formula with Phase Notation	Meaning
$Cl_2(g)$	Chlorine gas
$H_2O(\ell)$	Water in the liquid state as opposed to ice or steam
$NaCl(s)$	Sodium chloride as a solid
$NaCl(aq)$	A water solution of dissolved sodium chloride

An example of phase notation in an equation is

$$2HCl(aq) + Zn(s) \rightarrow ZnCl_2(aq) + H_2(g)$$

In words, this says that a water solution of hydrogen chloride (called hydrochloric acid) reacts with solid zinc to produce zinc chloride dissolved in water plus hydrogen gas.

Writing Ionic Equations

There are times that chemists choose to show only the substances that react in the chemical action. These are called *ionic* equations because they stress the reaction and production of ions. If we look at the preceding equation, we see that the complete cast of "actors" in this equation is:

$$2HCl(aq) \rightarrow 2H^+(aq) + 2Cl^-(aq)$$
$$Zn(s) \rightarrow Zn(s) \text{ particles}$$
$$ZnCl_2(aq) \rightarrow Zn^{2+}(aq) + 2Cl^-(aq)$$
$$H_2(g) \rightarrow H_2(g)$$

Writing the complete reaction using these results, we have

$$2H^+(aq) + 2Cl^-(aq) + Zn(s) \rightarrow Zn^{2+}(aq) + 2Cl^-(aq) + H_2(g)$$

Notice that nothing happened to the chloride ion. It appears the same on both sides of the equation. It is referred to as a spectator ion. In writing the simple ionic equation, spectator ions are omitted. So, the ionic equation is

$$2H^+(aq) + Zn(s) \rightarrow Zn^{2+}(aq) + H_2(g)$$

Chapter 4 Review

Write the formula or name:

1. $AgCl$ _____

2. $CaSO_4$ _____

3. $Al_2(SO_4)_3$ _____

4. NH_4NO_3 _____

5. $FeSO_4$ _____

6. Potassium chromate _____

7. Sodium fluoride _____

8. Magnesium sulfite _____

9. Copper(II) sulfate _____

10. Iron(III) chloride _____

11. Find the percentage of sulfur in H_2SO_4.

12. What are the empirical formula and the true formula of a compound composed of 85.7% C and 14.3% H with a true formula mass of 42 amu?

Find the oxidation numbers of the indicated atom in each of the following:

13. Cr in K_2CrO_4

14. S in $Na_2S_2O_3$

15. P in $(PO_4)^{3-}$

Answers

1. Silver chloride

2. Calcium sulfate

3. Aluminum sulfate

4. Ammonium nitrate

5. Ferrous sulfate or iron(II) sulfate

6. K_2CrO_4

7. NaF

8. $MgSO_3$

9. $CuSO_4$

10. $FeCl_3$

11. 32.65%

    H_2SO_4 is composed of

    $$\begin{array}{ll} 2\,H = \ \ 4 & \\ 1\,S = 32 & \\ 4\,O = \underline{64} & \\ \text{Total} = 98 & \end{array}$$

    Percentage of S:

    $$\frac{S = 32}{\text{Total} = 98} \times 100\% = 32.65 \text{ or } 33\%$$

12. CH_2 and C_3H_6

    $$C = 12\overline{)85.7\% \ C} \qquad H = 1\overline{)14.3\% \ H}$$
    $$\qquad \ \ 7.14 \qquad\qquad\qquad\qquad 14.3$$

To find the lowest ratio of the whole numbers:

$$7.14\overline{)7.14} \qquad 7.14\overline{)14.3}$$
$$1.0 \qquad\qquad 2.0$$

So the empirical formula is CH_2.

Since the formula mass is given as 42, the empirical formula CH_2, which totals 14 amu, divides into 42 three times. Therefore, the true formula is C_3H_6.

13. $Cr = +6$ because $2(K^{+1}) + 1(Cr^x) + 4(O^{-2})$ result in zero charge. So

$$2(+1) + 1(x) + 4(-2) = 0$$
$$2 + x \quad + (-8) \ = 0$$
$$x = +6$$

14. $S = +2$ because $2(Na^{+1}) + (S^x) + 3(O^{-2})$ results in zero charge. So

$$2(+1) + 2(x) + 3(-2) = 0$$
$$2 + 2x \quad + (-6) \ = 0$$
$$2x = +4$$
$$x = +2$$

15. P is +5 because $1(P^x) + 4(O^{-2}) = -3$

The equation for charges is

$$x + (-8) = -3$$
$$x = +5$$

Terms You Should Know

binary compound
coefficient
empirical formula
formula mass
molecular mass

percentage composition
polyatomic ion
ternary compound
true formula

Law of Definite Composition
Law of Multiple Proportions

PART IV:
THE STATES AND
PHASES OF MATTER

Chapter 5

GASES AND THE GAS LAWS

Introduction—Gases in the Environment

When we discuss gases today, the most pressing concern is the gases in our atmosphere. These are the gases that are held against the earth by the gravitational field. The mixture of gases in the air today has had 4.5 billion years in which to evolve. The earliest atmosphere must have consisted of volcanic discharges alone. Gases erupted by volcanoes today are mostly a mixture of water vapor, carbon dioxide, sulfur dioxide, and nitrogen, with almost no oxygen. If this is the same mixture that existed in the early atmosphere, various processes would have had to operate to produce the mixture we have today. One of these processes was condensation. As it cooled, much of the volcanic water vapor condensed to fill the earliest oceans. Chemical reactions would also have occurred. Some carbon dioxide would have reacted with the rocks of the earth's crust to form carbonate minerals, and some would have dissolved in the new oceans. Later, as primitive life capable of photosynthesis evolved in the seas, new marine organisms began producing oxygen. Almost all the free oxygen in the air today is believed to have formed this way. About 570 million years ago, the oxygen content of the atmosphere and oceans became high enough to permit marine life capable of respiration. Later, about 400 million years ago, the atmosphere contained enough oxygen for air-breathing animals to emerge from the seas.

The principal constituents of the atmosphere of the earth today are nitrogen (78%) and oxygen (21%). The gases in the remaining 1% are argon (0.9%), carbon dioxide (0.03%), varying amounts of water vapor, and trace amounts of hydrogen, ozone, methane, carbon monoxide, helium, neon, krypton, and xenon. Oxides and other pollutants added to the atmosphere by factories and automobiles have become a major concern because of their damaging effects in the form of acid rain (discussed further in Chapter 11). In addition, a strong possibility exists that the steady increase in atmospheric carbon dioxide, mainly attributed to fossil fuel combustion over the past century, may affect the earth's climate by causing a greenhouse effect, resulting in a steady rise in temperatures worldwide.

Studies of air samples show that up to 55 miles above sea level the composition of the atmosphere is substantially the same as at ground level; continuous stirring produced by atmospheric currents counteracts the tendency of the heavier gases to settle below the lighter ones. In the lower atmosphere ozone is normally present in extremely low concentrations. The atmospheric layer 12 to 30 miles up contains more ozone that is produced by the action of ultraviolet radiation from the sun. In this layer, however, the percentage of ozone is only 0.001 by volume. Human activity adds to the ozone concentration in the lower atmosphere where it can be a harmful pollutant.

The ozone layer became a subject of concern in the early 1970s when it was found that chemicals known as fluorocarbons, or chlorofluoromethanes, were rising into the atmosphere in large quantities because of their use as refrigerants and as propellants in aerosol dispensers. The concern centered on the possibility that these compounds, through the action of sunlight, could chemically attack and destroy stratospheric ozone, which protects the earth's surface from excessive ultraviolet radiation. As a result, US industries and the Environmental Protection Agency have phased out the use of certain chlorocarbons and fluorocarbons by the year 2000. There is still ongoing concern about both these environmental problems: the greenhouse effect and the deterioration of the ozone layer.

Some Representative Gases

Oxygen

Of the gases that occur in the atmosphere, the most important one to us is oxygen. Although it makes up only approximately 21% of the atmosphere by volume, the oxygen found on the earth is equal in weight to all the other elements combined. About 50% of the earth's crust (including the waters on the earth and the air surrounding it) is oxygen. (Note Figure 15.)

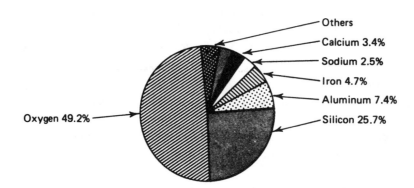

Figure 15.   Composition of Earth's Crust.

The composition of air varies slightly from place to place because it is a mixture of gases. The composition by volume is approximately as follows: nitrogen, 78%; oxygen, 21%; argon, 1%. There are also small amounts of carbon dioxide, water vapor, and trace gases.

Preparation of Oxygen

In 1774 an English scientist named Joseph Priestley discovered oxygen by heating mercuric oxide in an enclosed container with a magnifying glass. That mercuric oxide decomposes into oxygen and mercury can be expressed in an equation: $2HgO \rightarrow 2Hg + O_2$. After his discovery, Priestley visited one of the greatest of all scientists, Antoine Lavoisier, in Paris. As early as 1773 Lavoisier had carried out experiments involving burning, and they had caused him to doubt the phlogiston theory (that a substance called phlogiston was released when a substance burned; the theory went through several modifications before it was finally abandoned). By 1775 Lavoisier had demonstrated the true nature of burning and called the resulting gas "oxygen."

Today oxygen is usually prepared in the lab by heating an easily decomposed oxygen compound such as potassium chlorate ($KClO_3$). The equation for this reaction is

$$2KClO_3 + MnO_2 \rightarrow 2KCl + 3O_2(g) + MnO_2$$

The usual laboratory setup is shown in Figure 16.

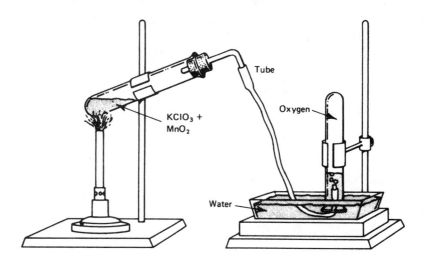

Figure 16. A Possible Laboratory Preparation of Oxygen.

In this preparation manganese dioxide (MnO_2) is often used. This compound is not used up in the reaction and can be shown to have the same composition as it had before the reaction occurred. The only effect it seems to have is that it lowers the temperature needed to decompose the $KClO_3$ and thus speeds up the reaction. Substances that behave in this manner are referred to as *catalysts*. Catalysts can also be used to slow down a reaction. The mechanism by which a catalyst acts is not completely understood in all cases, but it is known that in some reactions the catalyst does change its structure temporarily. Its effect is shown graphically in the reaction graphs in Figure 17.

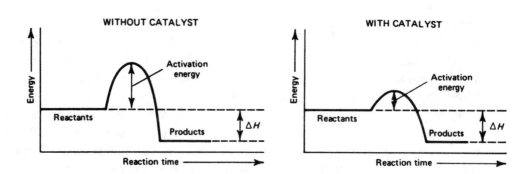

Figure 17. Reaction.

Notice that in the reaction with the catalyst the activation energy necessary to make the reaction take place is lower so that the reaction can take place at a more rapid rate.

Commercially, oxygen is prepared by either the fractional distillation of liquid air or the electrolysis of water. The fractional distillation of liquid air is typical of all fractional distillation processes in that it attempts to separate different components of a mixture in solution by taking advantage of the fact that each component has a distinct boiling point. By raising the temperature slowly and holding it constant when any boiling is taking place, the various components can be separated. With liquid air at −200°C, the temperature can be controlled so that nitrogen will boil off at −196°C. Then the oxygen is boiled off at −183°C.

Electrolysis of water will be explained more completely in Chapter 12, but essentially it consists of passing an electric current through water to cause it to decompose into hydrogen gas (H_2) and oxygen gas (O_2).

Oxidation

When oxygen combines with an element slowly so that *no* noticeable heat and light are given off, the process is referred to as slow oxidation. A common example of this is rusting of iron. When the combination of oxygen with an element occurs so rapidly that the amount of energy released is visible as light and felt as heat, the process is referred to as rapid oxidation or normal burning. It is not always necessary to have oxygen present for burning or combustion to take place. A jet of hydrogen gas will burn in an atmosphere of chlorine.

Properties of Oxygen

That oxygen is a gas under ordinary conditions of temperature and pressure, and that it is a gas that is colorless, odorless, tasteless, and slightly heavier than air, are all physical properties of this element. Oxygen is only slightly soluble in water, thus making it possible to collect the gas over water, as shown in Figure 16. Oxygen at −183°C changes to a pale blue liquid with magnetic properties. This attraction by a magnet (paramagnetism) is believed to be due to an unpaired group of electrons found in the *p* orbitals. The Lewis structure for oxygen is

Although oxygen will support combustion, as noted above, it will not burn. This is one of its chemical properties. The usual test for oxygen is to lower a glowing splint into the gas and see if the oxidation increases in its rate to reignite the splint. (*Note*: This is not the only gas that does this. N_2O reacts the same.)

Ozone

Ozone is another form of oxygen that contains three atoms in its molecular structure (O_3). Since ordinary oxygen and ozone differ in energy content and form, they have slightly different properties. They are called allotropic forms of oxygen. Ozone occurs, as explained in the introduction, in small quantities in the upper layers of the earth's atmosphere and can be formed in the lower atmosphere where high-voltage electricity in lightning passes through the air. This formation of ozone also occurs around machinery using high voltages. The reaction can be shown by the equation

$$3O_2 + \text{elec.} \rightarrow 2O_3$$

Because of its higher energy content, ozone is more reactive chemically than oxygen.

Hydrogen

Of all the 109 elements in the periodic chart, hydrogen is number 1 by reason of the fact that it is the lightest element known. Its lightness makes it useful in balloons, but care must be taken because of its extreme combustibility. Its ability to burn well makes it an important part of most fuels, and its powerful attraction to oxygen makes it a good reducing agent in some important industrial reactions. Hydrogen can be added to liquid fats, under the proper conditions, to make solid shortenings, or it can be added to certain fuels to improve their burning properties. Its isotopes were used in making the hydrogen bomb.

Preparation of Hydrogen

Although there is evidence of the preparation of hydrogen before 1766, Henry Cavandish was the first person to recognize this gas as a separate substance. He observed that whenever it burned it produced water. Lavoisier named it *hydrogen*, which means "water former."

Electrolysis of water, which is the process of passing an electric current through water to cause it to decompose, is one method of obtaining hydrogen. This is a widely used commercial method as well as a laboratory method.

Another method of producing hydrogen is to displace it from the water molecule by using a metal. To choose the metal you must be familiar with its activity with respect to hydrogen. The activities of common metals are shown in Table 8.

TABLE 8

ACTIVITY CHART OF METALS COMPARED TO HYDROGEN

Potassium	⎫	
Calcium	⎬ In cold water	
Sodium	⎭	
Magnesium	In hot water	
Aluminum	⎫	
Zinc		
Iron	⎬ In most dilute acids	
Tin	⎭	
HYDROGEN		
Copper	⎫	
Mercury	⎬ Not active enough	
Silver		
Gold	⎭	

(left margin vertical label: Increasing Activity ↑)

As noted in Table 8, the first three metals will react with water in the following manner:

Very active metal + water = Hydrogen + metal hydroxide

Using sodium as an example,

$$2Na + 2HOH \rightarrow H_2(g) + 2NaOH$$

With metals that react slower, a dilute acid reaction is needed to produce hydrogen in sufficient quantities to collect in the laboratory. This general equation is

Active metal + dilute acid → Hydrogen + salt of the acid

An example:

$$Zn + dil. H_2SO_4 \rightarrow H_2(g) + ZnSO_4$$

This equation shows the usual laboratory method of preparing hydrogen. Mossy zinc is used in a setup as shown in Figure 18. The acid is introduced down the thistle tube after the zinc is placed in the reacting bottle. In this sort of setup, you do not begin collecting the gas that bubbles out of the delivery tube for a few minutes so that the air in the system has a chance to be expelled and you can collect a rather pure volume of the gas generated.

In industry, hydrogen is produced by (1) the electrolysis of water, (2) passing steam over red-hot iron or through hot coke, or (3) decomposing natural gas (mostly methane, CH_4) with heat ($CH_4 + H_2O \rightarrow CO + 3H_2$).

Properties of Hydrogen

Hydrogen has the following important physical properties:

1. It is ordinarily a gas; colorless, odorless, and tasteless when pure.
2. It weighs 0.9 g/liter at 0°C and 1 atmosphere (atm) pressure. This is 1/14 as heavy as air.

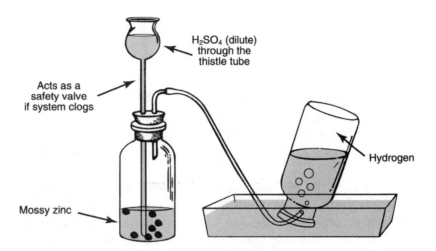

Figure 18.   Preparation of an Insoluble Gas by the Addition of a Liquid to the Other Reactant.

3. It is slightly soluble in water.
4. It becomes a liquid at a temperature of –240°C and a pressure of 13 atm.
5. It diffuses (moves from place to place in gases) more rapidly than any other gas. This property can be demonstrated as shown in Figure 19.

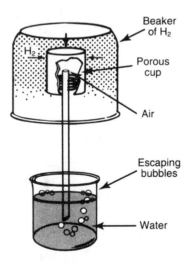

Figure 19.   Diffusion of Hydrogen.

Here the H_2 in the beaker that is placed over the porous cup diffuses faster through the cup than the air can diffuse out. Consequently, there is a pressure buildup in the cup, which pushes the gas out through the water in the lower beaker.

The chemical properties of hydrogen are:

1. It burns in air or in oxygen, giving off large amounts of heat. Its high heat of combustion makes it a good fuel.
2. It does not support ordinary combustion.
3. It is a good reducing agent in that it withdraws oxygen from many hot metal oxides.

General Characteristics of Gases

Measuring the Pressure of a Gas

Pressure is defined as force per unit area. With respect to the atmosphere, pressure is the result of the weight of a mixture of gases. This pressure, which is called *atmospheric pressure*, *air pressure*, or *barometric pressure*, is approximately equal to the weight of a kilogram mass on every square centimeter of surface exposed to it. This weight is about 10 newtons (N).

The pressure of the atmosphere varies with altitude. At higher altitudes, the weight of the overlying atmosphere is less, and so the pressure is less. Air pressure also varies somewhat with weather conditions as low- and high-pressure areas move with weather fronts. On the average, however, the air pressure at sea level can support a column of mercury 760 mm in height. This average sea-level air pressure is known as *normal atmospheric pressure*, also called *standard pressure*.

The instrument most commonly used for measuring air pressure is the *mercury barometer*. The diagram below shows how it operates. Atmospheric pressure is exerted on the mercury in the dish, and this in turn holds the column of mercury up in the tube. This column at standard pressure will measure 760 mm above the level of the mercury in the dish below.

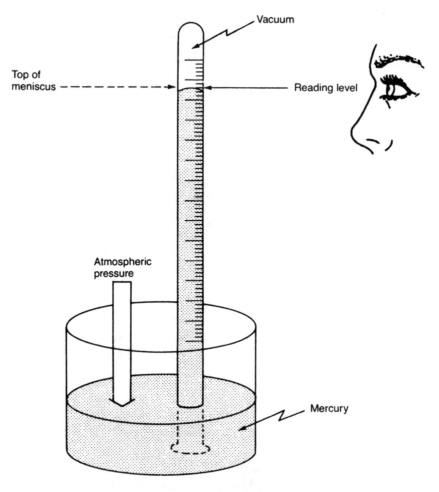

Mercury Barometer

In gas-law problems pressure may be expressed in various units. One standard atmosphere (1 atm) is equal to 760 mm Hg or 760 torr, a unit named for Evangelista Torricelli. In the SI system, the unit of pressure is the pascal (Pa), named in honor of the scientist of the same name. Standard pressure in pascals is 101,325 Pa or 101.325 kPa.

SUMMARY OF UNITS OF PRESSURE

Unit	Abbreviation	Unit Equivalent to 1 atm
Atmosphere	atm	1 atm
Millimeters of Hg	mm Hg	760 mm Hg
Torr	torr	760 torr
Inches of Hg	in. Hg	29.9 in. Hg
Pounds per square inch	lb/in.^{2} (psi)	14.7 lb/in.^{2}
Pascal	Pa	101,325 Pa

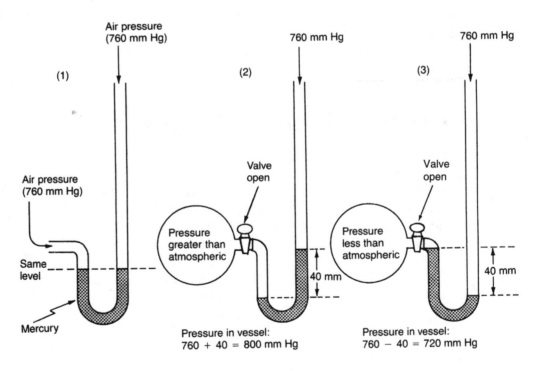

Manometer

A device similar to the barometer can be used to measure the pressure of a gas in a confined container. This apparatus, called a *manometer*, is illustrated above. A manometer is basically a U-tube containing mercury or some other liquid. When both ends are open to the air, as in (1) in the diagram, the level of the liquid will be the same on both sides since the same pressure is being exerted on both ends of the tube. In (2) and (3), a vessel is connected to one end of the U-tube. Now the height of the mercury column serves as a means of reading the pressure inside the vessel if the atmospheric pressure is known. When the pressure inside the vessel is the same as the atmospheric pressure outside, the levels of liquid are the same. When the pressure inside is greater than outside, the column of liquid will be higher on the side that is exposed to the air, as in (2). Conversely, when the pressure inside the vessel is less than the outside atmospheric pressure, the additional pressure will force the liquid to a higher level on the side near the vessel, as in (3).

Kinetic Molecular Theory

By indirect observations, the Kinetic Molecular Theory has been arrived at to explain the forces between molecules and the energy the molecules possess. There are three basic assumptions to the Kinetic Molecular Theory:

1. Matter in all its forms (solid, liquid, and gas) is composed of extremely small particles. In many cases these are called molecules. The space occupied by the gas particles themselves is ignored in comparison with the volume of the space they occupy.
2. The particles of matter are in constant motion. In solids, this motion is restricted to a small space. In liquids, the particles have a more random pattern but still are restricted to a kind of rolling over one another. In a gas, the particles are in continuous, random, straight-line motion.
3. When these particles collide with each other or with the walls of the container, there is no loss of energy.

Some Particular Properties of Gases

As the temperature of a gas is increased, its kinetic energy is increased, thereby increasing the random motion. At a particular temperature not all the particles have the same kinetic energy, but the temperature is a measure of the average kinetic energy of the particles. A graph of the various kinetic energies resembles a normal bell-shaped curve with the average found at the peak of the curve (see Figure 20).

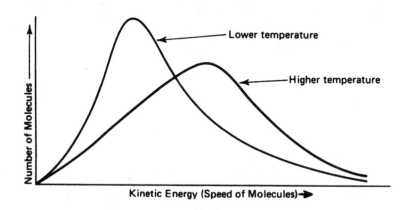

Figure 20.   Molecular Speed Distribution in a Gas at Different Temperatures.

When the temperature is lowered, the gas reaches a point at which the kinetic energy can no longer overcome the attractive forces between the particles (or molecules) and the gas condenses to a liquid. The temperature at which this condensation occurs is related to the type of substance the gas is composed of and the type of bonding in the molecules themselves. This relationship of bond type to condensation point (or boiling point) is pointed out in Chapter 3.

That gases are moving in a random motion so that they may move from one position to another is referred to as *diffusion.* You know that if a bottle of perfume is opened in one corner of a room, the perfume, that is, its molecules, will move or diffuse to all parts of the room in time. The rate of diffusion is the rate of the mixing of gases.

Effusion is the term used to describe the passage of a gas through a tiny orifice into an evacuated chamber. The rate of effusion measures the speed at which the gas is transferred into the chamber.

Gas Laws and Related Problems

Graham's Law

This law relates the rate at which a gas diffuses (or effuses) to the type of molecule in the gas. It can be expressed as follows:

The rate of diffusion of a gas is inversely proportional to the square root of its molecular mass.

Hydrogen, with the lowest molecular mass, can diffuse more rapidly than other gases under similar conditions. A sample problem:

Compare the rate of diffusion of hydrogen to that of oxygen under similar conditions.
The formula is

$$\frac{\text{Rate A}}{\text{Rate B}} = \frac{\sqrt{\text{Molecular mass of B}}}{\sqrt{\text{Molecular mass of A}}}$$

Let A be H_2 and B be O_2.

$$\frac{\text{Rate } H_2}{\text{Rate } O_2} = \frac{\sqrt{32}}{\sqrt{2}} = \frac{\sqrt{16}}{\sqrt{1}} = \frac{4}{1}$$

Therefore hydrogen diffuses four times as fast as oxygen.

In dealing with the gas laws, a student must know what is meant by standard conditions of temperature and pressure (STP). Standard pressure is defined as the height of mercury that can be held in an evacuated tube by 1 atm of pressure (14.7 psi). This is usually expressed as 760 mm Hg or 101.3 Pa. Standard temperature is defined as 273 K or absolute (which corresponds to 0°C).

Charles's Law

Jacques Charles, a French chemist of the early nineteenth century, discovered that when a gas under constant pressure is heated from 0°C to 1°C, it expands 1/273 of its volume. It contracts this amount when the temperature is dropped 1° to –1°C. Charles reasoned that if a gas at 0°C were cooled to –273°C (actually found to be –273.15°C), its volume would be zero. Actually, all gases are converted into liquids before this temperature is reached. By using the Kelvin scale to rid the problem of negative numbers, we can state Charles's Law as follows:

If the pressure remains constant, the volume of a gas varies directly as the absolute temperature.

Then, initial $\dfrac{V_1}{T_1}$ = final $\dfrac{V_2}{T_2}$ at constant pressure.

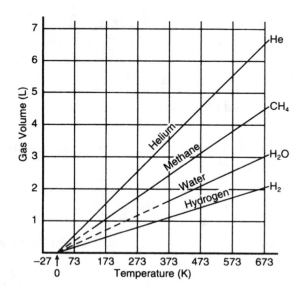

Plots of V versus T for representative gases.

Graphic Relationship—Charles's Law
The dashed lines represent extrapolation of the data into regions where the gas would become liquid or solid. Extrapolation shows that each gas, if it remained gaseous, would reach zero volume at 0 K or –273°C.

Example

The volume of a gas at 20°C is 500 mL. Find its volume at standard temperature if the pressure is held constant.

Solution:
Convert temperatures:

$$20°C = 20° + 273° = 293 \text{ K}$$
$$0°C = 0° + 273° = 273 \text{ K}$$

If you know that cooling a gas decreases its volume, then you know that 500 mL will have to be multiplied by a fraction (made up of the Kelvin temperatures) which has a smaller numerator than the denominator. So

$$500 \text{ mL} \times \frac{273}{293} = 465 \text{ mL}$$

Or you can use the formula

$$\frac{V_1}{T_1} = \frac{V_2}{T_2} \quad \text{so} \quad \frac{500 \text{ mL}}{293} = \frac{X \text{ mL}}{273}$$

$$X \text{ milliliters} = 465 \text{ mL}$$

Boyle's Law

Robert Boyle, a seventeenth-century English scientist, found that the volume of a gas decreases when the pressure on it is increased, and vice versa, when the temperature is held constant. Boyle's Law can be stated as follows:

If the temperature remains constant, the volume of a gas varies inversely as the pressure changes.

Then $P_1V_1 = P_2V_2$ at a constant temperature.

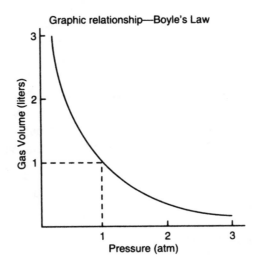

Graphic relationship—Boyle's Law

Example

Given the volume of a gas as 200 mL at 1.05 atm pressure. Calculate the volume of the same gas at 1.01 atm. The temperature is held constant.

Solution:

If you know that this *decrease* in pressure will cause an increase in the volume, then you know 200 mL must be multiplied by a fraction (made up of the two pressures) that has a larger numerator than the denominator. So

$$200 \text{ mL} \frac{1.05 \text{ atm}}{1.01 \text{ atm}} = 208 \text{ mL}$$

Or you can use the formula

$$P_1V_1 = P_2V_2$$

$$\text{So, } V_2 = V_1 \times \frac{P_1}{P_2}$$

$$V_2 = 200 \text{ mL} \times \frac{1.05 \text{ atm}}{1.01 \text{ atm}} = 208 \text{ mL}$$

Combined Gas Law

This is a combination of the two preceding gas laws and can be expressed as follows:

$$\frac{P_1V_1}{T_1} = \frac{P_2V_2}{T_2}$$

Example

The volume of a gas at 780 mm pressure and 30°C is 500 mL. What volume would the gas occupy at STP?

Solution:

Again you can use reasoning to determine the kind of fractions the temperatures and pressures must be to arrive at your answer. Since the pressure is going from 780 mm to 760 mm, the volume should increase. The fraction must then be $^{780}/_{760}$. Also, since the temperature is going from 30°C (303 K) to 0°C (273 K), the volume should decrease; this fraction must be $^{273}/_{303}$. So

$$500 \text{ mL} \times \frac{780}{760} \times \frac{273}{303} = 462 \text{ mL}$$

Or you can use the formula

$$\frac{P_1V_1}{T_1} = \frac{P_2V_2}{T_2}$$

$$\text{Solve for } V_2 = V_1 \times \frac{P_1}{P_2} \times \frac{T_2}{T_1}$$

$$V_2 = 500 \text{ mL} \times \frac{780 \text{ mm Hg}}{760 \text{ mm Hg}} \times \frac{273 \text{ K}}{303 \text{ K}} = 462 \text{ mL}$$

Pressure versus Temperature

At constant volume, the pressure of a given mass of gas varies directly with the absolute temperature.

Then $\dfrac{P_1}{T_1} \times \dfrac{P_2}{T_2}$ at constant volume.

Example

A steel tank contains a gas at 27°C and a pressure of 12 atm. Determine the gas pressure when the tank is heated to 100°C.

Solution:
Reasoning that an increase in temperature will cause an increase in pressure at constant volume, you know the pressure must be multiplied by a fraction that has a larger numerator than denominator. The fraction must be $^{373\,K}/_{300\,K}$. So

$$12\ \text{atm} \times \frac{373\,K}{300\,K} = 14.9\ \text{atm}$$

Or you can use the formula

$$\frac{P_1}{T_1} = \frac{P_2}{T_2}$$

$$P_2 = P_1 \times \frac{T_2}{T_1}$$

$$\text{So } P_2 = 12\ \text{atm} \times \frac{373\,K}{300\,K} = 14.9\ \text{atm}$$

Dalton's Law of Partial Pressures

When a gas is made up of a mixture of different gases, the total pressure of the mixture is equal to the sum of the partial pressures of the components; that is, the partial pressure of the gas is the pressure of the individual gas if it alone occupied the volume.

$$P_{\text{total}} = P_{\text{gas 1}} + P_{\text{gas 2}} + P_{\text{gas 3}} + \cdots$$

Example

A mixture of gases at 760 mm Hg pressure contains 65% nitrogen, 15% oxygen, and 20% carbon dioxide by volume. What is the partial pressure of each gas?

Solution:

$$0.65 \times 760 = 494\ \text{mm pressure } (N_2)$$
$$0.15 \times 760 = 114\ \text{mm pressure } (O_2)$$
$$0.20 \times 760 = 152\ \text{mm pressure } (CO_2)$$

If the pressure was given as 1 atmosphere (atm), you would substitute 1 atm for 760 mm Hg. The answers would be:

$$0.65 \times 1 \text{ atm} = 0.65 \text{ atm } (N_2)$$
$$0.15 \times 1 \text{ atm} = 0.15 \text{ atm } (O_2)$$
$$0.20 \times 1 \text{ atm} = 0.20 \text{ atm } (CO_2)$$

Corrections of Pressure

Correction of Pressure When a Gas Is Collected over Water

When a gas is collected over a volatile liquid, such as water, some of the water vapor is present in the gas and contributes to the total pressure. Assuming that the gas is saturated with water vapor for that temperature, you can find the partial pressure due to the water vapor in a table of such water vapor values. (See Table Ⓜ in the Tables for Reference section at the back of the book.) This vapor pressure, which depends on only the temperature, must be subtracted from the total pressure to find the partial pressure of the gas being measured.

Correction of Difference in the Height of the Fluid

When gases are collected in eudiometers (glass tube closed at one end), it is not always possible to get the level of the liquid inside the tube to equal the level on the outside. This deviation of levels must be taken into account when determining the pressure of the enclosed gas. There are then two possibilities: (1) When the level inside is higher than the level outside the tube, the pressure on the inside is less, by the height of fluid in excess, than the outside pressure. If the fluid is mercury, you simply subtract the difference from the outside pressure reading (also in height of mercury and in the same units) to get the corrected pressure of the gas. If the fluid is water, you must first convert the difference to an equivalent height of mercury by dividing the difference by 13.6 (since mercury is 13.6 times as heavy as water, the height expressed in Hg will be 1/13.6 the height of water). Again, care must be taken that this equivalent height of Hg is in the same units as the expression for the outside pressure before it is subtracted. This gives the corrected pressure for the gas in the eudiometer. (2) When the level inside is lower than the level outside the tube, a correction must be added to the outside pressure. If the difference in height between the inside and the outside is expressed in height of water, you must take 1/13.6 of this quantity to correct it to millimeters of mercury units. This relationship is shown in Figure 21. This quantity is then added to the expression of the outside pressure, which must also be in milliliters of mercury. If the tube contains mercury, then the difference between the inside and outside levels is merely added to the outside pressure to get the corrected pressure for the enclosed gas.

Example

Hydrogen gas is collected in a eudiometer tube over water. It is impossible to level the outside water with that in the tube, and so the water level inside the tube is 40.8 mm higher than that outside. The barometric pressure is 730 mm Hg. The water vapor pressure at the room temperature of 29°C is found in a handbook to be 30.0 mm. What is the pressure of the dry hydrogen?

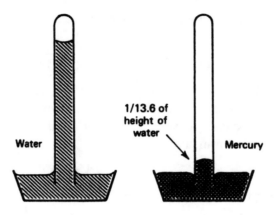

Figure 21.   Same Pressure Exerted on Both Liquids.

Solution:

Step 1:   To find the true pressure of the gas, we must first subtract the water level difference expressed in mm Hg: $^{40.8}/_{13.6}$ = 3 mm Hg, and so 730 mm − 3 mm = 727 mm total pressure of gases in the eudiometer.

Step 2:   Correcting for the partial pressure due to water vapor in the hydrogen, we subtract the vapor pressure (30.0 mm) from 727 mm and get our answer: 697 mm.

Ideal Gas Law

The laws discussed previously do not include the relationship of number of moles of a gas to the pressure, volume, and temperature of the gas. A law derived from the kinetic-molecular theory relates these variables. It is called the Ideal Gas Law and is expressed as

$$PV = nRT$$

P, V, and T retain their usual meanings, n stands for the number of moles of gas, and R represents the ideal gas constant.

Boyle's Law and Charles's Law are actually derived from the Ideal Gas Law. Boyle's Law applies when the number of moles and the temperature of the gas are constant. Then in $PV = nRT$, the number of moles, n, is constant; the gas constant, R, remains the same; and by definition T is constant. Therefore, $PV = k$. At the initial set of conditions of a problem P_1V_1 = a constant (k). At the second set of conditions, the terms on the right side of the equation are equal to the same constant, and so $P_1V_1 = P_2V_2$. This matches the Boyle's Law equation introduced earlier.

The same can also be done with Charles's Law, since $PV = nRT$ can be expressed with the variables on the left and the constants on the right.

$$\frac{V}{T} = \frac{nR}{P}$$

In Charles's Law the number of moles and the pressure are constant. Substituting k for the constant term, $^{nR}/_{P}$, we have

$$\frac{V}{T} = k$$

The expression relating two sets of conditions can be written as

$$\frac{V_1}{T_1} = \frac{V_2}{T_2}$$

To use the Ideal Gas Law in the form $PV = nRT$, the gas constant, R, must be determined. This can be done mathematically as shown in the following example.

One mole of oxygen gas is collected in the laboratory at a temperature of 24.0°C and a pressure of exactly 1 atm. The volume is 24.35 L. Find the value of R.

$$PV = nRT$$

Rearranging the equation to solve for R gives

$$R = \frac{PV}{nT}$$

Substituting the known values on the right, we have

$$R = \frac{1 \text{ atm} \times 24.35 \text{ L}}{1 \text{ mol} \times 297 \text{ K}}$$

Calculating R, we get

$$R = 0.0820 \frac{\text{L atm}}{\text{mol K}}$$

Once R is known, the Ideal Gas Law can be used to find any of the variables given the other three.

For example, calculate the pressure, at 16.0°C, of 1.00 g of hydrogen gas occupying 2.54 L.

Rearranging the equation to solve for P, we get

$$P = \frac{nRT}{V}$$

The molar mass of hydrogen is 2.00 g/mol, and so the number of moles in this problem is

$$\frac{1.00 \text{ g}}{2.00 \text{ g/mol}} = 0.500 \text{ mol}$$

Substituting the known values, we have

$$P = \frac{(0.500 \text{ mol})\left(0.0820 \frac{\text{L atm}}{\text{mol K}}\right) 289 \text{ K}}{2.54 \text{ L}}$$

Calculating the value, we get

$$P = 4.66 \text{ atm}$$

Another use of the Ideal Gas Law is to find the number of moles of a gas when P, T, and V are known.

For example, how many moles of nitrogen gas are there in 0.38 L of gas at 0°C and 380 mm Hg pressure?

Rearranging the equation to solve for n gives

$$n = \frac{PV}{RT}$$

Changing temperature to kelvins and pressure to atmospheres gives

$$T = 0° + 273 = 273 \text{ K}$$

$$P = \frac{380 \text{ mm Hg}}{760 \text{ mm Hg/atm}} = 0.5 \text{ atm}$$

Substituting in the equation, we have

$$n = \frac{(0.5 \text{ atm})(0.38 L)}{\left(0.220 \frac{L \text{ atm}}{\text{mol } K}\right)(273 K)} = 0.0085 \text{ mol}$$

$$n = 0.0085 \text{ mol of nitrogen gas}$$

Ideal Gas Deviations

In the use of the gas laws, we have assumed that the gases involved were "ideal" gases. This means that the molecules of the gas were not taking up space in the gas volume and that no intermolecular forces of attraction were serving to pull the molecules closer together. You will find that a gas behaves like an ideal gas at low pressures and high temperatures which move the molecules as far as possible from conditions that would cause condensation. In general, pressures below a few atmospheres cause most gases to exhibit sufficiently ideal properties for application of the gas laws with a reliability of a few percent or better. If, however, high pressures are used, the molecules will be forced into closer proximity to each other as the volume decreases until the attractive force between molecules becomes a factor. This factor decreases the volume and, therefore, the PV values at high-pressure conditions will be less than that predicted by the Ideal Gas Law where the product of PV remains a constant.

Examining what occurs at very low temperatures creates a similar situation. Again, the molecules, because they have slowed down at low temperatures, come into closer proximity with each other and begin to feel the attractive forces between them. This tends to make the gas volume smaller, and therefore causes the product PV to be lower than that expected in the ideal gas situation. So, under conditions of very high pressures and low temperatures, there will be deviations from the expected results of the Ideal Gas Law.

Chapter 5 Review

1. The most abundant element in the earth's crust is
   (A) sodium
   (B) oxygen
   (C) silicon
   (D) aluminum

2. A compound that can be decomposed to produce oxygen gas in the laboratory is
   (A) MnO_2
   (B) NaOH
   (C) CO_2
   (D) $KClO_3$

3. In the usual laboratory preparation equation for the reaction in question 2, what is the coefficient of the O_2?
   (A) 1
   (B) 2
   (C) 3
   (D) 4

4. In a graphic representation of the energy contents of reactants and resulting products, which would have a higher energy content in an exothermic reaction?
   (A) the reactants
   (B) the products
   (C) both the same
   (D) neither one

5. The process of separating components of a mixture by making use of the difference in their boiling points is called
   (A) destructive distillation
   (B) displacement
   (C) fractional distillation
   (D) filtration

6. When oxygen combines with an element to form a compound, the resulting compound is called a(n)
   (A) salt
   (B) oxide
   (C) oxidation
   (D) oxalate

7. According to the activity chart of metals, which metal would react most vigorously in a dilute acid solution?
   (A) zinc
   (B) iron
   (C) aluminum
   (D) magnesium

8. Graham's Law refers to
   (A) boiling points of gases
   (B) gaseous diffusion
   (C) gas compression problems
   (D) volume changes of gases when the temperature changes

9. When 200 mL of a gas at constant pressure is heated, its volume
   (A) increases
   (B) decreases
   (C) remains unchanged

10. When 200 mL of a gas at constant pressure is heated from 0°C to 100°C, the volume must be multiplied by
    (A) 0/100
    (B) 100/0
    (C) 273/373
    (D) 373/273

11. If you wish to find the corrected volume of a gas which was at 20°C and 1 atm pressure and conditions are changed to 0°C and 0.92 atm pressure, by what fractions should you multiply the original volume?

    (A) $\dfrac{293}{273} \times \dfrac{1}{0.92}$

    (B) $\dfrac{273}{293} \times \dfrac{0.92}{1}$

    (C) $\dfrac{273}{293} \times \dfrac{1}{0.92}$

    (D) $\dfrac{293}{273} \times \dfrac{0.92}{1}$

12. When the level of mercury inside a gas tube is higher than the level in the reservoir, you can find the correct pressure inside the tube by taking the outside pressure reading and _?_ the difference in the height of mercury.
    (A) subtracting
    (B) adding
    (C) dividing by 13.6
    (D) doing both (C) and (A)

13. If water is the liquid in question 12 instead of mercury, you can change the height difference to an equivalent mercury expression by
    (A) dividing by 13.6
    (B) multiplying by 13.6
    (C) adding 13.6
    (D) subtracting 13.6

14. Standard conditions are
    (A) 0°C and 14.7 mm
    (B) 32°C and 76 cm
    (C) 273°C and 760 mm
    (D) 4°C and 7.6 m

15. When a gas is collected over water, the pressure is corrected by
     (A) adding the vapor pressure of water
     (B) multiplying by the vapor pressure of water
     (C) subtracting the vapor pressure of water at that temperature
     (D) subtracting the temperature of the water from the vapor pressure

16. At 5 atm pressure and 70°C, how many moles are present in 1.5 L of O_2 gas?
     (A) 0.036
     (B) 0.267
     (C) 0.536
     (D) 1.60

Answers

1. **B**	5. **C**	8. **B**	11. **C**	14. **B**
2. **D**	6. **B**	9. **A**	12. **A**	15. **C**
3. **C**	7. **D**	10. **D**	13. **A**	16. **B**
4. **A**				

Terms You Should Know

atmospheric pressure
manometer
mercury barometer
oxidation
ozone

pascal
standard pressure
standard temperature
torr

Boyle's Law
Charles's Law
Combined Gas Law
Dalton's Law of Partial Pressures
Graham's Law
Ideal Gas Law
Kinetic Molecular Theory

Chapter 6

CHEMICAL CALCULATIONS (STOICHIOMETRY) AND THE MOLE CONCEPT

Solving Problems

This chapter deals with the solving of a variety of chemistry problems, which is often referred to as *stoichiometry*. Solving problems should be done in an organized manner, and it would be to your advantage to go back to the Introduction to this book and review the section called, "Problem Solving: A Thinking Skill." It describes a well-planned method for attacking the process of solving problems that you will find helpful in this chapter.

Two methods of solving the problems in this chapter are often given. It would be wise on your part to learn both the proportion method and the factor-label method so that you have more tools available for solving problems. Remember that no matter which method you use, you should still use the estimation process to verify the plausibility of the answer.

The Mole Concept

Providing a name for a quantity of things taken as a whole is common in everyday life. Some examples are a dozen, a gross, and a ream. Each of these represents a specific number of items and is not dependent on the commodity. A dozen eggs, oranges, or bananas will always represent 12 items.

In chemistry we have a unit that describes a quantity of particles. It is called the *mole* (abbreviated *mol*). A mole is 6.02×10^{23} particles. The particles can be atoms, molecules, ions, electrons, and so forth. Because particles are so small in chemistry, the mole is a very convenient unit. The number 6.02×10^{23} is often referred to as *Avogadro's number* in honor of the Italian scientist whose hypothesis led to its determination.

In 1811 Amedeo Avogadro made a far-reaching scientific assumption (hypothesis) that also bears his name. He stated that equal volumes of different gases contain equal numbers of molecules at the same pressure and temperature. The statement was called *Avogadro's hypothesis*. It means that, under the same conditions, the number of molecules of hydrogen in a 1-L container is exactly the same as the number of molecules of carbon dioxide or any other gas in a 1-L container even though the individual molecules of the different gases have different masses and size. Because of the substantiation of this hypothesis by much data since its inception, it's often referred to as *Avogadro's Law*.

Molar Mass and Moles

When the formula mass of an ionic compound is determined by the addition of its component atomic masses and expressed in grams, it is called the *gram-formula mass* or *molar mass*. An example is $CaCO_3$.

Example 1

$$
\begin{aligned}
1\ Ca &= 40 \\
1\ C &= 12 \\
3\ O &= 48 = 3 \times 16 \\
\hline
CaCO_3 &= 100 \text{ formula mass} \\
100\ g &= \text{gram-formula weight or molar mass}
\end{aligned}
$$

Solution:
The molar mass is equivalent to the mass of 1 *mol* of that material expressed in grams.

The term *molar mass* or *gram-molecular weight* (gmw) can also be used when it is known that the material is a molecular substance and not an ionic lattice like NaCl or NaOH.

Example 2

Find the mass of 1 mol of $KAL(SO_4)_2 \cdot 12\ H_2O$.

Solution:

$$
\begin{aligned}
1\ K &= 39 \\
1\ Al &= 27 \\
2(SO_4) &= 192 = 2(32 + 16 \times 4) \\
12\ H_2O &= 216 = 12(2 + 16) \\
\hline
1\ mol &= 474\ g
\end{aligned}
$$

Review Problems*

(Molar Mass and Moles)

Find the molecular mass or formula mass of the following:

1. HNO_3

2. $C_6H_{10}O_5$

3. H_2SO_4

4. KCl

5. $C_{12}H_{22}O_{11}$

   In some cases, there may be a question of the mass of 1 mol of an element if you are not told specifically what is referred to—the single atoms or a molecular state. An example of this situation may arise with hydrogen and other elements that form diatomic molecules. If you are asked the mass of 1 mole of hydrogen, you could say it is 1 g if you are dealing with single atoms of hydrogen. This 1 mole of hydrogen atoms contains 6.02×10^{23} atoms. In a similar fashion, 1 mole of hydrogen molecules (H_2) has a molar mass of 2 g. This 1 mole of hydrogen molecules contains 6.02×10^{23} molecules and, since each molecule is composed of two atoms, 12.04×10^{23} atoms.
   The other elements that exist as diatomic molecules are oxygen, nitrogen, fluorine, chlorine, bromine, iodine, and astatine.

Example

Find the mass of
(a) 1 mol of oxygen
(b) one molecule of oxygen
(c) 1 mol of oxygen atoms
(d) one atom of oxygen

Solution:
(a) Since the atomic mass is 16 g and the molecule is diatomic, 2×16 g = 32 g.
(b) There are 6.02×10^{23} molecules in 1 mole of this gas. 32 g $\div$ 6.02×10^{23} molecules = 5.32×10^{-23} g.
(c) Without the diatomic molecule structure, the atomic mass of oxygen expressed in grams is the answer. That is 16 g.
(d) The atomic molar mass, 16 g, is divided by 6.02×10^{23}. This gives 2.66×10^{-23} g.

   If you gave the same answer to the first two questions, you are confused about the mole concept. To say that a molecule of oxygen gas has the same mass as a mole of oxygen gas is the same as saying that an apple has the same mass as a dozen apples.

*After each type of problem in this chapter there are several problems to try to solve; therefore a list of review questions is not included at end of the chapter. Answers to Review problems are found at the end of the chapter. Page 127.

Mole Relationships

Since the mole is used often in chemistry to quantify the number of atoms, molecules, and several other items, it is important to know the relationships that exist and how to move from one to another. The following summarizes a number of these relationships:

When dealing with elements—

Moles of an element × molar mass (atomic) = mass of the element
Mass of an element / molar mass (atomic) = moles of the element

When dealing with compounds—

Moles of a compound × molar mass (molecular) = mass of the compound
Mass of a compound / molar mass (molecular) = moles of the compound

When dealing with the molecules of a compound—

Moles of molecules × 6.02×10^{23} = number of molecules
Number of molecules / 6.02×10^{23} = moles of molecules

When dealing with the atoms of elements—

Moles of atoms × 6.02×10^{23} = number of atoms
Number of atoms / 6.02×10^{23} = moles of atoms

Gas Volumes and Molar Mass

Because the volume of a gas may vary depending on the conditions of temperature and pressure, a standard is set for comparing gases. The standard conditions of temperature and pressure (abbreviated STP) are 0°C and 760 mm of mercury pressure.

The molecular mass of a gas expressed in grams and under standard conditions occupies 22.4 L. This is an important relationship to remember! The 22.4 L is referred to as the gram-molecular volume (gmv) or molar volume. Two scientists are associated with this relationship.

Gay-Lussac's Law states that when only gases are involved in a reaction, the volumes of the reacting gases and the volumes of the gaseous products are in a ratio to each other as small whole numbers. This law may be illustrated by the following cases:

Example 1

$$H_2(g) + Cl_2(g) \rightarrow 2HCl(g)$$

This balanced equation shows that

1 vol hydrogen + 1 vol chlorine = 2 vol hydrogen chloride

Example 2

$$2H_2(g) + O_2(g) \rightarrow 2H_2O(g)$$

This balanced equation shows that

2 vol hydrogen + 1 vol oxygen = 2 vol steam

Avogadro's Law, which explains Gay-Lussac's, states that equal volumes of gases under the same conditions of temperature and pressure contain equal numbers of molecules. This means that 1 mol of any gas at STP occupies 22.4 L, and so:

32 g O_2 at STP occupies 22.4 L.
2 g H_2 at STP occupies 22.4 L.
44 g CO_2 at STP occupies 22.4 L.
$2O_2$ (2 mol O_2) = 64 g = 44.8 L at STP.
$3H_2$ (3 mol H_2) = 6 g = 67.2 L at STP.

Density and Molar Mass

Since the density of a gas is usually given in g/L of gas at STP, we can use the molar volume to molar mass relationship to solve the following types of problems.

Solution:

Step 1: Write the balanced equation for the reaction.

$$2NaClO_3 \xrightarrow{\Delta} 2NaCl + 3O_2(g)$$

Step 2: Write the given quantity and the unknown quantity above the appropriate substances.

42.6 g x liters
$2NaClO_3 \xrightarrow{\Delta} 2NaCl + 3O_2$

Step 3: Calculate the equation mass or volume under the substances that have something indicated above them. Note that the units above and below *must* match.

42.6 g x liters
$2NaClO_3 \xrightarrow{\Delta} 2NaCl + 3O_2$
213 g   67.2 L
(2 × molar mass   (3 × 22.4 L)
   of $NaClO_3$)

Step 4: Form the proportion.

$$\frac{42.6 \text{ g}}{213 \text{ g}} = \frac{x \text{ liters}}{67.2 \text{ L}}$$

Step 5: Solve for x.

$$x = 13.4 \text{ L } O_2$$

This problem can also be solved using methods other than the *proportion method* shown above.

Another method to proceed from step 3 is called the *factor-label method*. The reasoning is this: Since the equation shows that 213 g of reactant produces 67.2 L of the required product, multiplying the given amount by this equality (so that the units of the answer will be correct) will give the same answer as above. So step 4 is

Step 4: $42.6 \text{ g NaClO}_3 \times \dfrac{67.2 \text{ L O}_2}{213 \text{ g NaClO}_3} = 13.4 \text{ L O}_2$

Still another method of solving this problem is called the *mole method*. Steps 1 and 2 are the same. Then step 3 is

Step 3:   Determine how many moles of substance are given.

$$42.6 \text{ g} \div \dfrac{106.5 \text{ g}}{1 \text{ mol NaClO}_3} = 0.4 \text{ mol NaClO}_3$$

The equation shows that 2 mol $NaClO_3$ makes 3 mol O_2. So 0.4 mol $NaClO_3$ yields

$$0.4 \text{ mol NaClO}_3 \times \dfrac{3 \text{ mol O}_2}{2 \text{ mol NaClO}_3} = 0.6 \text{ mol O}_2$$

Step 4:   Convert the moles of O_2 to liters.

$$0.6 \text{ mol O}_2 \times \dfrac{22.4 \text{ L O}_2}{1 \text{ mol O}_2} = 13.4 \text{ L O}_2$$

Example 2

Find the weight of $CaCO_3$ needed to produce 11.2 L of CO_2 when the calcium carbonate is reacted with hydrochloric acid.

Solution:

Step 1:   Write the balanced equation for the reaction.

$$CaCO_3 + 2HCl \rightarrow CaCl_2 + H_2O + CO_2$$

Step 2:   Write the given quantity and the unknown quantity above the appropriate substances.

x grams   11.2 L
$$CaCO_3 + 2HCl \rightarrow CaCl_2 + H_2O + CO_2$$

Step 3:   Calculate the equation mass or volume under substances that have something indicated above them.

x grams   11.2 L
$$CaCO_3 + 2HCl \rightarrow CaCl_2 + H_2O + CO_2$$
100 g   22.4 L

Proportion method

Form the proportion.

Step 4: $\dfrac{x \text{ grams}}{100 \text{ g}} = \dfrac{11.2 \text{ L}}{22.4 \text{ L}}$

Step 5:   Solve for x.

$x = 50 \text{ g CaCO}_3$

Factor-label method

Step 4: $11.2 \text{ L CO}_2 \times \dfrac{100 \text{ g CaCO}_3}{22.4 \text{ L CO}_2} = 50 \text{ g CaCO}_3$

Mole method

Step 4: $11.2 \text{ L CO}_2 \div \dfrac{22.4 \text{ L}}{1 \text{ mol}} = 0.5 \text{ mol CO}_2$

$0.5 \text{ mol CO}_2 \times \dfrac{100 \text{ g CaCO}_3}{1 \text{ mol CO}_2} = 50 \text{ g CaCO}_3$

Example 1

Find the molar mass of a gas when the density is given as 1.25 g/L.

Solution:

Since it is known that 1 mol of a gas occupies 22.4 L at STP, we can solve this problem by multiplying the mass of 1 L by 22.4 L/mol.

$$\dfrac{1.25 \text{ g}}{L} \times \dfrac{22.4 L}{1 \text{ mol}} = 28 \text{ g/mol}$$

Even if the weight given is not for 1 L, the same setup can be used.

Example 2

If 3 L of a gas weighs 2 g, find the molar mass.

Solution:

$$\dfrac{2 \text{ g}}{3 L} \times \dfrac{22.4 L}{1 \text{ mol}} = 14.9 \text{ g/mol}$$

You can also find the density of a gas if you know the molar mass. Since the molar mass occupies 22.4 L at STP, dividing the molar mass by 22.4 L will give you the mass per liter, or the density.

Example 3

Find the density of oxygen at STP.

Solution:
Oxygen is diatomic in its molecular structure.
  O_2 = molar mass of 2×16 or 32 g/mol
  32 g/mol occupies 22.4 L
Therefore

$$\frac{32 \text{ g}}{1 \text{ mol}} \div \frac{22.4 \text{ L}}{1 \text{ mol}} = \frac{32 \text{ g}}{1 \text{ mol}} \times \frac{1 \text{ mol}}{22.4 \text{ L}} = 1.43 \text{ g/L}$$

You can find the density of a gas, then, by dividing its molar mass by 22.4 L.

Mass-Volume Relationships

A typical *mass-volume problem*:

Example 1

How many liters of oxygen (STP) can you prepare from the decomposition of 42.6 g of sodium chlorate?
  The equation shows 1 mol $CaCO_3$ yields 1 mol CO_2, and so

$$0.5 \text{ mol } CO_2 \times \frac{1 \text{ mol } CaCO_3}{1 \text{ mol } CO_2} = 0.5 \text{ mol } CaCO_3$$

Coverting moles to grams:

$$0.5 \text{ mol } CaCO_3 \times \frac{100 \text{ g } CaCO_3}{1 \text{ mol } CaCO_3} = 50 \text{ g } CaCO_3$$

Review Problems*
(Mass-Volume)

6. What mass of water must be electrolyzed to obtain 20 L of oxygen?   Ans. 32.2 g

7. How many grams of aluminum will be completely oxidized by 44.8 L of oxygen?
                                                                Ans. 72 g Al

*Answer explanations are found at the end of the chapter. Page 127.

Mass-Mass Problems

A typical problem concerning just mass relationships is as follows:

Example 1

What mass of oxygen can be obtained from heating 100 g of potassium chlorate?

Step 1:   Write the balanced equation for the reaction.

$$2KClO_3 \rightarrow 2KCl + 3O_2$$

Step 2:   Write the given quantity and the unknown quantity above the appropriate substances.

100 g x grams
$2KClO_3 \rightarrow 2KCl +$ $3O_2$

Step 3:   Calculate the equation mass under the substances that have something indicated above them. Note that the units above and below *must* match.

100 g x grams
$2KClO_3 \rightarrow 2KCl +$ $3O_2$
245 g   96 g

Using the proportion method
Step 4:   Form the proportion.

$$\frac{100 \text{ g}}{245 \text{ g}} = \frac{x \text{ grams}}{96 \text{ g}}$$

Step 5:   Solve for x.

$$x = 39.3 \text{ g of } O_2$$

Using the factor-label method
From step 3 on you would proceed as follows:

Step 4:   The equation indicates that 245 g $KClO_3$ yields 96 g of O_2. Therefore multiplying the given quantity by a factor made up of these two quantities arranged appropriately so that the units of the answer remain uncancelled will give the answer to the problem.

$$100 \text{ g } \cancel{KClO_3} \times \frac{96 \text{ g } O_2}{245 \text{ g } \cancel{KClO_3}} = 39.3 \text{ g } O_2$$

Using the mole method
To solve this problem you would proceed as follows from step 2:

Step 3:   Determine how many moles of substance are given.

$$100 \text{ g } \cancel{KClO_3} \times \frac{1 \text{ mol } KClO_3}{122.5 \text{ g } \cancel{KClO_3}} = 0.815 \text{ mol } KClO_3$$

The equation shows that 2 mol of $KClO_3$ yields 3 mol of O_2. So 0.815 mol $KClO_3$ will yield

$$0.815 \cancel{\text{ mol } KClO_3} \times \frac{3 \text{ mol } O_2}{2 \cancel{\text{ mol } KClO_3}} = 1.22 \text{ mol } O_2$$

Step 4:   Convert the moles of O_2 to grams of O_2.

$$1.22 \cancel{\text{ mol } O_2} \times \frac{32 \text{ g } O_2}{1 \cancel{\text{ mol } O_2}} = 39 \text{ g } O_2$$

Example 2

What mass of potassium hydroxide is needed to neutralize 20 g of sulfuric acid?

Solution:
Step 1: $2KOH + H_2SO_4 \rightarrow K_2SO_4 + 2H_2O$

Step 2: $\quad\; x$ grams $\;\;$ 20 g
$\qquad\;\; 2KOH + H_2SO_4 \rightarrow K_2SO_4 + 2H_2O$

Step 3: $\quad\; x$ grams $\;\;$ 20 g
$\qquad\;\; 2KOH + H_2SO_4 \rightarrow K_2SO_4 + 2H_2O$
$\qquad\;\;\; 112 \text{ g} \qquad 98 \text{ g}$

Using the proportion method

Step 4: $\dfrac{x \text{ grams}}{112 \text{ g}} = \dfrac{20 \text{ g}}{98 \text{ g}}$

Step 5: $x = 22.8 \text{ g KOH}$

Using the factor-label method

Step 4: $20 \text{ g } \cancel{H_2SO_4} \times \dfrac{112 \text{ g KOH}}{98 \text{ g } \cancel{H_2SO_4}} = 22.8 \text{ g KOH}$

Using the mole method

Step 3: $20 \text{ g } \cancel{H_2SO_4} \times \dfrac{1 \text{ mol } H_2SO_4}{98 \text{ g } \cancel{H_2SO_4}} = 0.204 \text{ mol } H_2SO_4$

$$0.204 \text{ mol H}_2\text{SO}_4 \times \frac{2 \text{ mol KOH}}{1 \text{ mol H}_2\text{SO}_4} = 0.408 \text{ mol KOH}$$

Step 4: $0.408 \text{ mol KOH} \times \dfrac{56 \text{ g KOH}}{1 \text{ mol KOH}} = 22.8 \text{ g KOH}$

Review Problems*
(Mass-Mass)

8. What mass of manganese dioxide is needed to react with an excess of hydrochloric acid so that 100 g of chlorine is liberated?   Ans: 122.5 g

9. A 20-g sample of Mg is burned in 20 g of O_2. How much MgO is formed? (Hint: Determine which is in excess.)   Ans: 33.3 g MgO

Volume-Volume Problems

This type of problem involves only volume units and therefore can make use of Gay-Lussac's Law: When gases combine, they combine in simple whole number ratios. These simple numbers are the coefficients of the equation.

A typical problem concerning just volume relationships is as follows:

Example 1

What volume of NH_3 is produced when 22.4 L of nitrogen is combined with a sufficient quantity of hydrogen under the appropriate conditions?

Solution:
Step 1:   Write the balanced equation for the reaction.

   $N_2 + 3H_2 \rightarrow 2NH_3$

Step 2:   Write the given quantity and the unknown quantity above the respective substances.

   22.4 L x liters
   $N_2 + 3H_2 \rightarrow 2NH_3$

Step 3:   Set up a proportion using the coefficients of the substances that have something indicated above them as denominators.

   $$\frac{22.4 \text{ L}}{1 \text{ L}} = \frac{x \text{ liters}}{2 \text{ L}}$$

---

*Answer explanations are found at the end of the chapter. Page 127.

Step 4:   Solve for x.

$x = 44.8$ L

Using the factor-label method

From Step 2 on you would proceed as follows:

Step 3:   The equation shows that 1 vol of N_2 will yield 2 vol of NH_3. Therefore multiplying the given quantity by a factor made up of these two quantities appropriately arranged so that the units of the answer remain uncancelled will solve the problem.

$$22.4 \text{ L } N_2 \times \frac{2 \text{ L } NH_3}{1 \text{ L } N_2} = 44.8 \text{ L } NH_3$$

Using the mole method

To solve this problem you would proceed as follows:

Step 3:   The given quantity is converted to moles.

$$22.4 \text{ L } N_2 \times \frac{1 \text{ mol } N_2}{22.4 \text{ L } N_2} = 1 \text{ mol } N_2$$

The equation shows that 1 mol of N_2 will yield 2 mol of NH_3. Therefore using this relationship will yield

$$1 \text{ mol } N_2 \times \frac{2 \text{ mol } NH_3}{1 \text{ mol } N_2} = 2 \text{ mol } NH_3$$

Convert the moles of NH_3 to liters of NH_3.

$$2 \text{ mol } NH_3 \times \frac{22.4 \text{ L } NH_3}{1 \text{ mol } NH_3} = 44.8 \text{ L } NH_3$$

Example 2

What volume of SO_2 will result from the complete burning of pure sulfur in 8 L of oxygen until all the oxygen is used?

Solution:
Step 1: $S + O_2 \rightarrow SO_2$

Step 2:   8 L x liters
$S + O_2 \rightarrow SO_2$

Using the proportion method

Step 3: $\dfrac{8 \text{ L}}{1 \text{ L}} = \dfrac{x \text{ liters}}{1 \text{ L}}$

Step 4: $x = 8$ L of SO_2

Using the factor-label method

Step 3: $8 \text{ L } O_2 \times \dfrac{1 \text{ L } SO_2}{1 \text{ L } O_2} = 8 \text{ L } SO_2$

Using the mole method (not really practical in this case)

Step 3: $8 \text{ L } O_2 \times \dfrac{1 \text{ mol } O_2}{22.4 \text{ L } O_2} = 0.358 \text{ mol } O_2$

$0.358 \text{ mol } O_2 \times \dfrac{1 \text{ mol } SO_2}{1 \text{ mol } O_2} = 0.358 \text{ mol } SO_2$

Step 4: $0.358 \text{ mol } SO_2 \times \dfrac{22.4 \text{ L } SO_2}{1 \text{ mol } SO_2} = 8 \text{ L } SO_2$

Review Problems*
(Volume-Volume)

10. How many liters of hydrogen are necessary to react with sufficient chlorine to produce 12 L of hydrogen chloride?   Ans.: 6 L

11. How many liters of oxygen will be needed to burn 100 L of carbon monoxide?
                                                                            Ans.: 50 L

Problems with an Excess of One Reactant

It will not always be true that the amounts given in a particular problem are exactly in the proportion required for the reaction to use up all of the reactants. In other words, at times some of one reactant will be left over after the other has been used up. This is similar to the situation in which two eggs are required to mix with one cup of flour in a particular recipe, and you have four eggs and four cups of flour. Since two eggs require only one cup of flour, four eggs can use only two cups of flour and two cups of flour will be left over.

A chemical equation is very much like a recipe. Consider the following equation:

$$Zn + 2HCl \rightarrow ZnCl_2 + H_2$$

If you are given 65 g of zinc and 65 g of HCl, how many grams of hydrogen gas can be produced? Which reactant will be left over? How many grams of this reactant will not be consumed?

*Answer explanations are found at the end of the chapter. Page 128.

Step 1:   Set up the problem.

$$\begin{array}{cccc} 65\text{ g} & 65\text{ g} & & x\text{ grams} \\ \text{Zn} + & 2\text{HCl} \rightarrow & \text{ZnCl}_2 + & \text{H}_2 \\ 65\text{ g} & 73\text{ g} & & 2\text{ g} \end{array}$$

The given quantities are above the equation, and the equation masses are given beneath. To solve for the grams of hydrogen gas, the limiting reactant must be determined.

Step 2:   Compare the given quantities with the equation requirements.

Knowing that it takes 65 g of zinc to react with 73 g of hydrochloric acid, it is reasonable to surmise that, since there are only 65 g of hydrochloric acid, not all of the 65 g of zinc can be used. The limiting factor will be the amount of hydrochloric acid.

Step 3:   Solve for the quantity of product that will be produced using the amount of the limiting factor.

Now that we know that the limiting factor is the amount of hydrochloric acid (65 g), the equation can be solved. This means that the proportion is set up ignoring the 65 g of zinc.

$$\frac{65\text{ g}}{73\text{ g}} = \frac{x\text{ grams}}{2\text{ g}}$$

Solving for x, we get $x = 1.78$ g or 1.8 g of hydrogen produced.

Step 4:   To find the number of grams of zinc that were not consumed, a separate problem is necessary. The equation would be set up as follows:

$$\begin{array}{cccc} y\text{ grams} & 65\text{ g} & & \\ \text{Zn} & + 2\text{HCl} \rightarrow & \text{ZnCl}_2 + \text{H}_2 \\ 65\text{ g} & 73\text{ g} & & \end{array}$$

This gives the proportion

$$\frac{y\text{ grams}}{65\text{ g}} = \frac{65\text{ g}}{73\text{ g}}$$

$$y = \frac{65 \times 65}{73} = 57.8 \text{ or } 58 \text{ g}$$

Subtracting the 58 g of zinc that will be used from the original 65 g leaves 7 g of zinc that will not be consumed.

Often there are some equations on a chemistry test to either complete or complete and balance. The following review problems of representative equations are some you should be capable of writing and balancing.

Answers

1. 63

2. 162

3. 98

4. 74.5

5. 342

6. 32.2 g
    Equation:
    $$\begin{array}{cc} x \text{ grams} & 20 \text{ L} \\ 2H_2O \rightarrow & 2H_2 + O_2 \end{array}$$

    Calculate the equation mass and volume so that the units match the ones above the equation.

    $$\begin{array}{cc} x \text{ grams} & 20 \text{ L} \\ 2H_2O \rightarrow & 2H_2 + O_2 \end{array}$$

    $$\begin{array}{cc} (2 \times \text{molecular mass of water}) & (1 \times \text{gram-molecular volume}) \\ 36 \text{ g} & 22.4 \text{ L} \end{array}$$

    Then

    $$\frac{x \text{ grams}}{36 \text{ g}} = \frac{20 \text{ L}}{22.4 \text{ L}}$$
    $$x = 32.2 \text{ g}$$

7. 72 g
    $$\begin{array}{cc} x \text{ grams} & 44.8 \text{ L} \\ 4Al \quad + \quad 3O_2 & \rightarrow 2Al_2O_3 \end{array}$$

    Calculate the equation mass or volume so that the units match the ones above the equation.

    $$\begin{array}{ccc} x \text{ grams} & 44.8 \text{ L} & \\ 4Al \quad + & 3O_2 & \rightarrow 2Al_2O_3 \\ (4 \times 27 \text{ g}) & (3 \times 22.4 \text{ L}) & \\ 108 \text{ g} & 67.2 \text{ L} & \end{array}$$

    Then

    $$\frac{x \text{ grams}}{108 \text{ g}} = \frac{44.8 \text{ L}}{67.2 \text{ L}}$$
    $$x \text{ g} = 72 \text{ g}$$

8. 122.5 g
    The equation is

    $$\begin{array}{cc} x \text{ grams} & 100 \text{ g} \\ MnO_2 \quad + 4HCl \rightarrow & MnCl_2 + Cl_2 + 2H_2O \end{array}$$

Calculate the equation masses for the substances involved.

$$\underset{\substack{(1 \times 87 \text{ g}) \\ 87 \text{ g}}}{\overset{x \text{ grams}}{MnO_2}} + 4HCl \rightarrow MnCl_2 + \underset{\substack{(1 \times 71 \text{ g}) \\ 71 \text{ g}}}{\overset{100 \text{ g}}{Cl_2}} + 2H_2O$$

Then

$$\frac{x \text{ grams}}{87 \text{ g}} = \frac{100 \text{ g}}{71 \text{ g}}$$

$$x = 122.5 \text{ g}$$

9. 33.3 g
    The equation is

$$\underset{\substack{48 \text{ g}}}{\overset{20 \text{ g}}{2Mg}} + \underset{\substack{32 \text{ g}}}{\overset{20 \text{ g}}{O_2}} \rightarrow \underset{\substack{80 \text{ g}}}{\overset{x \text{ grams}}{2MgO}}$$

Since the equation masses indicate you must have more Mg than O_2, there is an excess of oxygen.

$$\frac{20 \text{ g Mg}}{48 \text{ g Mg}} = \frac{x \text{ g MgO}}{80 \text{ g MgO}}$$

$$x = 33.3 \text{ g MgO}$$

10. 6 L
    The equation is

$$\overset{x \text{ liters}}{H_2} + \overset{12 \text{ L}}{Cl_2} \rightarrow 2HCl$$

Set up the proportion using the coefficients of the substances that have something indicated above them as denominators.

$$\frac{x \text{ liters}}{1 \text{ L}} = \frac{12 \text{ L}}{2 \text{ L}}$$

$$x = 6 \text{ L}$$

11. 50 L
    The equation is

$$\overset{100 \text{ L}}{2CO} + \overset{x \text{ liters}}{O_2} \rightarrow 2CO_2$$

Set up the proportion using the coefficients of the substances that have something indicated above them as denominators.

$$\frac{100 \text{ L}}{2 \text{ L}} = \frac{x \text{ liters}}{1 \text{ L}}$$

$$x = 50 \text{ L}$$

Review Problems
(Balancing Equations)

12. zinc + sulfuric acid →

13. iron(III) chloride + sodium hydroxide → iron(III) hydroxide(s) + sodium chloride

14. aluminum hydroxide + sulfuric acid → aluminum sulfate + water

15. potassium + water →

16. magnesium + oxygen →

17. silver nitrate + copper → copper(II) nitrate + silver(s)

18. magnesium bromide + chlorine → magnesium chloride + bromine

Answers

12. zinc + sulfuric acid → zinc sulfate + hydrogen(g)

    $Zn + H_2SO_4 \rightarrow ZnSO_4 + H_2(g)$

13. $FeCl_3 + 3NaOH \rightarrow Fe(OH)_3(s) + 3NaCl$

14. $2Al(OH)_3 + 3H_2SO_4 \rightarrow Al_2(SO_4)_3 + 6H_2O$

15. potassium + water → potassium hydroxide + hydrogen(g)

    $2K + 2H_2O \rightarrow 2KOH + H_2(g)$

16. magnesium + oxygen → magnesium oxide

    $2Mg + O_2 \rightarrow 2MgO$

17. $2AgNO_3 + Cu \rightarrow Cu\ (NO_3)_2 + 2Ag(s)$

18. $MgBr_2 + Cl_2 \rightarrow MgCl_2 + Br_2$

Terms You Should Know

molar mass	mole
molar volume	STP

Avogadro's Law
Avogadro's number
Gay-Lussac's Law

Chapter 7

LIQUIDS, SOLIDS, AND PHASE CHANGES

Liquids

Importance of Intermolecular Interaction

In a liquid, the volume of the molecules and the intermolecular forces between them are much more important than in a gas. When you consider that in a gas the molecules constitute far less than 1% of the total volume, while in the liquid state the molecules constitute 70% of the total volume, it is clear that in a liquid the forces between molecules are more important. Because of this decreased volume and increased intermolecular interaction, a liquid expands and contracts only very slightly with a change in temperature and lacks the compressibility typical of gases.

Kinetics of Liquids

Even though the volume of space between molecules has decreased in a liquid and the mutual attraction forces between neighboring molecules can have great effects on the molecules, they are still in motion. This motion can be verified under a microscope when colloidal particles are suspended in a liquid. The particles' zigzag path, called *Brownian movement*, indicates molecular motion and supports the *Kinetic Molecular Theory*.

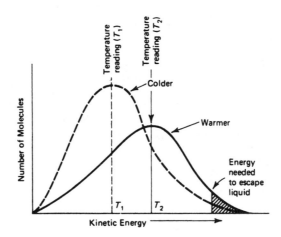

Figure 22. Distribution of the Kinetic Energy of Molecules.

Viscosity

Viscosity is the friction or resistance to motion that exists between the molecules of a liquid when they move past each other. It is logical that the stronger the attraction between the molecules of a liquid, the greater its resistance to flow—and thus the greater its viscosity. The viscosity of a liquid depends on its intermolecular forces. Because hydrogen bonds are such strong intermolecular forces, liquids with hydrogen bonds tend to have high viscosities. Water is strongly hydrogen-bonded and has a relatively high viscosity. You may have noticed how fast liquids without this trait, such as alcohol and gasoline, flow.

Surface Tension

Molecules at the surface of a liquid experience attractive forces downward, toward the inside of the liquid, and sideways, along the surface of the liquid. This is unlike the uniformly distributed attractive forces that molecules in the center of the liquid experience. This imbalance of forces at the surface of a liquid results in a property called *surface tension*. The uneven forces make the surface behave as if it had a tight film stretched across it. Depending on the magnitude of the surface tension of the liquid, the film is able to support the weight of a small object such as a razor blade or a needle. Surface tension also explains the beading of raindrops on the shiny surface of a car.

Increases in temperature increase the average kinetic energy of molecules and the rapidity of their movement. This is shown graphically in Figure 22. The molecules in the sample of cold liquid have, on the average, less kinetic energy than those in the warmer sample. Hence the temperature reading T_1 will be less than the temperature reading T_2. If a particular molecule gains enough kinetic energy when it is near the surface of a liquid, it can overcome the attractive forces of the liquid phase and escape into the gaseous phase. This is called a *change of phase*. When fast-moving molecules with high kinetic energy escape, the average energy of the remaining molecules is lower; hence the temperature is lowered.

Phase Equilibrium

Figure 23 shows water in a container enclosed by a bell jar. Observation of this closed system would show an initial small drop in the water level, but after some time the level would become constant. The explanation is that, at first, more energetic molecules near the surface are escaping into the gaseous phase faster than some of the gaseous water molecules are returning to the surface and possibly being caught by the attractive forces that will retain them in the liquid phase. After some time the rates of evaporation and condensation equalize. This is known as *phase equilibrium*.

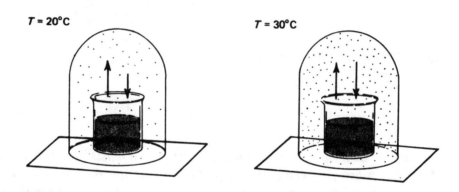

Figure 23. Closed System in Dynamic Equilibrium.

Note: Arrows show extent of evaporation (↑) and condensation (↓).

In a closed system like this, when opposing changes are taking place at equal rates, the system is said to have *dynamic equilibrium*. At higher temperatures, since the number of molecules at higher energies increases, the number of molecules in the liquid phase will be reduced and the number of molecules in the gaseous phase will be increased. The rates of evaporation and condensation, however, will again become equal.

The behavior of the system described above illustrates what is known as *Le Châtelier's Principle*. It is stated as follows: When a system at equilibrium is disturbed by the application of a stress (a change in temperature, pressure, or concentration), it reacts so as to minimize the stress and attain a new equilibrium position.

In the discussion above, if the 20°C system is heated to 30°C, the number of gas molecules will be increased while the number of liquid molecules will be decreased

$$\text{Heat} + H_2O(\ell) \rightleftharpoons H_2O(g)$$

The equation shifts to the right (any similar system that is endothermic shifts to the right when temperature is increased) until equilibrium is reestablished at the new temperature.

The molecules in the vapor that are in equilibrium with the liquid at a given temperature exert a constant pressure. This is called the *equilibrium vapor pressure* at that temperature.

Table Ⓜ in the Equations and Tables for Reference section at the back of this book gives the vapor pressure of water at various temperatures.

Boiling Point

The vapor pressure-temperature relation can be plotted on a graph for a closed system. (See Figure 24.) When a liquid is heated in an open container, the liquid and vapor are not in equilibrium and the vapor pressure increases until it becomes equal to the pressure above the liquid. At this point the average kinetic energy of the molecules is such that they are rapidly converted from the liquid to the vapor phase within the liquid as well as at the surface. The temperature at which this occurs is known as the *boiling point*.

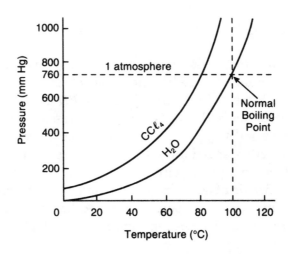

Figure 24. Vapor Pressure-Temperature Relationship for Carbon Tetrachloride and Water.

Critical Temperature and Pressure

There are conditions for particular substances when it is impossible for the liquid or gaseous phase to exist. Since the kinetic energy of a molecular system is directly proportional to the Kelvin temperature, it is logical to assume that there is a temperature at which the kinetic energy of the molecules is so great that the attractive forces between molecules are insufficient for the liquid phase to remain. The temperature above which the liquid phase of a substance cannot exist is called its *critical temperature*. Above its critical temperature, no gas can be liquefied regardless of the pressure applied. The minimum pressure required to liquefy a gas at its critical temperature is called its *critical pressure*.

Solids

Whereas particles in gases have the highest degree of disorder, the solid state has the most ordered system. Particles are fixed in a rather definite position and maintain a definite shape. Particles in solids do vibrate in position, however, and may even diffuse through the solid. (Example: Gold clamped to lead shows diffusion of some gold atoms into the lead over long periods of time.) Other solids do not show diffusion because of strong ionic or covalent bonds in network solids. (Examples: NaCl and diamond, respectively.)

When heated at certain pressures, some solids vaporize directly without passing through the liquid phase. This is called *sublimation*. Solids like solid carbon dioxide and solid iodine exhibit this property because of unusually high vapor pressure.

The temperature at which atomic or molecular vibrations of a solid become so great that the particles break free from fixed positions and begin to slide freely over each other in a liquid state is called the *melting point*. The amount of energy required at the melting point temperature to cause the change of phase to occur is called the *heat of fusion*. The amount of this energy depends on the nature of the solid and the type of bonds present.

Phase Diagrams

The simplest way to discuss a phase diagram is by an example, such as Figure 25.

A phase diagram ties together the effects of temperature and also pressure on the phase changes of a substance. In Figure 25 the line *BD* is essentially the vapor pressure curve for the liquid phase. Notice that at 760 mm (the pressure for 1 atm) the water will boil (change to the vapor phase) at 100°C (point *F*). However, if the pressure is raised, the boiling point temperature increases; and if the pressure is less than 760 mm, the boiling point decreases along the *BD* curve down to point *B*.

At 0°C the freezing point of water is found along the line *BC* at point *E* for pressure at 1 atm or 760 mm. Again, this point is affected by pressure along the line *BC*, so that if the pressure is decreased, the freezing point is slightly higher up to point *B* or 0.01°C.

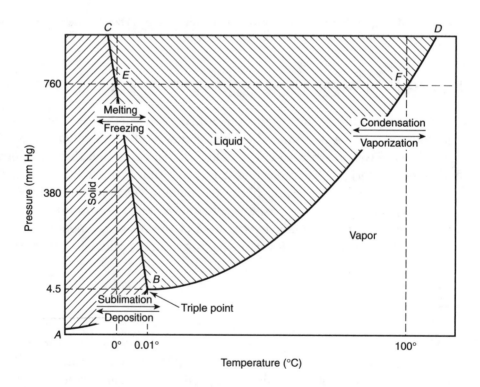

Figure 25. Partial Phase Diagram for Water (distorted somewhat to distinguish the triple point from the freezing point).

Point *B* represents the point at which the solid, liquid, and vapor phases can all exist at equilibrium. This point is known as the *triple point*. It is the only temperature and pressure at which three phases of a pure substance can exist in equilibrium with one another in a system containing only the pure substance.

Water

History of Water

Ancient philosophers like Aristotle regarded water as a basic element that typified all liquid substances. Scientists did not discard this view until the latter half of the eighteenth century. In 1781 the British chemist Henry Cavendish synthesized water by detonating a mixture of hydrogen and air. However, the results of his experiments were not clearly interpreted until 2 years later when the French chemist Antoine Laurent Lavoisier proved that water was not an element but a compound of oxygen and hydrogen. In a scientific paper presented in 1804, the French chemist Joseph Gay-Lussac and the German naturalist Alexander von Humboldt demonstrated jointly that water consisted of two volumes of hydrogen to one of oxygen, as expressed by the well-known formula H_2O.

It was not until 1932 that the American chemist Harold Clayton Urey discovered the presence of a small amount (1 part in 6000 parts) of so-called heavy water, or deuterium oxide (D_2O). Deuterium is the hydrogen isotope with an atomic number of 2. In 1951 the American chemist Aristid Grosse discovered that naturally occurring water also contained minute traces of tritium oxide (T_2O). Tritium is the hydrogen isotope with an atomic weight of 3.

Purification of Water

Water is so often involved in chemistry that it is important to have a rather complete understanding of this compound and its properties. Pure water has become a matter of national concern. Both the commercial methods of purification and the usual laboratory method of obtaining pure water, distillation, will be covered.

The process of distillation involves the evaporation and condensation of the water molecules. The usual apparatus for the distillation of any liquid is shown in Figure 26.

This method of purification will remove any substance that has a boiling point higher than that of water. It cannot remove dissolved gases or liquids that boil off before water. These substances will be carried over into the condenser and subsequently into the distillate.

Often naturally occurring water contains suspended and dissolved impurities that make it unsuitable for many purposes. Objectionable organic and inorganic materials are removed by such methods as screening and sedimentation to eliminate suspended materials. This treatment often involves the use of activated carbon to remove tastes and odors, filtration to remove insoluble solids, and chlorination or irradiation to kill infective microorganisms.

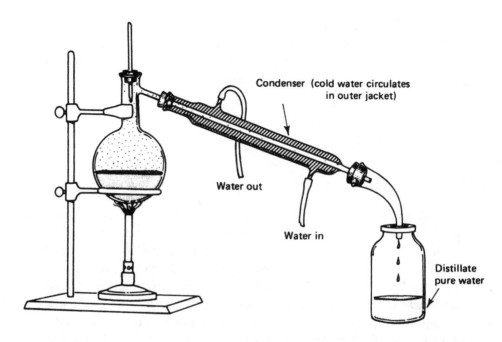

Figure 26. Distillation of Water.

Another method of water purification is aeration, or the saturation of water with air. In this method, water is brought into contact with air in such a manner as to produce maximum diffusion, usually by spraying the water into the air in fountains. Aeration removes odors and taste caused by decomposing organic matter, as well as industrial wastes such as phenols and volatile gases such as chlorine. It also converts dissolved iron and manganese compounds to insoluble hydrated oxides of the metals, which then settle out as precipitates.

When water contains bicarbonates of Ca, Mg, or Fe it is said to have carbonate hardness, formerly called "temporary hardness." Such hardness may be removed by boiling since heat causes the bicarbonate to decompose to a carbonate precipitate. This precipitate can cause detrimental scales in industrial boilers if not removed.

Natural waters are also apt to contain varying amounts of calcium sulfate, magnesium sulfate, calcium chloride, and magnesium chloride. Water in which these chemicals are dissolved is said to have noncarbonate hardness, formerly known as "permanent hardness." One way of softening water is to pass it through an artificial zeolite which replaces the "hard" positive ions (called cations) with "soft" ions of sodium. It is the "hard" cations that are responsible for "bathtub ring" by reacting with the soap to form an insoluble compound. Another effective method that gives pure water like that obtained by distillation is the use of organic deionizers (e.g., Dowex, Amberlite, and Zeolite). These organic resins exchange H+ ions for the cations, and OH− ions for the anions.

With the ever-increasing demands for fresh water, especially in arid and semiarid areas or very densely populated areas, much research has gone into finding efficient methods of removing salt from seawater and brackish water. The state of California is seriously considering desalinization as a possible solution to its growing water problems. In the United States, desalinization research is directed by the Bureau of Reclamation, Department of the Interior. Several processes are being developed to produce fresh water cheaply.

Three of the processes involve evaporation followed by condensation of the resultant steam and are known as multiple-effect evaporation, vapor-compression distillation, and flash evaporation. The last-named method, the most widely used, involves heating seawater and pumping it into lower-pressure tanks where it abruptly vaporizes into steam. The steam then condenses and is drawn off as pure water. In 1967 Key West, Florida, opened this type of plant and became the first city in the United States to draw its fresh water from the sea.

Other alternatives are freezing, reverse osmosis, and electrodialysis. The major obstacle to the use of any of these methods is cost. Using conventional fuels, plants with a capacity of 1 million gallons per day or less produce water at a cost of $1.00 or more per 1000 gallons. More than 500 such plants are in operation today. However, water from conventional sources, such as wells and reservoirs, is sold for less than 30 cents per 1000 gallons delivered to the home.

Composition of Water

Water can be analyzed, that is, broken into its components, by electrolysis. This process is discussed on page 303 and shows that its composition by volume is 2 parts of hydrogen to 1 part of oxygen. Water composition can also be arrived at by synthesis. Synthesis is the formation of a compound by uniting its components. Water can be made by mixing hydrogen and oxygen in a eudiometer over mercury and passing an igniting spark through the mixture. Again the ratio of combination is found to be 2 parts hydrogen to 1 part oxygen. In a steam-jacketed eudiometer, which keeps the water formed in the gas phase, 2 vol of hydrogen combine with 1 vol of oxygen to form 2 vol of steam. Another interesting method is the Dumas experiment pictured in Figure 27.

Data obtained show that the ratio of hydrogen to oxygen combined to form water is 1 : 8 by mass. This means that 1 g of hydrogen combines with 8 g of oxygen to form 9 g of water.

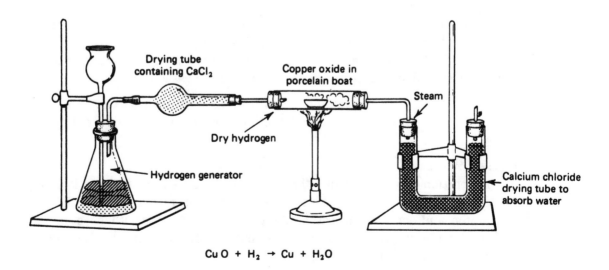

$$Cu\,O + H_2 \rightarrow Cu + H_2O$$

Figure 27. Synthesis of Water.

Some sample problems involving the composition of water are shown below.

Example 1

(By Mass)

An electric spark is passed through a mixture of 12 g of hydrogen and 24 g of oxygen in the eudiometer setup shown. Find the number of grams of water formed and the number of grams of gas left uncombined.

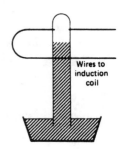

Solution:

Since water forms in a ratio of $1:8$ by mass, to use up the oxygen (which by inspection will be the limiting factor since it has enough hydrogen present to react completely) we need only 3 g of hydrogen.

3 g of hydrogen + 24 g of oxygen = 27 g of water

This leaves 12 g – 3 g = 9 g of hydrogen uncombined.

Example 2

(By Volume)

A mixture of 8 mL of hydrogen and 200 mL of air is placed in a steam-jacketed eudiometer, and a spark is passed through the mixture. What will be the total volume of gases in the eudiometer?

Solution:

Since this is a combination by volume, 8 mL of hydrogen require 4 mL of oxygen. (Ratio $H_2 : O_2$ by volume = 2 : 1.)

The 200 mL of air is approximately 21% oxygen. This will more than supply the needed oxygen and leave 196 mL of the air uncombined.

The 8 mL of hydrogen and 4 mL of oxygen will form 8 mL of steam since the eudiometer is steam-jacketed and keeps the water formed in the gaseous state.

(Ratio by volume of hydrogen : oxygen : steam = 2 : 1 : 2)
TOTAL VOLUME = 196 mL of air + 8 mL of steam = 204 mL

Heavy Water

A small portion of water is called "heavy" water because it contains an isotope of hydrogen, deuterium (symbol D) rather than ordinary hydrogen nuclei. Deuterium has a nucleus of one proton and one neutron rather than just one proton. Another isotope of hydrogen is tritium. Its nucleus is composed of two neutrons and one proton. Both of these isotopes have had use in the nuclear energy field.

Hydrogen Peroxide

The prefix *per-* indicates that this compound contains more than the usual oxide. Its formula is H_2O_2. It is a well-known bleaching and oxidizing agent. Its electron-dot formula is shown in Figure 28.

Figure 28. Hydrogen Peroxide.

Properties and Uses of Water

Water has been used in the definition of various standards.

1. For mass—1 mL (or 1 cc) of water at 4°C is 1 g
2. For heat— (a) The heat needed to raise 1 g of water 1°C on the Celsius scale = 1 cal; 1000 cal = 1 kcal
   (b) the heat needed to raise 1 lb of water 1°C on the Fahrenheit scale = 1 British thermal unit (Btu)
3. Degree of heat—the freezing point of water = 0°C, 32°F, 273 K
   —the boiling point of water = 100°C, 212°F, 373 K

Water Calorimetry Problems

A calorimeter is a container well insulated from outside sources of heat or cold so that most of its heat is contained in the vessel. If a very hot object is placed in a calorimeter containing some ice crystals, we can find the final temperature of the mixture mathematically and check it experimentally. To do this, certain behaviors must be understood. Ice changing to water and then to steam is not a continuous and constant change of temperature as time progresses. In fact, the chart would look as shown in Figure 29.

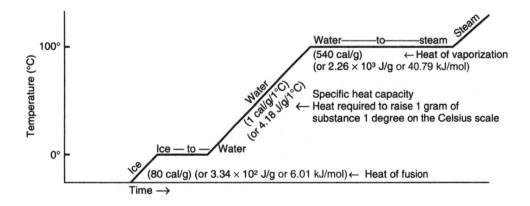

Figure 29. Changing Ice to Steam.

From this graph, you see that heat is being used at 0°C and 100°C to change the state of water but not its temperature. One gram of ice at 0°C needs 80 cal or 3.34×10^2 J to change to water at 0°C. This is called its *heat of fusion*. Likewise, energy is being used at 100°C to change water to steam, not to change the temperature. One gram of water at 100°C absorbs 540 cal or 2.26×10^3 J of heat to change to 1 g of steam at 100°C. This is called its *heat of vaporization*. This energy being absorbed at the plateaus in the curve is being used to break up the bonding forces between molecules by increasing their potential energy content so that a specific change of state can occur.

Example 1

What quantity of ice at 0°C can be melted by 100 J of heat?

Solution:
Heat to fuse (melt) a substance = heat of fusion of the substance × mass of the substance. This can be expressed by the following formula where, q is used to denote heat measurement made in a calorimeter:

$$q = m \text{ (mass)} \times C \text{ (heat of fusion)}$$

Solving for m, we get

$$m = q/C$$
$$= 100 \text{ J}/3.34 \times 10^2 \text{ J/g}$$
$$= 29.9 \times 10^{-2} \text{ g or } 0.299 \text{ g of ice melted}$$

Because heat is absorbed in this melting, this is an endothermic action.

Example 2

How much heat is needed to change 100 g of ice at 0°C to steam at 100°C?
To melt 100 g of ice at 0°C:

Solution:
Use m (mass) × C (heat of fusion) = q (quantity of heat).

$$100 \text{ g} \times \frac{80 \text{ cal}}{1 \text{ g}} = 8000 \text{ cal} = 8 \text{ kcal}$$

To heat 100 g of water from 0°C to 100°C:
Use m (mass) × ΔT × specific heat = q (quantity of heat).

$$100 \text{ g} \times \overbrace{(100° - 0°)°C}^{\text{temperature change}} \times \frac{1 \text{ cal}}{1 \text{ g} \times 1°C} = 10{,}000 \text{ cal} = 10 \text{ kcal}$$

To vaporize 100 g of water at 100°C to steam at 100°C:
Use m (mass) × heat of vaporization = q (quantity of heat).

$$100 \text{ g} \times \frac{540 \text{ cal}}{1 \text{ g}} = 54{,}000 \text{ cal} = 54 \text{ kcal}$$

$$\text{Total heat} = 8 + 10 + 54 = 72 \text{ kcal}$$

MOST HEAT
OF VAPORIZATION

Using the factor-label system to express the answer in joules, we have

$$72 \text{ kcal} \times \frac{1000 \text{ cal}}{1 \text{ kcal}} \times \frac{4.18 \text{ J}}{1 \text{ cal}} = 3.01 \times 10^5 \text{ J}$$

Reactions of Water with Anhydrides

Anhydrides are certain oxides that react with water to form two classes of compounds—acids and bases.

Many metal oxides react with water to form bases such as sodium hydroxide, potassium hydroxide, and calcium hydroxide. For this reason, they are called *basic anhydrides*. The common bases are water solutions that contain the hydroxyl (OH^-) ion. Some common examples are

$$Na_2O + H_2O \rightarrow 2NaOH, \text{ sodium hydroxide}$$
$$CaO + H_2O \rightarrow Ca(OH)_2, \text{ calcium hydroxide}$$
In general, then: Metal oxide + H_2O → Metal hydroxide

In a similar manner, nonmetallic oxides react with water to form an acid such as sulfuric acid, carbonic acid, or phosphoric acid. For this reason, they are referred to as *acid anhydrides*. The common acids are water solutions containing hydrogen ions (H^+). Some common examples are

$$CO_2 + H_2O \rightarrow H_2CO_3, \text{ carbonic acid}$$
$$SO_3 + H_2O \rightarrow H_2SO_4, \text{ sulfuric acid}$$
$$P_2O_5 + 3H_2O \rightarrow 2H_3PO_4, \text{ phosphoric acid}$$
In general, then: Nonmetallic oxide + H_2O → Acid

Polarity and Hydrogen Bonding

Water is different from most liquids in that it reaches its greatest density at 4°C and then begins to expand from 4°C to 0°C (which is its freezing point). When it freezes at 0°C, its volume expands by about 9%. Most liquids contract as they cool and change state to a solid because their molecules have less energy, move more slowly, and are closer together. This abnormal behavior can be explained as follows. X-ray studies of ice crystals show that H_2O molecules are bound into large molecules in which each oxygen atom is connected through *hydrogen bonds* to four other oxygen atoms as shown in Figure 30.

This is a rather wide open structure, which accounts for the low density of ice. As heat is applied and melting begins, this structure begins to collapse, but not all the hydrogen bonds are broken. The collapsing increases the density, but the remaining bonds keep the structure from completely collapsing. As the heat is absorbed, the kinetic energy of the molecules breaks more of these bonds as the temperature goes from 0° to 4°C. At the same time this added kinetic energy tends to distribute the molecules farther apart. At 4°C these opposing forces are in balance—thus the greatest density. Above 4°C the increasing molecular motion again causes a decrease in density since it is the dominant force and offsets the breaking of any more hydrogen bonds.

Figure 30. Study of Ice Crystal.

Note: δ indicates partial charge.

This behavior of water can be explained by studying the water molecule itself. The water molecule is composed of two hydrogen atoms bonded by a polar covalent bond to one oxygen atom.

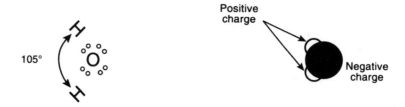

Because of the polar nature of the bond, the molecule exhibits the charges shown in the above drawing. It is this polar charge that causes the polar bonding discussed in Chapter 3 as the hydrogen bond. This bonding is stronger than the usual molecular attraction called van der Waals forces.

Solubility

Water is often referred to as "the universal solvent" because of the number of common substances that dissolve in it. When substances are dissolved in water to the extent that no more will dissolve at that temperature, the solution is said to be *saturated*. The substance dissolved is called the *solute*, and the dissolving medium is called a *solvent*. To give an accurate statement of a substance's solubility, three conditions are mentioned: the amount of solute, the amount of solvent, and the temperature of the solution. Since the solubility varies for each substance and for different temperatures, a student must be acquainted with the use of solubility curves such as those shown in Figure 31. A more complete table is given as Table Ⓑ in the Equations and Tables for Reference section.

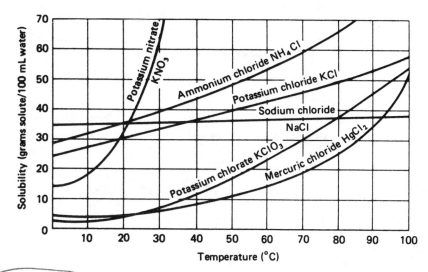

Figure 31. Solubility Curves.

These curves show the number of grams of solute that will dissolve in 100 g of water over a temperature of 0° to 100°C. Take, for example, the very lowest curve at 0°C. This curve shows the number of grams of $KClO_3$ that will dissolve in 100 g of water over a temperature range of 0 to 100°C. To find the solubility at any particular temperature, for example, at 50°C, follow the vertical line up from 50°C until it crosses the curve. At that point, place a ruler so that it is horizontal across the page and take the reading on the vertical axis. This point happens to be slightly below the 20-g mark, or 18 g. This means that 18 g of $KClO_3$ will dissolve in 100 g of water at 50°C.

Type of Problem Using the Solubility Curve

A solution contains 20 g of $KClO_3$ in 200 g of H_2O at 80°C. How many more grams of $KClO_3$ can be dissolved to saturate the solution at 90°C?

Reading the graph at 90° and up to the graph line for $KClO_3$, we find that 100 g of H_2O can dissolve 48 g. And so 200 g can hold (2×48) g or 96 g. Therefore 96 g − 20 g = 76 g $KClO_3$ can be added to the solution.

General Rules of Solubility

All nitrates, acetates, and chlorates are soluble.

All common compounds of sodium, potassium, and ammonium are soluble.

All chlorides are soluble except those of silver, mercury(ious), and lead. (Lead chloride is noticeably soluble in hot water.)

All sulfates are soluble except those of lead, barium, strontium, and calcium. (Calcium sulfate is slightly soluble.)

The normal carbonates, phosphates, silicates, and sulfides are insoluble except those of sodium, potassium, and ammonium.

All hydroxides are insoluble except those of sodium, potassium, ammonium, calcium, barium, and strontium.

Some general trends of solubility are shown in the chart below.

	Temperature Effect	Pressure Effect
Solid	Solubility usually increases with temperature increase	Little effect
Gas	Solubility usually decreases with temperature increase	Solubility varies in direct proportion to the pressure applied to it. *Henry's Law*

Factors That Affect Rate of Solubility

The following procedures increase the rate of solubility.

Pulverizing—Increases surface exposed to solvent

Stirring—Brings more solvent that is unsaturated into contact with solute

Heating—Increases molecular action and gives rise to mixing by convection currents. (This heating affects the solubility as well as the rate of solubility.)

Summary of Types of Solutes and Relationship of Type to Solubility

Generally speaking, solutes are most likely to dissolve in solvents with similar characteristics; that is, ionic and polar solutes dissolve in polar solvents, and nonpolar solutes dissolve in nonpolar solvents.

It should also be mentioned that polar molecules that do not ionize in aqueous solution (e.g., sugar, alcohol, glycerol) have molecules as solute particles; polar molecules that partially ionize in aqueous solution (e.g., ammonia, acetic acid) have a mixture of molecules and ions as solute particles; and polar molecules that completely ionize in aqueous solution (e.g., hydrogen chloride, hydrogen iodide) have ions as solute particles.

Water Solutions

In order to make molecules or ions of another substance go into solution, water molecules must overcome the forces that hold these molecules or ions together. The mechanism of the actual process is complex. To make sugar molecules go into solution, the water molecules cluster around the sugar molecules, pull them off, and disperse, forming the solution.

For an ionic crystal such as salt, the water molecules orient themselves around the ions (which are charged particles) and again must overcome the forces holding the ions together. Since the water molecule is polar, this orientation around the ion is an attraction of the polar ends of the water molecule. For example:

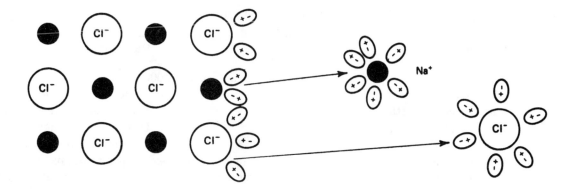

Once surrounded, the ion is insulated to an extent from other ions in solution because of the *dipole* property of water. The water molecules that surround the ion differ in number for various ions, and the whole group is called a *hydrated ion*. In general, as stated in the preceding section, polar substances and ions dissolve in polar solvents, and nonpolar substances such as fats dissolve in nonpolar solvents such as gasoline. The process of going into solution is *exothermic* if energy is released in the process, and *endothermic* if energy from the water is used up to a greater extent than energy is released in freeing the particle.

When two liquids are mixed and they dissolve in each other, they are said to be completely *miscible*. If they separate and do not mix, they are said to be *immiscible*.

Two molten metals may be mixed and allowed to cool. This gives a "solid solution" called an *alloy*.

Continuum of Water Mixtures

Figure 32 shows the general sizes of the particles found in a water mixture.

The basic difference between a colloid and a suspension is in the diameter of the particles dispersed. All the boundaries marked are only the general ranges in which the distinctions among solutions, colloids, and suspensions are usually made.

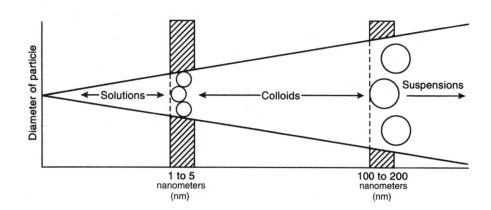

Figure 32. Size of Particles in Water Mixture.

The characteristics of water mixtures are as follows:

Solutions	Colloids	Suspensions
.... 1 nm		1000 nm ....
Clear; may have color		Cloudy; opaque color
Particles do not settle		Particles settle on standing
Particles pass through ordinary filter paper		Do not pass through ordinary filter paper
Particles pass through membranes	Do not pass through semipermeable membranes like animal bladders, cellophane, and parchment, which have very small pores*	
Particles are not visible	Visible in ultramicroscope	Visible with microscope or naked eye
	Show Brownian movement	No Brownian movemen

*Separation of a solution from a colloidal dispersion through a semipermeable membrane is called *dialysis*.

When a bright light is directed at right angles to the stage of an ultramicroscope, the individual reflections of colloidal particles can be seen. They can be observed to be following a random zigzag path. This is explained as follows: The molecules in the dispersing medium are in motion and continuously bumping into the colloidal particles, causing them to change direction in a random fashion. This motion is called *Brownian movement* after the scientist who first observed it.

Expressions of Concentration

There are general terms and very specific terms used to express the concentration of a solution. The general terms and their definitions are

Dilute—Small amount of solute is dispersed in the solvent.

Concentrated—Large amount of solute is dissolved in the solvent.

Saturated—The solution is holding all the solute possible at that temperature. This is not a static condition; that is, some solute particles are exchanging places with some of the undissolved particles, but the total solute in solution remains the same. This is an example of equilibrium.

Unsaturated—More solute can go into solution at that temperature. The solvent has further capacity to hold more solute.

Supersaturated—Sometimes a saturated solution at a higher temperature can be carefully cooled so that the solute does not have a chance to come out of solution. At a lower temperature, then, the solution will be holding more solute in solution than it should for saturation and is said to be supersaturated. As soon as the solute particles are jarred or a "seed" particle is added to the solution to act as a nucleus, they rapidly come out of solution so that the solution reverts to the saturated state.

It is interesting to note that the words *saturated* and *concentrated* are NOT synonymous. In fact, it is possible to have a saturated dilute solution when the solute is only slightly soluble and a small amount of it makes the solution saturated but its concentration is still dilute.

The more specific terms used to describe concentration are mathematically calculated.

Percentage concentration is based on the percent of solute in the solution by mass. The general formula is

$$\frac{\textbf{No. of grams of solute}}{\textbf{No. of grams of solution}} \times \textbf{100\%} = \textbf{\% concentration}$$

Example

How many grams of NaCl are needed to prepare 200 g of a 10% salt solution?

Solution:

$$10\% \text{ of } 200 \text{ g} = 20 \text{ g of salt}$$

You could also solve the problem using the above formula and solving for the unknown quantity.

$$\frac{x \text{ grams solute}}{200 \text{ g solution}} \times 100\% = 10\%$$

$$x \text{ grams solute} = \frac{10}{100} \times 200 = 20 \text{ g of solute}$$

Using Specific Gravity in Solutions

Specific gravity is the ratio of the mass of a substance to the mass of an equal volume of water. Often chemistry handbooks give the specific gravity of solutions. These in turn can be used to determine the mass of a solute in solution. Specific gravity has no units, but since the density of water is 1.0 g/mL, specific gravity is numerically equal to the density in grams per milliliter. For example, 7.9 is the specific gravity of iron. This means that the density of iron is 7.9 g/mL.

This relationship can be used in solving problems of concentration that use percentage composition and a known specific gravity.

Example

Suppose that a solution of hydrogen chloride gas dissolved in water (hydrochloric acid) has a concentration of 30% HCl by mass. How much solute is there in 100 mL of this solution if the specific gravity of the solution is 1.15? (This means that its density is 1.15 g/mL.)

Solution:
First, determine the mass of 100 mL of the solution:

$$\text{Mass} = \text{Density} \times \text{volume}$$
$$\text{Mass} = 1.15 \text{ g/mL} \times 100 \text{ mL} = 115 \text{ g}$$

Then, determine what mass of HCl is dissolved in 115 g of the solution:

$$\text{Mass HCl} = 30\% \times \text{mass of solution}$$
$$\text{Mass HCl} = 0.30 \times 115 \text{ g} = 34.5 \text{ g}$$

So, there are 34.5 g of HCl gas dissolved in 100 mL of this solution.

The next two expressions depend on the fact that if the formula mass of a substance is expressed in grams, it is called a *gram-formula mass* (gfm), *molar mass*, or 1 *mol*. *Gram-molecular mass* can be used in place of gram-formula mass when the substance is really of molecular composition and not ionic like NaCl or NaOH. The definitions and examples are:

Molarity   (abbreviated M) is defined as the number of moles of a substance dissolved in 1 L of solution.

Example 1

A 1 M H_2SO_4 solution has 98 g of H_2SO_4 (its gram-formula mass) in 1 L of the solution.

Solution:
This can be expressed as a formula, so that

$$\text{Molarity} = \frac{\text{No. of moles of solute}}{1 \text{ L of solution}}$$

If the molarity (M) and the volume of a solution are known, the mass of the solute can be determined by using the above equations and solving for the number of moles of solute and then multiplying this number by the molar mass.

Example 2

How many grams of NaOH are dissolved in 200 mL of solution if its concentration is 1.50 M?

Solution:

$$M = \frac{\text{No. of moles of solute}}{1 \text{ L of solution}}$$

Solving for number of moles, we have

No. of moles of solute $= M \times$ volume in liters of solution
No. of moles of NaOH $= 1.50$ $M \times$ volume in liters of solution
No. of moles of NaOH $= 1.5$ mol/L $\times 0.200$ L $= 0.30$ mol of NaOH

The molar mass of NaOH $= 23 + 16 + 1 = 40$ g of NaOH.

$$0.30 \; \cancel{mol \; NaOH} \times \frac{40 \text{ g of NaOH}}{1 \; \cancel{mol \; NaOH}} = 12 \text{ g of NaOH}$$

Molality   (abbreviated m) is defined as the number of moles of the solute dissolved in 1000 g of solvent.

Example 1

A 1 m solution of H_2SO_4 has 98 g of H_2SO_4 dissolved in 1000 g of water. This, you will notice, gives a total volume greater than 1 L, whereas the molar solution had 98 g in 1 L of solution since

$$\text{Molality } (m) = \frac{\text{Moles of solute}}{1000 \text{ g of solvent}}$$

Example 2

Suppose that 0.25 mol of sugar are dissolved in 500 g of water. What is the molality of this solution?

Solution:

$$m = \frac{\text{Moles of solute}}{1000 \text{ g of solvent}}$$

If there are 0.25 mol in 500 g of H_2O, then there are 0.50 mol in 1000 g of H_2O. So,

$$m = \frac{0.50 \text{ mol of sugar}}{1000 \text{ g of } H_2O} \quad \text{or} \quad 0.50$$

Normality   (abbreviated N) is used less frequently to express concentration and depends on knowledge of what a *gram-equivalent mass* is. This can be defined as the amount of a substance that reacts with or displaces 1 g of hydrogen or 8 g of oxygen. A simple method of determining the number of equivalents in a formula is to count the number of hydrogens or find the total positive oxidation numbers since each 1+ charge can be replaced by a hydrogen.

In H_2SO_4 (gram-formula mass = 98 g), there are 2 hydrogens, and so the gram-equivalent mass will be

$$\frac{98 \text{ g}}{2} = 49 \text{ g}$$

In Al_2SO_4 (molar mass = 150 g), there are 2 aluminums, each of which has a +3 oxidation number, making a total of +6. The gram-equivalent mass will be

$$\frac{150 \text{ g}}{6} = 49 \text{ g}$$

This idea of gram-equivalent mass is used in the definition of this means of expressing concentration, namely, the normality of a solution.

Normality is defined as the number of gram-equivalent masses of solute in 1 L of solution.

Example

If 49 g of H_2SO_4 is mixed with enough H_2O to make 500 mL of solution, what is the normality?

Solution:
Since normality is expressed for a liter, we must double the expression to 98 g of H_2SO_4 in 1000 mL of solution. To find the normality we find the number of gram-equivalents in a liter of solution. So

$$\frac{98\text{ g (no. of grams 1 L of solution)}}{49\text{ g (gram-equivalent mass)}} = 2N \text{ (normal)}$$

Dilution

Since the expression of molarity gives the quantity of solute per volume of solution, the amount of solute dissolved in a given volume of solution is equal to the product of the concentration and the volume. Hence 0.5 L of 2 M solution contains

$$M \times V = \text{amount of solute (in moles)}$$

$$\frac{2 \text{ mol}}{\cancel{L}} \times 0.5\cancel{L} = 1 \text{ mol (of solute in 0.5 L)}$$

Notice that volume units must be identical.

If you dilute a solution with water, the amount or number of moles of solute present remains the same, but the concentration changes. So you can use the expression

$$\begin{array}{cc} \text{Before} & \text{After} \\ M_1V_1 & = M_2V_2 \end{array}$$

This expression is useful in solving problems involving dilution.

Example

If you wish to make 1 L of solution that is 6 M into a 3 M solution, how much water must be added?

Solution:

$$M_1V_1 = M_2V_2$$
$$6\,M \times 1\,L = 3\,M \times ?\,L$$

Solving this expression:

? liters = 2 L. This is the total volume of the solution after dilution and means that 1 L of water has to be added to the original volume of 1 L to get a total of 2 L for the dilute solution volume.

An important use of the molarity concept is in the solution of *titration* problems, which are covered in Chapter 11 along with pH expressions of concentration for acids.

Colligative Properties of Solutions

Colligative properties are properties that depend primarily on the concentration of particles and not the type of particle. There is usually a direct relationship between the concentration of particles and the effect recorded.

The vapor pressure of an aqueous solution is always lowered by the addition of more solute. From the molecular standpoint, it is easy to see that there are fewer molecules of water per unit volume in the liquid, and therefore fewer molecules of water in the vapor phase are required to maintain equilibrium. The concentration in the vapor drops and so does the pressure that molecules exert. This is shown graphically in Figure 33.

Notice that the effects of this change in vapor pressure are registered in the freezing point and the boiling point. The freezing point is lowered, and the boiling point is raised, in direct proportion to the number of particles of solute present. For water solutions, the concentration expression that expresses this relationship is molality (m), that is, the number of moles of solute per kilogram of solvent. For molecules that do not dissociate, it has been found that a 1 m solution freezes at –1.86°C and boils at 100.51°C. A 2 m solution would then freeze at twice this lowering, or –3.72°C, and boil at twice the 1 m increase of 0.51°C, or 101.02°C.

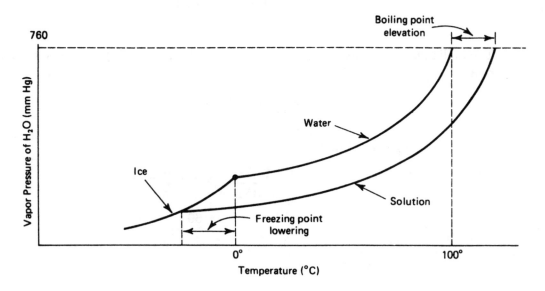

Figure 33. Diagram of Freezing Point Lowering—Boiling Point Rise.

The table below summarizes the colligative effect for aqueous solutions.

Type	Concentration (m)	Examples	Moles of Particles	Freezing Point (°C)	Boiling Point (°C)
Molecular nonionizing	1	Sugar, urea	1	−1.86	100.51
Molecular completely ionized	1	HCl	2	−3.72	101.02
Ionic, completely dissociated	1	NaCl	2	−3.72	101.02
Ionic, completely dissociated	1	$CaCl_2$ $Cu(NO_3)_2$	3	−5.58	101.53

Example 1

A 1.50-g sample of urea is dissolved in 105.0 g of water and produces a solution that boils at 100.12°C. From the data, what is the molecular mass of urea?

Solution:
Since this property is related to the molality, then

$$\frac{1.5 \text{ g}}{105.0 \text{ g}} = \frac{x \text{ grams}}{1000 \text{ g}}$$

$$x = 14.28 \text{ g in 1000 g of water}$$

The boiling point change is 0.12°C, and since each mole of particles causes a 0.51° increase, then

$$0.12°C \div \frac{0.51°C}{\text{mol}} = 0.235 \text{ mol}$$

Then 14.28 g = 0.235 mol and

$$\frac{14.28 \text{ g}}{0.235 \text{ mol}} = \frac{x \text{ grams}}{1 \text{ mol}}$$

$$x = 60.8 \text{ g/mol}$$

Example 2

Suppose that there are two water solutions, one of glucose (molar mass = 180), and the other of sucrose (molar mass = 342). Each contains 50 g of solute in 1000 g (1 kg) of water. Which has the higher boiling point? The lower freezing point?

Solution:

The molality of each of these nonionizing substances is found by dividing the number of grams of solute by the molecular mass.

$$\text{Glucose: } \frac{50 \text{ g}}{180 \text{ g/mol}} = 0.278 \text{ mol}$$

$$\text{Sucrose: } \frac{50 \text{ g}}{342 \text{ g/mol}} = 0.146 \text{ mol}$$

Both are in 1000 g of water, and so their respective molalities are 0.278 m and 0.146 m. Since the freezing point and boiling point are colligative properties, the effect depends only on the concentration. Because the glucose has a higher concentration, it will have a higher boiling point and a lower freezing point. The respective boiling points are

$$0.278 \text{ } m \times \frac{0.51°\text{C rise}}{1 \text{ } m} = 0.14 °\text{C rise or } 100.14°\text{C for glucose}$$

and $0.146 \text{ } m \times \dfrac{0.51°\text{C rise}}{1 \text{ } m} = 0.07°\text{C}$ rise in boiling point or 100.07°C for sucrose.

The lowering of the freezing point is

$$0.278 \text{ } m \times \frac{-1.86°\text{C drop}}{1 \text{ } m} = -0.52°\text{C drop below } 0°\text{C for glucose}$$

$$0.146 \text{ } m \times \frac{-1.86°\text{C drop}}{1 \text{ } m} = -0.27°\text{C drop below } 0°\text{C for sucrose}$$

Using a solute that is an ionic solid that completely ionizes in an aqueous solution introduces a consideration of the number of particles present in the solution. Notice in the previous chart that a 1-m solution of NaCl yields a solution with 2 mol of particles.

This is because

NaCl(aq)	→ Na$^+$	+ Cl$^-$
1 mol of ionic	1 mol of	1 mol of
sodium chloride salt	Na$^+$ ions	Cl$^-$ ions

So a 1-m solution of salt has 2 mol of ion particles in 1000 g of solvent. The colligative property of lowering the freezing point and raising the boiling point depends primarily on the concentration of particles and not the type of particles.

In a 1-m solution of NaCl there is 1 mol of sodium chloride salt dissolved in 1000 g of water, and so there is a total of 2 mol of ions released—1 mol of Na$^+$ ions and 1 mol of Cl$^-$ ions. Because this provides 2 mol of particles, it will cause a 2 × −1.86°C (drop caused by 1 mol) = −3.72°C drop in the freezing point and likewise a 2 × 0.51°C (rise caused by 1 mol) = 1.02°C rise in the boiling point or a boiling point of 101.02°C.

The chart also shows that CaCl$_2$ releases 3 mol of particles from 1 mol of CaCl$_2$ dissolved in 1000 g of water. Note that its effect is to lower the freezing point 3 × −1.86°C or to −5.58°C. The boiling point rise is also three times the molal rise of 0.51°C and gives a boiling point of 101.53°C.

This explains the use of salt on icy roads in the winter and the increased effectiveness of calcium chloride per mole of solute. The use of glycols in antifreeze solutions in automobile radiators is based on this same concept.

Crystallization

Many substances form a repeated pattern structure as they come out of solution. The structure is bounded by plane surfaces that make definite angles with each other to produce a geometric form called a *crystal*. The smallest portion of the crystal lattice that is repeated throughout the crystal is called the *unit cell*. The kinds of unit cells are shown in Figure 34. The crystal structure can also be classified by its internal axis, as shown in Figure 35.

A substance that holds a definite proportion of water in its crystal structure is called a *hydrate*. The formulas of hydrates show this water in the following manner: $CuSO_4 \cdot 5H_2O$; $CaSO_4 \cdot 2H_2O$; and $Na_2CO_3 \cdot 10H_2O$. (The · is read as "with.") When these crystals are heated gently, the water of hydration can be forced out of the crystal and the structure collapses into an anhydrous (without water) powder. The dehydration of hydrated $CuSO_4$ serves as a good example since the hydrated crystals are deep blue because of the water molecules present with the copper ions. When this water is removed, the structure crumbles into the anhydrous white powder. Some hydrated crystals, such as magnesium sulfate (Epsom salt), lose the water of hydration on exposure to air at ordinary temperatures. They

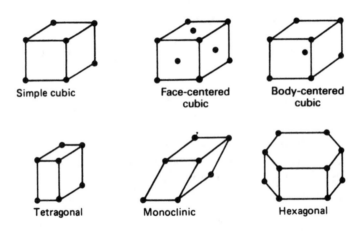

Figure 34. Kinds of Unit Cells.

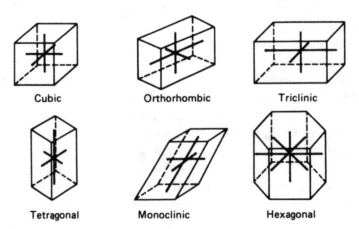

Figure 35. Crystal Structures Classified by Internal Axis.

are said to be *efflorescent*. Other hydrates, such as magnesium chloride and calcium chloride, absorb water from the air and become wet. They are said to be *deliquescent*, or *hydroscopic*. This property explains why calcium chloride is often used as a drying agent in laboratory experiments.

Chapter 7 Review

1. Distillation of water cannot remove
   (A) volatile liquids like alcohol
   (B) dissolved salts
   (C) suspensions
   (D) precipitates

2. The ratio in water of hydrogen to oxygen by mass is
   (A) 1 : 9
   (B) 2 : 1
   (C) 1 : 2
   (D) 1 : 8

3. Decomposing water by an electric current will give a ratio of hydrogen to oxygen by volume equal to
   (A) 1 : 9
   (B) 2 : 1
   (C) 1 : 2
   (D) 1 : 8

4. If 10 g of ice melts at 0°C, the total quantity of heat absorbed is
   (A) 10 cal
   (B) 80 cal
   (C) 800 cal
   (D) 8000 cal

5. To heat 10 g of water from 4°C to 14°C requires
   (A) 10 cal
   (B) 14 cal
   (C) 100 cal
   (D) 140 cal

6. The abnormally high boiling point of water in comparison to similar compounds is due primarily to
   (A) van der Waals forces
   (B) polar covalent bonding
   (C) dipole insulation
   (D) hydrogen bonding

7. A metallic oxide placed in water would most likely yield a(n)
   (A) acid
   (B) base
   (C) metallic anhydride
   (D) basic anhydride

8. A solution can be *both*
   (A) dilute and concentrated
   (B) saturated and dilute
   (C) saturated and unsaturated
   (D) supersaturated and saturated

9. The solubility of a solute must indicate
   (A) the temperature of the solution
   (B) the quantity of solute
   (C) the quantity of solvent
   (D) all of these

10. A foam is an example of
    (A) a gas dispersed in a liquid
    (B) a liquid dispersed in a gas
    (C) a solid dispersed in a liquid
    (D) a liquid dispersed in a liquid

11. When another crystal was added to a water solution of the same substance, the crystal seemed to remain unchanged. Its particles were
    (A) going into an unsaturated solution
    (B) exchanging places with others in the solution
    (C) causing the solution to become supersaturated
    (D) not going into solution in this static condition

12. A 10% solution of NaCl means that in 100 g of solution there is
    (A) 5.85 g NaCl
    (B) 58.5 g NaCl
    (C) 10 g NaCl
    (D) 94 g of H_2O

13. The gram-equivalent mass of $AlCl_3$ is its gram-formula mass divided by
    (A) 1
    (B) 3
    (C) 4
    (D) 6

14. The molarity of a solution made by placing 98 g of H_2SO_4 in sufficient water to make 500 mL of solution is
    (A) 0.5
    (B) 1
    (C) 2
    (D) 3
    (E) 4

15. If 684 g of sucrose (molecular mass = 342 g) is dissolved in 2000 g of water (essentially 2 L), what will be the freezing point of this solution?
    (A) –0.51°C
    (B) –1.86°C
    (C) –3.72°C
    (D) –6.58°C

Answers

1. **(A)**

2. **(D)**

3. **(B)**

4. **(C)**
   Each gram of ice that dissolves at 0°C absorbs 80 cal. So, 10 g × 80 cal/g = <u>800 cal</u>.

5. **(C)**
   By definition, 1 cal is the amount of heat needed to heat 1 g of water 1° on the Celsius scale.
   To heat 10 g from 4° to 14° would require

$$\underset{\text{Temperature difference}}{10\ g \times \ \overbrace{(14°-4°)}} = \underline{100\ cal}$$

6. **(D)**

7. **(B)**

8. **(B)**

9. **(D)**

10. **(A)**

11. **(B)**

12. **(C)**
    A 10% solution contains 10 g of solute per 100 g of solution because

$$10\% \times 100\ g = \underline{100\ g}$$

13. **(B)**

14. **(C)**
    98 g of H_2SO_4 = 1 mol H_2SO_4
    $\qquad$ 500 mL = 0.5 L
    So 1 mol in 0.5 L would be 2 mol in 1 L or a 2-L solution.

15. **(B)**

$$684\ g\ of\ sucrose = \frac{684}{342\ g/mol} = 2\ mol\ of\ sucrose$$

$$\frac{2\ mol}{2000\ g\ H_2O} = 1\ m\ solution$$

So the freezing point is lowered only –1.86°C.

Terms You Should Know

anhydride
boiling point
Brownian movement
colligative property
critical pressure
critical temperature
crystal
dynamic equilibrium
endothermic
exothermic
heat of fusion
heat of vaporization
"heavy" water
hydrate
Le Châtelier's Principle

melting point
molality
molarity
normality
phase equilibrium
polarity
solute
solvent
specific gravity
sublimation
surface tension
van der Waals forces
viscosity
water desalination

PART V:
CHEMICAL
REACTIONS

CHEMICAL REACTIONS AND THERMOCHEMISTRY

Types of Reactions

The many kinds of reactions you may encounter can be placed in four basic categories: combination, decomposition, single replacement, and double replacement.

The first type, *combination*, can also be called *synthesis*. This means the formation of a compound from the union of its elements. Some examples of this type are

$$Zn + S \rightarrow ZnS \quad \text{Zinc Sulfide}$$
$$2H_2 + O_2 \rightarrow 2H_2O \quad \text{WATER}$$
$$C + O_2 \rightarrow CO_2 \quad \text{CARBON DIOXIDE}$$

The second type of reaction, *decomposition*, can also be referred to as *analysis.* This means the breakdown of a compound to release its components as individual elements or other compounds. Some examples of this type are

$$2H_2O \rightarrow 2H_2 + O_2 \text{ (electrolysis of water)}$$
$$C_{12}H_{22}O_{11} \rightarrow 12C + 11H_2O$$
$$2HgO \rightarrow 2Hg + O_2$$

The third type of reaction is called *single replacement* or *single displacement*. This type can best be shown by examples in which one substance is displacing another. Some examples are

$$Fe + CuSO_4 \rightarrow FeSO_4 + Cu$$
$$Zn + H_2SO_4 \rightarrow ZnSO_4 + H_2(g)$$
$$Cu + 2AgNO_3 \rightarrow Cu(NO_3)_2 + 2Ag$$

The last type of reaction is called *double replacement* or *double displacement* because there is an actual exchange of "partners" to form new compounds. Some examples of this are

$$AgNO_3 + NaCl \rightarrow AgCl + NaNO_3$$

Sulfuric Acis + $H_2SO_4 + 2NaOH \rightarrow Na_2SO_4 + 2H_2O$ (neutralization)

Sodium Hydroxide $CaCO_3 + 2HCl \rightarrow H_2CO_3 + CaCl_2$

CARBONIC (unstable)

ACID $\rightarrow H_2O + CO_2(g)$

Predicting Reactions

One of the most important topics of chemistry deals with the reasons why reactions take place. Taking each of the above types of reactions, let us see how a prediction can be made concerning how the reaction gets the driving force to make it occur.

Combination (Synthesis)

The best source of information in predicting a chemical combination is the heat of formation table. A *heat of formation* table gives the number of calories evolved or absorbed when a mole (gram-formula mass) of the compound in question is formed by the direct union of its elements. In this book a positive number indicates that heat is absorbed, and a negative number that heat is evolved. It makes some difference whether the compounds formed are in the solid, liquid, or gaseous state. Unless otherwise indicated (g = gas, ℓ = liquid), the compounds are in the solid state. The values given are in kilocalories (1 kcal = 1000 cal). One calorie is the amount of heat needed to raise the temperature of 1 g of water 1° on the Celsius scale. The symbol ΔH is used to indicate the heat of formation.

If the heat of formation is a large number preceded by a minus sign, the combination is likely to occur spontaneously and the reaction is exothermic. If, on the other hand, the number is small and negative or is positive, heat will be needed to get the reaction to go at any noticeable rate.

Example 1

$$Zn + S \rightarrow ZnS + 48.5 \text{ kcal} \qquad \Delta H = -48.5 \text{ kcal}$$

Solution:
This means that 1 mol of zinc (65 g) reacts with 1 mol of sulfur (32 g) to form 1 mol of zinc sulfide (97 g) and releases 48.5 kcal of heat.

EXOTHERMIC

Example 2

$$Mg + \tfrac{1}{2}O_2 \rightarrow MgO + 143.84 \text{ kcal} \qquad \Delta H = -143.84 \text{ kcal}$$

Solution:

Indicates that the formation of 1 mol of magnesium oxide requires 1 mol of magnesium and $\frac{1}{2}$ mol of oxygen with the release of 143.84 kcal of heat. Notice the use of the fractional coefficient for oxygen. If the equation had been written with the usual whole-number coefficients, 2 mol of magnesium oxide would have been released.

$$2Mg + O_2 \rightarrow 2MgO + 2(-143.84) \text{ kcal}$$

Since, by definition, the heat of formation is given for the formation of *one* mole, this latter thermal equation shows 2(−143.84) kcal released.

Example 3

$$H_2(g) + \tfrac{1}{2}O_2(g) \rightarrow H_2O(\ell) + 68.32 \text{ kcal} \quad \Delta H = -68.32 \text{ kcal}$$

Solution:

In combustion reactions the heat evolved when 1 mol of a substance is completely oxidized is called the *heat of combustion* of the substance. So, in the equation

$$C(g) + O_2(g) \rightarrow CO_2(g) + 94.05 \text{ kcal} \quad \Delta H = -94.05 \text{ kcal}$$

ΔH is the heat of combustion of carbon. Because the energy of a system is conserved during chemical activity, the same equation could be arrived at by adding the following equations:

$$
\begin{array}{ll}
C(s) + \tfrac{1}{2}O_2(g) \rightarrow CO(g) & \Delta H = -26.41 \text{ kcal} \\
CO(g) + \tfrac{1}{2}O_2(g) \rightarrow CO_2(g) & \Delta H = -67.64 \text{ kcal} \\
\hline
C(s) + O_2(g) \rightarrow CO_2(g) & \Delta H = -94.05 \text{ kcal}
\end{array}
$$

Decomposition (Analysis)

The prediction of decomposition reactions uses the same source of information, the heat of formation table. If the heat of formation is a high exothermic (ΔH is negative) value, the compound will be difficult to decompose since this same quantity of energy must be returned to the compound. A low heat of formation indicates decomposition would not be difficult, such as the decomposition of mercuric oxide with $\Delta H = -21.68$ kcal/mol

$$2HgO \rightarrow 2Hg + O_2 \quad \text{(Priestley's method of preparation)}$$

A high positive heat of formation indicates extreme instability of a compound, and it can decompose explosively.

Single Replacement

A prediction of the feasibility of this type of reaction can be based on a comparison of the heat of formation of the original compound and that of the compound to be formed. For example, in a reaction of zinc with hydrochloric acid, the 2 mol of HCl have $\Delta H = 2 \times -22.06$ kcal.

$$Zn + 2HCl \rightarrow ZnCl_2 + H_2(g) \quad \textit{Note}: \Delta H = 0 \text{ for elements}$$
$$2 \times -22.06$$
$$-44.12 \text{ kcal} \neq -99.40 \text{ kcal}$$

and the zinc chloride has $\Delta H = -99.40$. This comparison leaves an excess of 55.28 kcal of heat given off, and so the reaction would occur.

In the next example, $-220.5 - (-184.0) = -36.5$ excess kilocalories to be given off as the reaction occurs

$$Fe + CuSO_4 \rightarrow FeSO_4 + Cu$$
$$-184.0 \text{ kcal} \qquad -220.5 \text{ kcal}$$

Another simple way of predicting single replacement reactions is to check the relative positions of the two elements in the electromotive chart (Table 10). If the element that is to replace the other in the compound is higher on the chart, the reaction will occur. If it is below, there will be no reaction.

Some simple examples of this are the following reactions: *p 214*

In predicting the replacement of hydrogen by zinc in hydrochloric acid, reference to the electromotive chart shows that zinc is above hydrogen. This reaction will occur

$$Zn + 2HCl \rightarrow ZnCl_2 + H_2(g)$$

In fact, all the metals above hydrogen in the chart on page 214 will replace hydrogen in an acid solution. If a metal such as copper is chosen, there will be no reaction.

$$Cu + HCl \rightarrow No \text{ reaction}$$

The determination of these replacements using a quantitative method is covered in Chapter 12.

Double Replacement

For double-replacement reactions to go to completion, that is, proceed until the supply of one of the reactants is exhausted, one of the following conditions must be present: an insoluble precipitate is formed, a nonionizing substance is formed, or a gaseous product is given off.

To predict the formation of an insoluble precipitate, you should have some knowledge of the solubilities of compounds. Table 9 gives some general solubility rules. (A table of solubilities could also be used as reference.)

An example of this type of reaction is given in its complete ionic form.

$$(K^+ + Cl^-) + (Ag^+ + NO_3^-) \rightarrow AgCl(s) + (K^+ + NO_3^-)$$

The silver ions combine with the chloride ions to form an insoluble precipitate, silver chloride. If the reaction had been like

$$(K^+ + Cl^-) + (Na^+ + NO_3^-) \rightarrow K^+ + NO_3^- + Na^+ + Cl^-$$

only a mixture of the ions would have been shown in the final solution.

TABLE 9

SOLUBILITIES OF COMPOUNDS

Soluble	Except
Na^+ ⎫	
NH_4^+ ⎬ compounds	
K^+ ⎭	
Acetates	
Bicarbonates	
Chlorates	
Chlorides.................................Ag^+, Hg^+, Pb ($PbCl_2$, sol. in hot water)	
Nitrates	
Sulfates ....Ba, Ca (slight), Pb	

Insoluble
Carbonates, phosphates ....Na, NH_4, K compounds
Sulfides, hydroxides................Na, NH_4, K, Ba, Ca

Another reason for a reaction of this type to go to completion is the formation of a non-ionizing product such as water. This weak electrolyte keeps its component ions in molecular form and thus removes the possibility of reversing the reaction. All neutralization reactions are of this type. BASE + ACID = H_2O

$$(H^+ + Cl^-) + (Na^+ + OH^-) \rightarrow H_2O + Na^+ + Cl^-$$

This example shows the ions of the reactants, hydrochloric acid and sodium hydroxide, and the nonelectrolyte product, water with sodium and chloride ions in solution. Since the water does not ionize to any extent, the reverse reaction cannot occur.

The third reason for double displacement to occur is the evolution of a gaseous product. An example of this is calcium carbonate reacting with hydrochloric acid

$$CaCO_3 + 2HCl \rightarrow CaCl_2 + H_2O + CO_2(g)$$

Another example of a compound that evolves a gas in sodium sulfite with an acid is

$$Na_2SO_3 + 2HCl \rightarrow 2NaCl + H_2O + SO_2(g)$$

In general, acids with carbonates or sulfites are good examples of this type of equation.

Hydrolysis Reactions

Hydrolysis reactions are the opposite of neutralization reactions. In hydrolysis the salt and water react to form an acid and a base. For example, if sodium chloride is placed in solution, this reaction occurs to some degree:

$$(Na^+ + Cl^-) + H_2O \rightarrow (Na^+ + OH^-) + (H^+ + Cl^-)$$

In this hydrolysis reaction the same number of hydrogen ions and of hydroxide ions is released, and so the solution is neutral. This is because sodium hydroxide is a strong base and hydrochloric acid is a strong acid. (There is a chart of acid and base strengths on page 452 to use as a reference.) Because they are both classified as strong, it means that they essentially exist as ions in solutions. Therefore, there is an excess of neither hydrogen nor hydroxide ions in the solution and it will test neutral. So, the salt of a strong acid and strong base forms a neutral solution when dissolved in water. However, if Na_2CO_3 is dissolved, we have

$$(2Na^+ + CO_3^{2-}) + 2H_2O \rightarrow (2Na^+ + 2OH^-) + H_2CO_3$$

The H_2CO_3 is written together because it is a slightly ionized acid or, in other words, a weak acid. Since the hydroxide ions are free in the solution, the solution is basic. Notice that here it is the salt of a strong base and a weak acid that forms a basic solution. This generalization is true for this type of salt.

If we use the salt of a strong acid and a weak base, the reaction will be

$$(Zn^{2+} + 2Cl^-) + H_2O \rightarrow (2H^+ + 2Cl^-) + Zn(OH)_2$$

In this case the hydroxide ions are held in the weakly ionizing compound while the hydrogen ions are free to make the solution acidic. In general then, the salt of a strong acid and a weak base forms an acid solution by hydrolysis.

The fourth possibility is that of a salt of a weak acid and a weak base dissolving in water. An example is ammonium carbonate, $(NH_4)_2CO_3$, which is the salt of a weak base and a weak acid. The hydrolysis reaction is

$$(NH_4)_2CO_3 + 2H_2O \rightarrow 2NH_4OH + 2H_2CO_3$$

Both the ammonium hydroxide, NH_4OH, and the carbonic acid, H_2CO_3, are written as nonionized compounds because they are classified as a weak base and a weak acid, respectively. Therefore, a salt of a weak acid and a weak base forms a neutral solution since neither hydrogen ion nor hydroxide ion will be present in excess.

Entropy

In many of the preceding predictions of reactions, we used the concept that reactions will occur when they result in the lowest possible energy state.

There is, however, another driving force for reactions that relates to their state of disorder or of randomness. This state of disorder is called *entropy*. A reaction is also driven, then, by a need for a greater degree of disorder. An example is the intermixing of gases in two connected flasks when a valve is opened to allow the two previously isolated gases to travel between the two flasks. Since temperature remains constant throughout the process, the total heat content cannot have changed to a lower energy level, and yet the gases will become evenly distributed in the two flasks. The system has thus reached a higher degree of disorder or entropy. A quantitative treatment of entropy is given on page 191.

Thermochemistry

In general, all chemical reactions either liberate or absorb heat. The origin of chemical energy lies in the position and motion of atoms, molecules, and subatomic particles. The total energy possessed by a molecule is the sum of all the forms of potential and kinetic energy associated with it.

The energy changes in a reaction are due, to a large extent, to the changes in potential energy that accompany the breaking of chemical bonds in reactants to form new bonds in products.

The molecule may also have rotational, vibrational, and translational energy, along with some nuclear energy sources. All these make up the total energy of molecules. In beginning chemistry, the greatest concern in reactions is the electronic energy involved in the making and breaking of chemical bonds.

Since it is virtually impossible to measure the total energy of molecules, the energy change is usually the experimental data that we deal with in reactions. This change in quantity of energy is known as the change in *enthalpy* (heat content) of the chemical system and is symbolized by ΔH.

Changes in Enthalpy

Changes in enthalpy for exothermic and endothermic reactions can be shown graphically.

Notice that the ΔH for an endothermic reaction is positive, while that for an exothermic reaction is negative. It should be noted also that changes in enthalpy are always independent of the path taken to change a system from the initial state to the final state.

Since the quantity of heat absorbed or liberated during a reaction varies with the temperature, scientists have adopted 25°C and 1 atm pressure as the *standard state* condition for reporting heat data. A superscript zero on ΔH (i.e., ΔH^0) indicates that the corresponding process is carried out under standard conditions. The *standard enthalpy of formation* (ΔH_f^0) of a compound is defined as the change in enthalpy that accompanies the formation of 1 mol of a compound from its elements with all substances in their standard states at 25°C. This is called the *molar heat of reaction*.

To calculate the enthalpy of a reaction, it is necessary to write an equation for the reaction. The standard enthalpy change, ΔH, for a given reaction is usually expressed in kilocalories and depends on how the equation is written. For example, here is the equation for the reaction of hydrogen with oxygen written in two ways

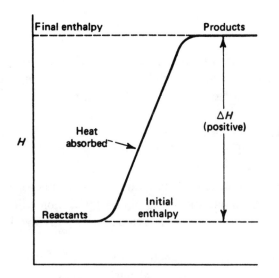

Course of Endothermic Reaction

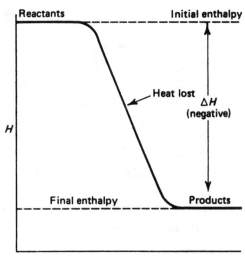

Course of Exothermic Reaction

$$H_2(g) + \tfrac{1}{2}O_2(g) \rightarrow \quad H_2O(g) \qquad \Delta H_f^0 = -57.8 \text{ kcal}$$

$$2H_2(g) + O_2(g) \rightarrow \quad 2H_2O(g) \qquad \Delta H_f^0 = -115.6 \text{ kcal}$$

Experimentally, ΔH_f^0 for the formation of 1 mol of $H_2O(g)$ is -57.8 kcal. Since the second equation represents the formation of 2 mol of $H_2O(g)$, the quantity is twice -57.8 or -115.6 kcal. It is assumed that the initial and final states are measured at 25°C and 1 atm, although the reaction occurs at a higher temperature.

Example

How much heat is liberated when 40.0 g of $H_2(g)$ reacts with excess $O_2(g)$?

Solution:
The reaction equation is

$$H_2(g) + \tfrac{1}{2}O_2(g) \rightarrow H_2O(g) \qquad \Delta H_f^0 = -57.8 \text{ kcal}$$

This represents 1 mol or 2 g of H(g) forming 1 mol of $H_2O(g)$

$$40 \text{ g} \times \frac{1 \text{ mol}}{2 \text{ g}} = 20 \text{ mol of hydrogen}$$

Since each mole gives off -57.8 kcal, then

$$20 \text{ mol} \times \frac{-57.8 \text{ kcal}}{1 \text{ mol}} = -1156 \text{ kcal}$$

Notice that the physical state of each participant must be given since the phase changes involve energy changes.

Combustion reactions produce a considerable amount of energy in the form of light and heat when a substance is combined with oxygen, The heat released by the complete combustion of one mole of a substance is called the *heat of combustion* of the substance. Heat of combustion is defined in terms of one mole of reactant, whereas heat of formation is defined in terms of one mole of product. All substances are in their standard state. The general enthalpy notation, ΔH, applies to heats of reaction, but with the addition of a subscripted c, ΔH_c, refers specifically to heat of combustion.

Additivity of Reaction Heats and Hess's Law

Chemical equations and ΔH^0 values may be manipulated algebraically. Finding the ΔH for the formation of vapor from liquid water shows how this can be done.

$$H_2(g) + \tfrac{1}{2}O_2(g) \rightarrow H_2O(g) \qquad \Delta H_f^0 = -57.8 \text{ kcal}$$
$$H_2(g) + \tfrac{1}{2}O_2(g) \rightarrow H_2O(\ell) \qquad \Delta H_f^0 = -68.3 \text{ kcal}$$

Since we want the equation for $H_2O(\ell) \rightarrow H_2O(g)$, we can reverse the second equation. This changes the sign of ΔH.

$$H_2O(\ell) \rightarrow H_2O(g) + \tfrac{1}{2}O_2(g) \qquad \Delta H_f^0 = 68.3 \text{ kcal}$$

Adding

$$H_2(g) + \tfrac{1}{2}O_2(g) \rightarrow H_2O(g) \qquad \Delta H_f^0 = -57.8 \text{ kcal}$$

yields

$$H_2O(\ell) + H_2(g) + \tfrac{1}{2}O_2(g) \rightarrow$$
$$H_2(g) + \tfrac{1}{2}O_2(g) + H_2O(g) \qquad \Delta H_f^0 = 10.5 \text{ kcal}$$

Simplification gives a net equation of

$$H_2O(\ell) \rightarrow H_2O(g) \qquad \Delta H_f^0 = 10.5 \text{ kcal}$$

The principle underlying the preceding calculations is known as *Hess's Law of Heat Summation*. This principle states that when a reaction can be expressed as the algebraic sum of two or more other reactions, the heat of the reaction is the algebraic sum of the heats of these reactions. This is based upon the *First Law of Thermodynamics*, which, simply stated, says that the total energy of the universe is constant and cannot be created or destroyed.

These laws allow calculations of ΔH values that cannot be easily determined experimentally. An example is determination of the ΔH of CO from the ΔH_f^0 of CO_2.

$$C(s) + \tfrac{1}{2}O_2(g) \rightarrow CO(g) \qquad \Delta H = ?$$

The calculation of ΔH can be done in kcal/mol units or kJ/mol. The solution is shown using both units.

$$C(s) + O_2(g) \rightarrow CO_2(g) \qquad \Delta H_f^0 = -94.0 \text{ kcal or } -393.5 \text{ kJ/mol}$$

$$CO(g) + \tfrac{1}{2}O_2(g) \rightarrow CO_2(g) \qquad \Delta H_f^0 = -67.6 \text{ kcal or } -283.0 \text{ kJ/mol}$$

The equation wanted is

$$C(s) + \tfrac{1}{2}O_2(g) \rightarrow CO(g)$$

To get this, we reverse the second equation and add it to the first

$$C(s) + O_2(g) \rightarrow CO_2(g) \qquad \Delta H_f^0 = -94.0 \text{ kcal or } -393.5 \text{ kJ/mol}$$

$$CO_2(g) \rightarrow \tfrac{1}{2}O_2(g) + CO(g) \qquad \Delta H_f^0 = +67.6 \text{ kcal or } +283.0 \text{ kJ/mol}$$

Addition yields

$$C(s) + \tfrac{1}{2}O_2(g) \rightarrow CO(g) \qquad \Delta H_f^0 = -26.4 \text{ kcal or } -110.5 \text{ kJ/mol}$$

The relationship can be shown schematically as below.

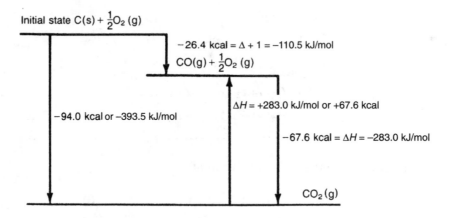

Some commonly used standard heats of formation (enthalpy), designated as ΔH_f^0, are listed in Table Ⓔ in the Equations and Tables for Reference section.

An alternative (and easier) method of calculating enthalpies is based on the concept that $\Delta H_{reaction}^0$ is equal to the difference between the total enthalpy of the reactants and that of the products. This can be expressed as

$$\Delta H_{reaction}^0 = \Sigma \text{ (sum of) } \Delta H_f^0 \text{ (products)} - \Sigma \text{ (sum of) } \Delta H_f^0 \text{ (reactants)}$$

Example 1

Calculate $\Delta H^0_{reaction}$ for the decomposition of sodium chlorate.

Solution:

$$NaClO_3(s) \rightarrow NaCl(s) + \tfrac{3}{2}O_2(g)$$

Step 1:   Obtain ΔH^0_f for all substances.

$$NaClO_3(s) = -85.7 \text{ kcal/mol}$$
$$NaCl(s) \quad = -98.2 \text{ kcal/mol}$$
$$O_2(g) \qquad = 0 \text{ kcal/mol (all elements} = 0)$$

Step 2:   Substitute these values in the equation.

$$\Delta H^0_{reaction} = \Sigma \Delta H^0_f \text{ (products)} - \Sigma \Delta H^0_f \text{ (reactants)}$$
$$\Delta H_{reaction} = -98.2 - (-85.7) \text{ kcal}$$
$$\Delta H_{reaction} = -12.5 \text{ kcal}$$

Example 2

Calculate $\Delta H_{reaction}$ for this oxidation of ammonia:

$$4NH_3(g) + 5O_2(g) \rightarrow 6H_2O(g) + 4NO(g)$$

Solution:
The individual ΔH^0_f values are

$$4NH_3 \quad = 4(-11 \text{ kcal}) \quad = -44.9 \text{ kcal}$$
$$5O_2(g) \quad = 0$$
$$6H_2O(g) = 6(-57.8 \text{ kcal}) = -346.8 \text{ kcal}$$
$$4NO(g) \quad = 4(21.6 \text{ kcal}) \quad = 86.4 \text{ kcal}$$

Substituting these in the $\Delta H_{reaction}$ equation gives

$$\Delta H_{reaction} = [-346.8 \text{ } (6H_2O) + 86.4 \text{ } (4NO)] - [-44.0 \text{ } (4NH_3) + 0 \text{ } (5O_2)]$$
$$\Delta H_{reaction} = -216.4 \text{ kcal}$$

Bond Dissociation Energy

The same principle of additivity applies to bond energies. Experimentation has found average bond energies for particular bonds; some of the more common are shown in the chart.

Bond	Energy (kcal/mol)	Bond	Energy (kcal/mol)
H—H	104	O=O	119
C—H	99	O—H	111
C—C	83	C—O	86
C=C	146	C=O	177
C≡C	200	H—F	135
O—O	35	H—Cl	103
N—N	39	H—Br	87
Cl—Cl	58	H—I	71

Example

Find the bond energy, H_{be}, for ethane.

Solution:

Ethane = H—C—C—H and has 6 C—H bonds and 1 C—C bond.

$$H_{be} = 6E_{C—H} + 1E_{C—C} = 6(99 \text{ kcal}) + 1(83 \text{ kcal}) = 677 \text{ kcal}$$

Enthalpy from Bond Energies

A reaction's enthalpy can be approximated through the summation of bond energies. An example is the reaction

$$H_2(g) + Cl_2(g) = 2HCl(g)$$

Bonds Broken	Heat Energy Absorbed (kcal)	Bonds Formed	Heat Energy Evolved (kcal)
H—H	104.2		
Cl—Cl	57.8	2H—Cl	206.0
Total	162.0		206.0

The difference between heat evolved and heat absorbed is 206.0 − 162.0 = 44.0 kcal. This is for 2 mol of HCl, and so −44 kcal divided by 2 = −22 kcal/mol. This answer is very close to the experimentally determined ΔH_f^0.

Chapter 8 Review

1. A synthesis reaction will occur spontaneously if the heat of formation of the product is
    (A) large and negative
    (B) small and negative
    (C) large and positive
    (D) small and positive

2. The reaction of aluminum with dilute H_2SO_4 can be classified as
    (A) synthesis
    (B) decomposition
    (C) single replacement
    (D) double replacement

3. For a metal atom to replace another kind of metallic ion in a solution, the metal atom must be
    (A) a good oxidizing agent
    (B) higher on the electromotive chart than the metal in solution
    (C) lower on the electromotive chart than the metal in solution
    (D) equal in activity to the metal in solution

4. One reason for a double-displacement reaction to go to completion is that
    (A) a product is soluble
    (B) a product is given off as a gas
    (C) the products can react with each other
    (D) the products are miscible

5. Hydrolysis will give an acid reaction when which of these is placed in solution with water?
    (A) Na_2SO_4
    (B) K_2SO_4
    (C) $NaNO_3$
    (D) $Cu(NO_3)_2$

6. A salt derived from a strong base and a weak acid will undergo hydrolysis and give a solution that will be
    (A) basic
    (B) acid
    (C) neutral
    (D) volatile

7. Enthalpy is an expression for the
    (A) heat content
    (B) energy state
    (C) reaction rate
    (D) activation energy

8. The ΔH_f^0 of a reaction is recorded for
    (A) 0°C
    (B) 25°C
    (C) 100°C
    (D) 200°C

9. The property of being able to add enthalpies is based on the
   (A) Law of Conservation of Heat
   (B) First Law of Thermodynamics
   (C) Law of Constants
   (D) Law of $E = mc^2$

10. If $\Delta H_{reaction}$ is –100 kcal/mol, it indicates the reaction is
    (A) endothermic
    (B) unstable
    (C) in need of a catalyst
    (D) exothermic

Answers

1. **A**	3. **B**	5. **D**	7. **A**	9. **B**
2. **C**	4. **B**	6. **A**	8. **B**	10. **D**

Terms You Should Know

combination reaction
decomposition reaction
double-replacement reaction
enthalpy
entropy

heat of combustion
hydrolysis reaction
molar heat of formation
reaction mechanism
single replacement reaction
standard enthalpy of formation

First Law of Thermodynamics
Hess's Law of Heat Summation

RATES OF CHEMICAL REACTIONS

Measurements of Reaction Rates

The measurement of reaction rate is based on the rate of appearance of a product or disappearance of a reactant. It is usually expressed in terms of a change in concentration of one of the participants per unit time.

Experiments have shown that for most reactions the concentrations of all participants change most rapidly at the beginning of the reaction, that is, the concentration of the product shows the greatest rate of increase, and the concentrations of the reactants the highest rate of decrease, at this point. This means that the rate of a reaction changes with time. Therefore a rate must be identified with a specific time.

Factors Affecting Reaction Rates

Five important factors control the rate of a chemical reaction. These are summarized below.

1. ***The nature of the reactants.*** In chemical reactions, some bonds are broken and others are formed. Therefore, the rates of chemical reactions should be affected by the nature of the bonds in the reacting substances. For example, reactions between ions in an aqueous solution may take place in a fraction of a second. Thus, the reaction between silver nitrate and sodium chloride is very fast. The appearance of the white silver chloride precipitate is immediate. In reactions where many covalent bonds must be broken, reactions usually take place slowly at room temperatures. The decomposition of hydrogen peroxide into water and oxygen happens slowly at room temperature. In fact, it takes about 17 minutes (min) for half the peroxide in a 0.50 M solution to decompose.

2. ***The surface area exposed.*** Since most reactions depend on the reactants coming into contact, the surface exposed proportionally affects the rate of the reaction. Consider what happens to the surface area exposed when a cube is sliced in half.

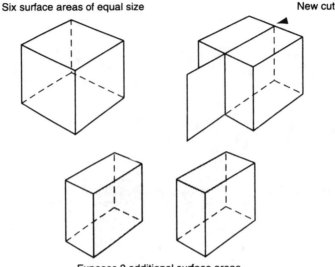

Six surface areas of equal size   New cut

Exposes 2 additional surface areas
or a total of 8 surface areas of equal area.

There is an increase of $33\frac{1}{3}\%$ of surface area that can now become involved in a chemical reaction. This is why reactants that depend on surface reactions are often introduced into reaction vessels as powders. It also explains the danger of a dust explosion if the reactant is flammable.

3. ***The concentrations.*** The reaction rate is usually proportional to the concentrations of the reactants. The usual dependence of the reaction rate on the concentration of the reactants can be simply explained by theorizing that if there are more molecules or ions of the reactant in the reaction area, then there is a greater chance that more reactions will occur. This idea is further developed in the collision theory described below.

4. ***The temperature.*** A temperature increase of 10°C above room temperature usually causes the reaction rate to double or triple. The basis for this generality is that as the temperature increases, the average kinetic energy of the particles involved increases. This means that the particles move faster and have a greater probability of hitting another reactant particle and, because they have more energy, causing an effective collision that results in the chemical reaction that forms the product substance.

5. ***The catalyst.*** A substance that increases the rate of chemical reaction without itself undergoing any permanent chemical change. The catalyst provides an alternative pathway by which the reaction can proceed and in which the activation energy is lower. It thus increases the rate at which the reaction comes to completion or equilibrium. Generally, the term is used for a substance that increases the reaction rate (a positive catalyst). Some reactions can be slowed down by negative catalysts.

Collision Theory of Reaction Rates

This theory makes the assumption that, for a reaction to occur, there must be collisions between the reacting species. This means that the rate of reaction depends on two factors: the number of collisions per unit time, and the fraction of these collisions that are successful because enough energy is involved.

There is a definite relationship between the concentrations of the reactants and the number of collisions. Graphically, the reaction of

$$A + B \rightarrow AB$$

can be examined as the concentrations are changed.

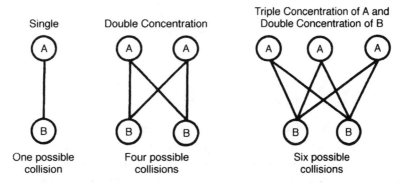

This shows that the number of collisions, and consequently the rate of reaction, are proportional to the product of the concentrations. Simply stated, then, the rate of reaction is directly proportional to the concentrations.

Activation Energy

Often a reaction rate may be increased or decreased by affecting the activation energy, that is, the energy necessary to cause a reaction to occur. This is shown graphically below.

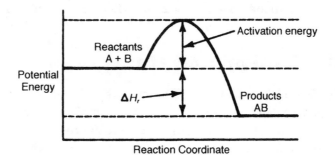

A *catalyst* is a substance that is introduced into a reaction to either speed up or slow down the reaction. This is accomplished by changing the amount of activation energy needed. The effect of a catalyst used to speed up a reaction can be shown as

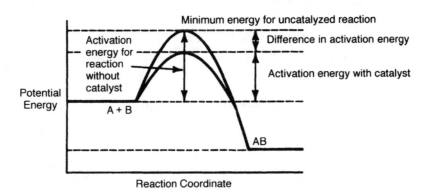

Reaction Rate Law

The relationship between the rate of a reaction and the masses (expressed as concentrations) of the reacting substances is summarized in the *Law of Mass Action*. It states that the rate of a chemical reaction is proportional to the product of the concentrations of the reactants. For a general reaction between A and B, represented by

$$aA + bB \rightarrow \cdots$$

the rate law expression is

$$r \propto [A]^a[B]^b$$

or, inserting a constant of proportionality that mathematically changes the expression to an equality, we have

$$r = k[A]^a[B]^b$$

Here k is called the *specific rate* constant for the reaction at the temperature of the reaction. The exponents a and b may be added to give the total reaction order. For example

$$H_2(g) + I_2(g) \rightarrow 2HI(g)$$
$$r = k[H_2]^1[I_2]^1$$

The sum of the exponents is $1 + 1 = 2$, and therefore we have a second-order reaction.

Reaction Mechanism and Rates of Reaction

Under the factors listed as affecting the rates of reaction, it was stated that the reaction rate is *usually* proportional to the concentration of the reactants. The reason for this is that some reactions do not occur directly between the reactants but may go through intermediate steps to get to the final product. The series of steps by which the reacting particles rearrange themselves to form the products of a chemical reaction is called the *reaction mechanism*. For example

Step 1:	A + B	→ I$_1$ (fast)
Step 2:	A + I$_1$	→ I$_2$ (slow)
Step 3:	C + I$_2$	→ D (fast)
Net equation:	2A + B + C → D	

Notice that the reactions in Steps 1 and 3 occur relatively fast compared to the reaction in Step 2. Now suppose that we increase the concentration of C. This will make the reaction in Step 3 go faster, but it will have little effect on the speed of the overall reaction since Step 2 is the rate-determining step. If, however, the concentration of A is increased, then the overall reaction rate will increase because it will speed up Step 2. Knowing the reaction mechanism provides the basis for predicting the effect of a concentration change of a reactant on the overall rate of reaction. Another way of determining the effect of concentration changes is to do actual experimentation.

Chapter 9 Review

1. List the five factors that affect the rate of a reaction:

    (1) _____

    (2) _____

    (3) _____

    (4) _____

    (5) _____

2. The addition of a catalyst to a reaction
    (A) changes the enthalpy
    (B) changes the entropy
    (C) changes the nature of the products
    (D) changes the activation energy

3. An increase in concentration
    (A) is related to the number of collisions directly
    (B) is related to the number of collisions inversely
    (C) has no effect on the number of collisions

4. At the beginning of a reaction, the reaction rate for the reactants is
    (A) largest, then decreasing
    (B) largest and remains constant
    (C) smallest, then increasing
    (D) smallest and remains constant

5. The reaction rate law applied to the reaction $aA + bB \rightarrow AB$ gives the expression
    (A) $r \propto [A]^b[B]^a$
    (B) $r \propto [AB]^a[A]^b$
    (C) $r \propto [B]^a[AB]^b$
    (D) $r \propto [A]^a[B]^b$

Answers

1. (1) Nature of the reactants
   (2) Surface area exposed
   (3) Concentrations
   (4) Temperature
   (5) Presence of a catalyst

2. (**D**)

3. (**A**)

4. (**A**)

5. (**D**)

Terms You Should Know

catalyst   Law of Mass Action
collision theory

Chapter 10

CHEMICAL EQUILIBRIUM

Reversible Reactions and Equilibrium

In some reactions no product is formed to allow the reaction to go to completion; that is, the reactants and products can still interact in both directions. This can be shown as

$$A + B \rightleftharpoons C + D$$

The double arrow indicates that C and D can react to form A and B, while A and B react to form C and D.

The reaction is said to have reached *equilibrium* when the forward reaction rate is equal to the reverse reaction rate. Notice that this is a dynamic condition, *not* a static one, although in appearance the reaction *seems* to have stopped. An example of an equilibrium is a crystal of copper sulfate in a saturated solution of copper sulfate. Although to the observer the crystal seems to remain unchanged, there is actually an equal exchange of crystal material with the copper sulfate in solution. As some solute comes out of solution, an equal amount goes into solution.

To express the rate of reaction in numerical terms, we can use the *Law of Mass Action*, discussed in Chapter 9, which states: The rate of a chemical reaction is proportional to the product of the concentrations of the reacting substances. The concentrations are expressed in moles of gas per liter of volume or moles of solute per liter of solution. Suppose, for example, that 1 mol/L of gas A_2 (diatomic molecule) is allowed to react with 1 mol/L of another diatomic gas, B_2, and they form gas AB; let R be the rate for the forward reaction forming AB. The bracketed symbols $[A_2]$ and $[B_2]$ represent the concentrations in moles per liter for these diatomic molecules. Then $A_2 + B_2 \rightarrow 2AB$ has the rate expression

$$R \propto [A_2] \times [B_2]$$

where $\propto$ is the symbol for "proportional to." When $[A_2]$ and $[B_2]$ are both 1 mol/L, the reaction rate is a certain constant value (k_1) at a fixed temperature.

$$R = k_1 \qquad (k_1 \text{ is called the rate constant})$$

For any concentrations of A and B, the reaction rate is

$$R = k_1 \times [A_2] \times [B_2]$$

If $[A_2]$ is 3 mol/L and $[B_2]$ is 2 mol/L, the equation becomes

$$R = k_1 \times 3 \times 2 = 6k_1$$

The reaction rate is six times the value for a 1 mol/L concentration of both reactants.

At the fixed temperature of the forward reaction, AB molecules are also decomposing. If we designate this reverse reaction as R', then, since

$$2AB \ (or \ AB + AB) \rightarrow A_2 + B_2$$

two molecules of AB must decompose to form one molecule of A_2 and one of B_2. Thus the reverse reaction in this equation is proportional to the square of the molecular concentration of AB.

$$R' \propto [AB] \times [AB]$$
$$or \ R' \propto [AB]^2$$
$$and \ R' \propto k_2 \times [AB]^2$$

where k_2 represents the rate of decomposition of AB at the fixed temperature. Both reactions can be shown

$$A_2 + B_2 \rightleftharpoons 2AB \qquad \text{(note double arrows)}$$

When the first reaction begins to produce AB, some AB is available for the reverse reaction. If the initial condition is only the presence of A_2 and B_2 gases, then the forward reaction will occur rapidly to produce AB. As the concentration of AB increases, the reverse reaction will increase. At the same time, the concentrations of A_2 and B_2 will be decreasing and consequently the forward reaction rate will decrease. Eventually the two rates will be equal, that is, $R = R'$. At this point, equilibrium has been established, and

$$k_1[A_2] \times [B_2] = k_2[AB]^2$$

or

$$\frac{k_1}{k_2} = \frac{[AB]^2}{[A_2] \times [B_2]} = K_{eq}$$

The convention is that k_1 (forward reaction) is placed over k_2 (reverse reaction) to obtain this expression. Then k_1/k_2 can be replaced by K_{eq}, which is called the *equilibrium constant* for this reaction under the particular conditions.

In another general example, like

$$aA + bB \rightleftharpoons cC + dD$$

the reaction rates can be expressed as

$$R = k_1[A]^a \times [B]^b$$
$$R' = k_2[C]^c \times [D]^d$$

Note that the values of k_1 and k_2 are different, but that each is a constant for the conditions of the reaction. At the start of the reaction, [A] and [B] will be at their greatest values, and R will be large; [C], [D], and R' will be zero. Gradually R will decrease, and R' will become equal. At this point the reverse reaction is forming the original reactants just as rapidly as they are being used up by the forward reaction. Therefore no further change in R, R', or any of the concentrations will occur.

If we set R' equal to R, we have

$$k_2 \times [\text{C}]^c \times [\text{D}]^d = k_1 \times [\text{A}]^a \times [\text{B}]^b$$

or

$$\frac{[\text{C}]^c \times [\text{D}]^d}{[\text{A}]^a \times [\text{B}]^b} = \frac{k_1}{k_2} = K_{eq}$$

This process of two substances A and B reacting to form products C and D, and the reverse, can be shown graphically to visualize what happens as equilibrium is established. This hypothetical equilibrium reaction is described by the following general equation.

$$a\text{A} + b\text{B} \rightleftharpoons c\text{C} + d\text{D}$$

At the beginning (at time t_0), the concentrations of C and D are zero and those of A and B are maximum. The graph below shows that over time the rate of the forward reaction decreases as A and D are used up. Meanwhile, the rate of the reverse reaction increases as C and D are formed. When these two reaction rates become equal (at time t_1), equilibrium is established. The individual concentrations of A, B, C, and D no longer change if the conditions remain the same. To an observer, it appears that all reaction has stopped, when in fact both reactions are occurring at the same rate.

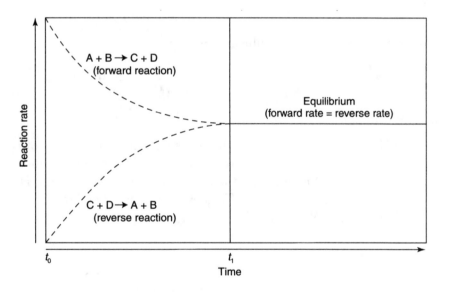

Rate Comparison for the Equilibrium Reaction System

We see that for the given reaction and the given conditions, K_{eq} is a constant, called the *equilibrium constant*. If K_{eq} is large, this means that equilibrium will not occur until the concentrations of the original reactants are small and those of the products large. A small value of K_{eq} means that equilibrium occurs almost at once and relatively little product is produced.

The equilibrium constant, K_{eq}, has been determined experimentally for many reactions, and the values are printed in chemical handbooks. (See Tables ① and ⓙ in the Equations and Tables for Reference section.)

Suppose we find the K_{eq} for reacting H_2 and I_2 at 490°C to be equal to 45.9. Then the equilibrium constant for the reaction

$$H_2 + I_2 \rightleftharpoons 2HI \text{ at } 490°C$$

is

$$K_{eq} = \frac{[HI]^2}{[H_2][I_2]} = 45.9$$

Example

Three moles of H_2 and 3 mol of I_2 are introduced into a 1-L box at a temperature of 490°C. Find the concentration of each substance in the box when equilibrium is established.

Solution:
Initial conditions:

$$[H_2] = 3 \text{ mol/L}$$
$$[I_2] = 3 \text{ mol/L}$$
$$[HI] = 0 \text{ mol/L}$$

The reaction proceeds to equilibrium and

$$K_{eq} = \frac{[HI]^2}{[H_2][I_2]} = 45.9$$

At equilibrium, then,

$\quad\quad [H_2] = (3 - x)$ moles/liter (where x is the number of moles of H_2 that are in the form of HI at equilibrium)

$\quad\quad [I_2] = (3 - x)$ moles/liter (the same x is used since 1 mol of H_2 requires 1 mol of I_2 to react to form 2 mol of HI)

$\quad\quad [HI] = 2x$ moles/liter

and so

$$K_{eq} = \frac{(2x)^2}{(3-x)(3-x)} = 45.9$$

If

$$\frac{(2x)^2}{(3-x)^2} = 45.9$$

then taking the square root of each side gives

$$\frac{2x}{3-x} = 6.77$$

Solving for x,

$$x = 2.32$$

Substituting this x value into the concentration expressions at equilibrium, we have

$$[H_2] = (3 - x) = 0.68 \text{ mol/L}$$
$$[I_2] \;\; = (3 - x) = 0.68 \text{ mol/L}$$
$$[HI] = 2x \quad\;\; = 4.64 \text{ mol/L}$$

The crucial step in this type of problem is setting up the concentration expressions from your knowledge of the equation. Suppose that this problem had been as follows.

Find the concentrations at equilibrium for the same conditions as in the preceding example except that only 2 mol of HI are injected into the box.

$$[H_2] = 0 \text{ mol/L}$$
$$[I_2] \;\; = 0 \text{ mol/L}$$
$$[HI] = 2 \text{ mol/L}$$

At equilibrium

$$[HI] = (2 - x) \text{ moles/liter}$$

(For every mole of HI that decomposes, only 0.5 mol of H_2 and 0.5 mol of I_2 are formed.)

$$[H_2] = 0.5x$$
$$[I_2] = 0.5x$$

$$K_{eq} = \frac{(2-x)^2}{(x/2)^2} = 45.9$$

Solving for x gives

$$x = 0.456$$

Then, substituting into the equilibrium conditions,

$$[HI] = 2 - x = 1.54 \text{ mol/L}$$

$$[I_2] = \tfrac{1}{2}x = 0.228 \text{ mol/L}$$

$$[H_2] = \tfrac{1}{2}x = 0.228 \text{ mol/L}$$

Le Châtelier's Principle

A general law, *Le Châtelier's Principle*, can be used to explain the results of applying any change of condition (stress) on a system in equilibrium. It states that if a stress is placed upon a system in equilibrium, the equilibrium is displaced in the direction that counteracts the effect of the stress. An increase in concentration of a substance favors the reaction that uses up that substance and lowers its concentration. A rise in temperature favors the reaction that absorbs heat and so tends to lower the temperature. These ideas are further developed below.

Effects of Changing Conditions

Effect of Changing Concentrations

When a system at equilibrium is disturbed by adding or removing one of the substances (thus changing its concentration), all the concentrations will change until a new equilibrium point is reached with the same value of K_{eq}.

If the concentration of a reactant in the forward reaction is increased, the equilibrium is displaced to the right, favoring the forward reaction. If the concentration of a reactant in the reverse reaction is increased, the equilibrium is displaced to the left. Decreases in concentration will produce effects opposite to those produced by increases.

Effect of Temperature on Equilibrium

If the temperature of a given equilibrium reaction is changed, the reaction will shift to a new equilibrium point. If the temperature of a system in equilibrium is raised, the equilibrium is shifted in the direction that absorbs heat. Note that the shift in equilibrium as a result of temperature change is actually a change in the value of the equilibrium constant. This is different from the effect of changing the concentration of a reactant; when concentrations are changed, the equilibrium shifts to a condition that maintains the same equilibrium constant.

Effect of Pressure on Equilibrium

A change in pressure affects only equilibria in which a gas or gases are reactants or products. Le Châtelier's Principle can be used to predict the direction of displacement. If it is assumed that the total space in which the reaction occurs is constant, the pressure will depend on the total number of molecules in that space. An increase in the number of molecules will increase pressure; a decrease in the number of molecules will decrease pressure. If the pressure is increased, the reaction that will be favored is the one that will lower the pressure, that is, decrease the number of molecules.

An example of the application of these principles is the Haber process of making ammonia. The reaction is

$$N_2 + 3H_2 \rightleftharpoons 2NH_3 + \text{heat (at equilibrium)}$$

If the concentrations of the nitrogen and hydrogen are increased, the forward reaction is increased. At the same time, if the ammonia produced is removed by dissolving it in water, the forward reaction is again favored.

Since the reaction is exothermic, the addition of heat must be considered with care. Increasing the temperature causes an increase in molecular motion and collisions, thus allowing the product to form more readily. At the same time, the equilibrium equation shows that the reverse reaction is favored by the increased temperature, and so a compromise temperature of about 500°C is used to get the best yield.

An increase in pressure will cause the forward reaction to be favored since the equation shows that four molecules of reactants form two molecules of products. This effect tends to reduce the increase in pressure by the formation of more ammonia.

Equilibria in Heterogeneous Systems

The examples so far have been of systems made up of only gaseous substances. The expression of the K_{eq} of systems is changed with the presence of other phases.

Equilibrium Constant for Systems Involving Solids

If the experimental data for this reaction are studied:

$$CaCO_3(s) \rightleftharpoons CaO(s) + CO_2(g)$$

it is found that at a given temperature an equilibrium is established in which the concentration of CO_2 is constant. It is also true that the concentrations of the solids have no effect on the CO_2 concentration as long as both solids are present. Therefore the K_{eq}, which would conventionally be written as

$$K_{eq} = \frac{[CaO][CO_2]}{[CaCO_3]}$$

can be modified by incorporating the concentrations of the two solids. This can be done since the concentration of solids is fixed by the density. The K_{eq} becomes a new constant, K

$$K = [CO_2]$$

Any heterogeneous reaction involving gases does not include the concentrations of pure solids. As another example, K for the reaction.

$$NH_4Cl(s) \rightleftharpoons NH_3(g) + HCl(g)$$

is

$$K = [NH_3][HCl]$$

Acid Ionization Constants

When a weak acid that does not ionize completely is placed in solution, an equilibrium is reached between the acid molecule and its ions. The mass action expression can be used to derive an equilibrium constant, called the acid *ionization constant*, for this condition. For example, an acetic acid solution ionizing is shown as

$$HC_2H_3O_2 + H_2O \rightleftharpoons H_3O^+ + C_2H_3O_2^-$$

$$K = \frac{[H_3O^+][C_2H_3O_2^-]}{[HC_2H_3O_2][H_2O]}$$

The concentration of water in mol/L is found by dividing the mass of 1 L of water (which is 1000 g at 4°C) by its gram-molecular mass, 18 g, giving H_2O a value of 55.6 mol/L. Since this number is so large compared to the other numbers involved in the equilibrium constant, it is practically constant and is incorporated into a new equilibrium constant designated K_a. Then the new expression is

$$K_a = \frac{[H_3O^+][C_2H_3O_2^-]}{[HC_2H_3O_2]}$$

Ionization constants have been found experimentally for many substances and are listed in chemical tables. The ionization constants of ammonia and acetic acid are about 1.8×10^{-5}. For boric acid $K_a = 5.8 \times 10^{-10}$, and for carbonic acid $K_a = 4.3 \times 10^{-7}$.

If the concentrations of the ions present in the solution of a weak electrolyte are known, the value of the ionization constant can be calculated. Also, if the value of K_i is known, the concentrations of the ions can be calculated.

A small value for K_a means that the concentration of the un-ionized molecule must be relatively large compared to the ion concentrations. Conversely, a large value for K_a means that the concentrations of ions are relatively high. Therefore the smaller the ionization constant of an acid, the weaker the acid. Thus, for the three acids referred to above, the ionization constants show that the weakest of these is boric acid, and the strongest, acetic acid. It should be remembered that, in all cases where ionization constants are used, the electrolytes must be weak in order to be involved in ionic equilibria.

Ionization Constant of Water

Since water is a very weak electrolyte, its ionization constant can be expressed as

$$2H_2O \rightleftharpoons H_3O^+ + OH^-$$

(Equilibrium constant) $K = \dfrac{[H_3O^+][OH^-]}{[H_2O]^2}$ (since $[H_2O]^2$ remains relatively constant, it is incorporated into K_w)

(Ionization constant) $K_w = [H_3O^+][OH^-] = 1 \times 10^{-14}$ at 25°C

From this expression, we see that for distilled water $[H_3O^+] = [OH^-] = 1 \times 10^{-7}$. Therefore the pH, which is $-\log[H_3O^+]$, is

$$pH = -\log[1 \times 10^{-7}]$$
$$pH = -[-7] = 7 \text{ for a neutral solution}$$

A pH of less than 7 is acid, and a pH of greater than 7 is basic. (More information on pH is given on page 199.)

Example

(This sample incorporates the entire discussion of ionization constants, including finding the pH.)

Calculate (a) the $[H_3O^+]$, (b) the pH, and (c) the percentage dissociation for 0.100 M acetic acid at 25°C. The symbol K_a is used for the ionization of acids. K_a for $HC_2H_3O_2$ is 1.8×10^{-5}.

Solution:
(a) For this reaction

$$H_2O(\ell) + HC_2H_3O_2(\ell) \rightleftharpoons H_3O^+(aq) + C_2H_3O_2^-(aq)$$

and

$$K_a = \frac{[H_3O^+][C_2H_3O_2^-]}{[HC_2H_3O_2]} = 1.8 \times 10^{-5}$$

Let x = number of moles/liter of $HC_2H_3O_2$ that dissociate and reach equilibrium. Then

$$[H_3O^+] = x, \; [C_2H_3O_2^-] = x, \; [HC_2H_3O_2] = 0.1 - x$$

Substituting in the expression for K_a gives

$$K_a = 1.8 \times 10^{-5} = \frac{(x)(x)}{0.10 - x}$$

Since weak acids, like acetic acid, at concentrations of $0.01 \, M$ or greater dissociate very little, the equilibrium concentration of the acid is very nearly equal to the original concentration; that is,

$$0.10 - x \cong 0.10$$

Therefore, the expression can be changed to

$$1.8 \times 10^{-5} = \frac{(x)(x)}{0.10}$$

$$x^2 = 1.8 \times 10^{-6}$$
$$x^2 = 1.3 \times 10^{-3} = [H_3O^+]$$

(b) Substituting this result into the pH expression gives

$$pH = -\log[H_3O^+] = -\log[1.3 \times 10^{-3}]$$
$$= 3 - \log 1.3$$
$$= 2.9$$

(c) The percentage of dissociation of the original acid may be expressed as

$$\% \text{ dissociation} = \frac{\text{mol/L that dissociate}}{\text{original concentration}}(100)$$

$$\% \text{ dissociation} \times \frac{1.3 \times 10^{-3}}{1.0 \times 10^{-1}}(100) = 1.3\%$$

Solubility Products

A saturated solution of an ionic solid has been defined as an equilibrium condition between the solute and its ions. For example

$$AgCl \rightleftharpoons Ag^+ + Cl^-$$

The equilibrium constant is

$$K = \frac{[Ag^+][Cl^-]}{[AgCl]}$$

Since the concentration of the solute remains constant for that temperature, the $[AgCl]$ is incorporated into K to give K_{sp}, called the *solubility product constant*:

$$K_{sp} = [Ag^+][Cl^-] = 1.2 \times 10^{-10} \text{ at } 25°C$$

This setup can be used to solve problems in which the ionic concentrations are given and the K_{sp} is to be found or the K_{sp} is given and the ionic concentrations are to be determined.

Example 1

By experimentation it is found that a saturated solution of $BaSO_4$ at 25°C contains 3.9×10^{-5} mol/L of Ba^{2+} ions. Find the K_{sp} of this salt.

Solution:
Since $BaSO_4$ ionizes into equal numbers of Ba^{2+} and SO_4^{2-}, the barium ion concentration will equal the sulfate ion concentration. So the solution is

$$BaSO_4 \rightleftharpoons Ba^{2+} + SO_4^{2-}$$
$$K_{sp} = [Ba^{2+}][SO_4^{2-}]$$

Therefore

$$K_{sp} = (3.9 \times 10^{-5})(3.9 \times 10^{-5}) = 1.5 \times 10^{-9}$$

Example 2

If the K_{sp} of radium sulfate, $RaSO_4$, is 4×10^{-11}, calculate its solubility in pure water.

Solution:
Let x = moles of $RaSO_4$ that dissolve per liter of water. Then, in the saturated solution,

$$[Ra^{2+}] = x \text{ mol/L}$$
$$[SO_4^{2-}] = x \text{ mol/L}$$
$$RaSO_4(s) \rightleftharpoons Ra^{2+} + SO_4^{2-}$$
$$[Ra^{2+}][SO_4^{2-}] = K_{sp} = 4 \times 10^{-11}$$

Let $x = [Ra^{2+}]$ and $[SO_4^{2-}]$. Then

$$(x)(x) = 4 \times 10^{-11}$$
$$x = 6 \times 10^{-6} \text{ mol/L}$$

Thus the solubility of $RaSO_4$ is 6×10^{-6} mol/L of water, giving a solution 6×10^{-6} M in Ra^{2+} and 6×10^{-6} M in SO_4^{2-}.

Example 3

In some cases the solubility products of solutions can be used to predict the formation of a precipitate.

Suppose we have two solutions. One contains 1.00×10^{-3} mol of silver nitrate, $AgNO_3$, per liter. The other solution contains 1.00×10^{-2} mol of sodium chloride, $NaCl$, per liter. If 1 L of each of these solutions is mixed to make a 2-L mixture, will a precipitate of $AgCl$ form?

Solution:
In the $AgNO_3$ solution, the concentrations are

$$[Ag^+] = 1.00 \times 10^{-3} \text{ mol/L} \quad \text{and} \quad [NO_3^-] = 1.00 \times 10^3 \text{ mol/L}$$

In the NaCl solution, the concentrations are

$$[Na^+] = 1 \times 10^{-2} \text{ mol/L} \quad \text{and} \quad [Cl^-] = 1.00 \times 10^{-2} \text{ mol/L}$$

When 1 L of each of these solutions is mixed to form a total volume of 2 L, the concentrations will be halved.

In the mixture then the initial concentrations will be

$$[Ag^+] = 0.50 \times 10^{-3} \quad \text{or} \quad 5.0 \times 10^{-4} \text{ mol/L}$$
$$[Cl^-] = 0.50 \times 10^{-2} \quad \text{or} \quad 5.0 \times 10^{-3} \text{ mol/L}$$

In the K_{sp} of AgCl,

$$[Ag^+][Cl^-] = [5.0 \times 10^{-4}][5.0 \times 10^{-3}]$$
$$[Ag^+][Cl^-] = 2.5 \times 10^{-7} \quad \text{or} \quad 2.5 \times 10^{-6}$$

This is far greater than 1.7×10^{-10} which is the K_{sp} of AgCl. These concentrations cannot exist, and Ag^+ and Cl^- will combine to form solid AgCl precipitate. Only enough Ag^+ ions and Cl^- ions will remain to make the product of the respective ions concentration equal 1.7×10^{-10}.

Common Ion Effect

When a reaction has reached equilibrium and an outside source adds more of one of the ions that is already in solution, the result is to cause the reverse reaction to occur at a faster rate and reestablish the equilibrium. This is called the *common ion effect*. For example, in the equilibrium reaction

$$NaCl(s) \rightleftharpoons Na^+ + Cl^-$$

the addition of concentrated HCl ($12\,M$) adds H^+ and Cl^- both in a concentration of $12\,M$. This increases the concentration of the Cl^- and disturbs the equilibrium. The reaction will shift to the left and cause some solid NaCl to come out of solution.

The "common" ion is the one already present in an equilibrium before a substance is added that increases the concentration of that ion. The effect is to reverse the solution reaction and to decrease the solubility of the original substance, as shown in the above example.

Factors Related to the Magnitude of K

Relation of Minimum Energy (Enthalpy) to Maximum Disorder (Entropy)

Some reactions are said to go to completion because the equilibrium condition is achieved when practically all the reactants have been converted to products. At the other extreme, some reactions reach equilibrium immediately, with very little product being formed. These two examples are representative of very large K values and very small K values, respectively. There are essentially two driving forces that control the extent of a reaction and determine when equilibrium will be established. These are the drive to the lowest heat content, or *enthalpy*, and the drive to the greatest randomness or disorder, which is called *entropy*. Reactions with negative ΔH values (enthalpy or heat content) are exothermic, and reactions with positive ΔS values (entropy or randomness) proceed to greater randomness.

The *Second Law of Thermodynamics* states that the entropy of the universe increases for any spontaneous process. This means that the entropy of a system may increase or decrease but that, if it decreases, then the entropy of the surroundings must increase to a greater extent so that the overall change in the universe is positive. In other words,

$$\Delta S_{universe} = \Delta S_{system} + \Delta S_{surroundings}$$

The following is a list of conditions in which ΔS is positive for the system:

1. When a gas is formed from a solid, for example,

$$CaCO_3(s) \rightarrow CaO(s) + CO_2(g)$$

2. When a gas is evolved from a solution, for example,

$$Zn(s) + 2H^+(aq) \rightarrow H_2(g) + Zn^{2+}(aq)$$

3. When the number of moles of gaseous product exceeds the number of moles of gaseous reactant, for example,

$$2C_2H_6(g) + 7O_2(g) \rightarrow 4CO_2(g) + 6H_2O(g).$$

4. When crystals dissolve in water, for example,

$$NaCl(s) \rightarrow Na^+(aq) + Cl^-(aq).$$

Looking at specific examples, we find that in some cases endothermic reactions occur when the products provide greater randomness or positive entropy. This reaction is an example:

$$CaCO_3(s) \rightleftharpoons CaO(s) + CO_2(g)$$

The production of the gas, and thus greater entropy, might be expected to take this reaction almost to completion. However, this is not true because another force is hampering this reaction. It is the absorption of energy, and thus the increase in enthalpy, as the $CaCO_3$ is heated.

The equilibrium condition, then, at a particular temperature, is a compromise between the increase in entropy and the increase in enthalpy of the system.

The Haber process of making ammonia is another example of this compromise of driving forces that affect the establishment of an equilibrium. In the reaction

$$N_2(g) + 3H_2(g) \rightleftharpoons 2NH_3(g) + heat$$

the forward reaction to reach the lowest heat content and thus release energy cannot go to completion because the force to maximum randomness is driving the reverse reaction.

Change in Free Energy of a System—Gibbs Equation

There factors can be combined in an equation that summarizes the change of *free energy* in a system. This is designated as ΔG. The relationship is

$$\Delta G = \Delta H - T\,\Delta S \qquad (T \text{ is temperature in kelvins})$$

and is called the Gibbs Free-Energy Equation. The sign of ΔG can be used to predict the spontaneity of a reaction at constant temperature and pressure. If ΔG is negative, the reaction is (probably) spontaneous; if ΔG is positive, the reaction is improbable; and if ΔG is 0, the system is at equilibrium and there is no net reaction.

The ways in which the factors in the equation affect ΔG are shown in this table

ΔH	ΔS	Will It Happen?	Comment
Exothermic (−)	+	Yes	No exceptions
Exothermic (−)	−	Probably	At low temperature
Endothermic (+)	+	Probably	At high temperature
Endothermic (+)	−	No	No exceptions

This drive to achieve a minimum of free energy may be interpreted as the driving force of a chemical reaction.

Chapter 10 Review

1. For the reaction $A + B \rightleftharpoons C + D$, the equilibrium constant can be expressed as

   (A) $K_{eq} = \dfrac{[A][B]}{[C][D]}$

   (B) $K_{eq} = \dfrac{[C][B]}{[A][D]}$

   (C) $K_{eq} = \dfrac{[C][D]}{[A][B]}$

   (D) $K_{eq} = \dfrac{C \cdot D}{A \cdot B}$

2. The concentrations in an expression of the equilibrium constant are given in
   (A) mol/mL
   (B) g/L
   (C) gram-equivalents/L
   (D) mol/L

3. In the equilibrium expression for the reaction $BaSO_4 \rightleftharpoons Ba^{2+} + SO_4^{2-}$ K_{sp} is equal to

   (A) $[Ba^{2+}] + [SO_4^{2-}]$

   (B) $\dfrac{[Ba^{2+}][SO_4^{2-}]}{BaSO_4}$

(C) $\dfrac{[Ba^{2+}][SO_4{}^{2-}]}{[BaSO_4]}$

(D) $\dfrac{[BaSO_4]}{[Ba^{2+}][SO_4{}^{2-}]}$

4. The K_w of water is equal to
   (A) 1×10^{-7}
   (B) 1×10^{-17}
   (C) 1×10^{-14}
   (D) 1×10^{-1}

5. The pH of a solution that has a hydrogen ion concentration of 1×10^{-4} mol/L is
   (A) 4
   (B) –4
   (C) 10
   (D) –10

6. The pH of a solution that has a hydroxide ion concentration of 1×10^{-4} mol/L is
   (A) 4
   (B) –4
   (C) 10
   (D) –10

7. A small value for K_{eq}, the equilibrium constant, indicates that
   (A) the concentration of the un-ionized molecules must be relatively small compared to the ion concentrations
   (B) the concentration of the ionized molecules must be larger than the ion concentrations
   (C) the substance ionizes to a large degree
   (D) the concentration of the un-ionized molecules must be relatively large compared to the ion concentrations

8. In the Haber process for making ammonia, an increase in pressure favors
   (A) the forward reaction
   (B) the reverse reaction
   (C) neither reaction

9. A change in which of these conditions will change the K of an equilibrium?
   (A) temperature
   (B) pressure
   (C) concentration of reactants
   (D) concentration of products

10. If $Ca(OH)_2$ is dissolved in a solution of NaOH, its solubility, compared to that in pure water, is
    (A) increased
    (B) decreased
    (C) unaffected

Answers

1. **C**	3. **A**	5. **A**	7. **D**	9. **A**
2. **D**	4. **C**	6. **C**	8. **A**	10. **B**

Terms You Should Know

acid ionization constant
common ion effect
enthalpy
entropy
equilibrium

Le Châtelier's Principle
Second Law of Thermodynamics

equilibrium constant
Haber process
Gibbs free-energy equation
solubility product constant

Chapter 11

ACIDS, BASES, AND SALTS

Definitions and Properties

Acids

There are some characteristic properties by which an acid may be defined. The most important are

Water solutions of acids conduct electricity. This conduction depends on their degree of ionization. A few acids ionize almost completely, while others ionize only to a slight degree. The table below indicates some common acids and their degrees of ionization.

DEGREE OF IONIZATION OF COMMON ACIDS

Completely or Nearly Completely Ionized	Moderately Ionized	Slightly Ionized
Nitric	Oxalic	Hydrofluoric
Hydrochloric	Phosphoric	Acetic
Sulfuric	Sulfurous	Carbonic
Hydriodic		Hydrosulfuric
Hydrobromic		(Most others)

Acids will react with metals that are more active than hydrogen ions (see chart on page 214) to liberate hydrogen. (Some acids are also strong oxidizing agents and will not release hydrogen. Somewhat concentrated nitric acid is such an acid.)

Acids have the ability to change the color of indicators. Some common indicators are litmus and phenolphthalein. Litmus is a dyestuff obtained from plants. When litmus is added to an acidic solution, or paper impregnated with litmus is dipped into an acid, the neutral purple color changes to pink-red. Phenolphthalein is red in a basic solution and becomes colorless in a neutral or acid solution.

Acids react with bases so that the properties of both are lost to form water and a salt.   This is called *neutralization*. The general equation is:

$$\text{Acid} + \text{base} \rightarrow \text{salt} + \text{water}$$

An example is

$$Mg(OH)_2 + H_2SO_4 \rightarrow MgSO_4 + 2H_2O$$

If an acid is known to be a weak solution, you might taste it and note the sour taste.

Acids react with carbonates to release carbon dioxide.   An example $CaCO_3 + 2HCl$ $\rightarrow CaCl_2 + H_2CO_3$ (unstable and decomposes)

$$1 \rightarrow H_2O + CO_2(g)$$

The most common theory used in first-year chemistry is the *Arrhenius Theory*, which states that an acid is a substance that yields hydrogen ions in an aqueous solution. Although we speak of hydrogen ions in solution, they are really not separate ions but become attached to the oxygen of the polar water molecule to form the H_3O^+ ion (the hydronium ion). So it is really this hydronium ion we are concerned with in an acid solution.

The general reaction for the dissociation of an acid, HX, is commonly written

$$HX \rightleftharpoons H^+ + X^-$$

To show the formation of the hydronium ion, H_3O^+, the complete equation is

$$HX + H_2O \rightleftharpoons H_3O^+ + X^-$$

A list of common acids and their formulas is given in Chapter 4, Table 7, with an explanation of the naming procedures for acids preceding the table.

Bases

Bases may also be defined by some operational definitions based on experimental observations. Some of the important ones are

Bases are conductors of electricity in an aqueous solution.   Their degree of conduction depends on their degree of ionization. See the table of common bases below and their degree of ionization.

DEGREE OF IONIZATION OF COMMON BASES

Completely or Nearly Completely Ionized	Slightly Ionized
Potassium hydroxide	Ammonium hydroxide
Sodium hydroxide	(All others)
Barium hydroxide	
Strontium hydroxide	
Calcium hydroxide	

Bases cause a color change in indicators.   Litmus changes from red to blue in a basic solution, and phenolphthalein turns pink from its colorless state.

Bases react with acids to neutralize each other and form a salt and water.

Bases react with fats to form a class of compounds called soaps.   Earlier generations used this method to make their own soap.

Aqueous solutions of bases feel slippery, and the stronger bases are very caustic to the skin.

The Arrhenius Theory

Defines a base as a substance that yields hydroxide ions (OH^-) in an aqueous solution.
 Some of the common bases have common names. The list below gives some of these:

Sodium hydroxide—Lye, caustic soda
Potassium hydroxide—Caustic potash
Calcium hydroxide—Slaked lime, hydrated lime, limewater
Ammonium hydroxide—Ammonia water, household ammonia

Much of the sodium hydroxide produced today comes from the Hooker cell electrolysis apparatus. We have already discussed the electrolysis apparatus used for the decomposition of water. When an electric current is passed through a salt water solution, hydrogen, chlorine, and sodium hydroxide are the products. The formula for this equation is

$$2NaCl + 2HOH \xrightarrow[\text{energy}]{\text{electrical}} H_2(g) + Cl_2(g) + 2NaOH$$

Broader Acid-Base Theories

Besides the common Arrhenius Theory of acids and bases discussed for aqueous solutions, two other theories, the Brønsted Theory and the Lewis Theory, are widely used.

The Brønsted Theory (1923)

Defines acids as proton donors, and bases as proton acceptors. This definition agrees with the aqueous solution definition of an acid giving up hydrogen ions in solution but goes beyond to other cases as well.
 An example of this is dry HCl gas reacting with ammonia gas to form the white solid NH_4Cl.

$$HCl(g) + NH_3(g) \rightarrow NH_4^+(s) + Cl^-(s)$$

The HCl is the proton donor or acid, and the ammonia is a Brønsted base which accepts the proton.

Conjugate Acids and Bases

In an acid-base reaction, the original acid gives up its proton to become a conjugate base. In other words, after losing its proton, the remaining ion is capable of gaining a proton, thus qualifying as a base. The original base accepts a proton, and so it now is classified as a conjugate acid since it can release this newly acquired proton and thus behave like an acid.

Some examples of this are

$$\text{HCl} + \text{NH}_3 \longrightarrow \text{NH}_4^+ + \text{Cl}^-$$

Acid   Base   Conjugate acid   Conjugate base

* * * *

$$\text{HC}_2\text{H}_3\text{O}_2 + \text{H}_2\text{O} \rightleftharpoons \text{H}_3\text{O}^+ + \text{C}_2\text{H}_3\text{O}_2^-$$

Acid   Base   Conjugate acid   Conjugate base

The Lewis Theory (1916)

The Lewis Theory of acids and bases is defined in terms of the electron-pair concept. It is probably the most generally useful concept of acids and bases. According to the Lewis definition, an acid is an electron-pair acceptor; and a base is an electron-pair donor. An example of this is the formation of ammonium ions from ammonia gas and hydrogen ions

x hydrogen electrons
o nitrogen electrons

Notice that the hydrogen ion is in fact accepting the electron pair of the ammonia, and so it is a Lewis acid. The ammonia is donating its electron pair, and so it is a Lewis base.

Acid Concentration Expressed as pH

Frequently, acid and base concentrations are expressed by means of the pH system. The pH can be defined as $-\log[\text{H}^+]$, where $[\text{H}^+]$ is the concentration of hydrogen ions expressed in mol/L. The logarithm is the exponent of 10 when the number is written in the base 10. For example,

$$100 = 10^2 \text{ and so logarithm of 100, base } 10 = 2$$
$$10{,}000 = 10^4 \text{ and so logarithm of 10,000, base } 10 = 4$$
$$0.01 = 10^{-2} \text{ and so logarithm of 0.01, base } 10 = -2$$

The logarithms of more complex numbers can be found in a logarithm table. An example of a pH problem is

Find the pH of a 0.1-M solution of HCl.

Step 1: Since HCl ionizes almost completely into H^+ and Cl^-, $[H^+] = 0.1$ mol/L.

Step 2: By definition,

$$pH = -\log[H^+]$$

So

$$pH = -\log[10^{-1}]$$

Step 3: The logarithm of 10^{-1} is -1.
So

$$pH = -(-1)$$

Step 4: The pH then is 1.

Since water has a normal H^+ concentration of 10^{-7} mol/L because of the slight ionization of water molecules, the water pH is 7 when it is neither acid nor base. The normal pH range is from 1 to 14.

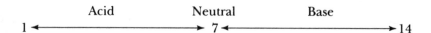

The pOH is the negative logarithm of the hydroxide ion concentration

$$pOH = -\log[OH^-]$$

If the concentration of the hydroxide ion is $10^{-9}\,M$, then the pOH of the solution is 9. From the equation

$$[H^+][OH^-] = 1.0 \times 10^{-14} \text{ at } 25°C$$

the following relationship can be derived

$$pH + pOH = 14.00$$

In other words, the sum of the pH and pOH of an aqueous solution at 25°C must always equal 14.00. For example, if the pOH of a solution is 9.00, then its pH must be 5.00.

Example

What is the pOH of a solution whose pH is 3.0?

Solution:
Substituting the 3.0 for pH in the expression

$$pH + pOH = 14.0$$
$$3.0 + pOH = 14.0$$
$$pOH = 14 - 3.0 = 11.0$$

Indicators

Some indicators can be used to determine pH because of their color changes somewhere along this pH scale. Some common indicators and their respective color changes are given below.

Indicator	pH Range of Color Change *(mixture of 2 colors)*	Color below Lower pH	Color above Higher pH
Methyl orange	3.1–4.4	Red	Yellow
Bromthymol blue	6.0–7.6	Yellow	Blue
Litmus	4.5–8.3	Red	Blue
Phenolphthalein	8.3–10.0	Colorless	Red

The following is an example of how to read this chart. At pH values below 4.5, litmus is red; above 8.3, it is blue. Between these values, it is a mixture of the two colors.

In choosing an indicator for a titration, we need to consider if the solution formed when the endpoint is reached has a pH of 7. Depending on the type of acid and base used, the resulting hydrolysis of the salt formed may cause it to be slightly acidic, slightly basic, or neutral. If the titration is of a strong acid and a strong base, the endpoint will be at pH 7 and practically any indicator can be used. This is because the addition of 1 drop of either reagent will change the pH at the endpoint by about 6 units. For titrations of strong acids and weak bases, we need an indicator, such as methyl orange, that changes color between 3.1 and 4.4 in the acid region. In the titration of a weak acid and a strong base, we should use an indicator that changes in the basic range. Phenolphthalein is a suitable choice for this type of titration since it changes color in the pH 8.3 to 10.0 range.

Volumetric Analysis—Titration

Knowledge of the concentrations of solutions and the reactions they take part in can be used to determine the concentrations of "unknown" solutions or solids. The use of volume measurement in solving these problems is called *volumetric analysis*.

A common example of a volumetric analysis uses acid-base reactions. If you are given a base of known concentration (standard solution), let us say 0.10 M NaOH, and you want to determine the concentration of an HCl solution, you could *titrate* the solutions in the following manner.

First, introduce a measured quantity, 25.0 mL, of the NaOH into a flask by using a pipet or buret. Next, introduce 2 drops of a suitable indicator. Since NaOH and HCl are considered a strong base and strong acid, respectively, an indicator that changes color in the middle pH range would be appropriate. Litmus solution would be one choice. It is blue in a basic solution but changes to red when the solution becomes acidic. Slowly introduce the HCl until the color change occurs. This is called the *endpoint*. This is the point at which enough acid solution is added to neutralize all the base in solution. It is also referred to as the *equivalence point*.

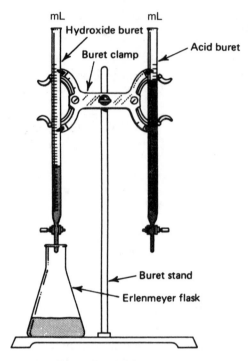

Buret Setup for Titration

Suppose 21.5 mL is needed to produce the color change; the reaction that occurs is

$$H^+(aq) + OH^-(aq) \rightarrow H_2O$$

until all the OH^- is neutralized, and then the excess H^+ causes the litmus paper to change color. To solve the question of the concentration of the NaOH we use

$$M_{acid} \times V_{acid} = M_{base} \times V_{base}$$

Substituting the known amounts into this equation gives

$$xM_{acid} \times 21.5 \text{ mL} = 0.1\,M \times 25.0 \text{ mL}$$
$$x = 0.116\,M$$

In choosing an indicator for a titration, we need to consider whether the solution formed when the endpoint is reached has a pH of 7. Depending on the types of acids and bases used, the resulting hydrolysis of the salt formed may cause the solution to be slightly acidic, slightly basic, or neutral. If a strong acid and a strong base are titrated, the endpoint will be at pH 7, and practically any indicator can be used because adding a drop of either reagent will change the pH at the endpoint by about 6 units. For the titration of strong acids and weak bases, we need an indicator, such as methyl orange, that changes color between pH 3.1 and 4.4 in the acid region. When titrating a weak acid and a strong base, we should use an indicator that changes in the basic range. Phenolphthalein is a suitable choice for this type of titration since it changes color in the pH 8.3 to 10.0 range.

The process of the neutralization reaction can be represented by a titration curve like the one below, which shows the titration of a strong acid with a strong base.

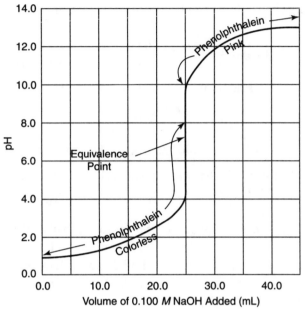

Titration of a Strong Acid with a Strong Base

Example 1

Find the concentration of acetic acid in vinegar if 21.6 mL of 0.20 M NaOH is needed to titrate a 25-mL sample of the vinegar.

Solution:
Using the equation $M_{acid} \times V_{acid} = M_{base} \times V_{base}$, we have

$$xM \times 25 \text{ mL} = 0.2 \times 21.6 \text{ mL}$$
$$x = 0.173 \, M_{acid}$$

Another type of titration problem involves a solid and a titrated solution.

Example 2

A solid mixture contains NaOH and NaCl. If a 0.100-g sample of this mixture required 10 mL of 0.100 M HCl to titrate the sample to its endpoint, what is the percent of NaOH in the sample?

Solution:
Since

$$\text{Molarity} = \frac{\text{No. of moles}}{\text{Liter of solution}}$$

then

$$M \times V = \frac{\text{No. of moles}}{\text{Liter of solution}} \times \text{liter of solution} = \text{No. of moles}$$

Substituting the HCl information into the equation, we have

$$\frac{0.100 \text{ mol}}{\text{Liter of solution}} \times 0.010 \text{ L of solution} = 0.001 \text{ mol}$$
(*Note*: This is 10 mL expressed in liters)

Since 1 mol of HCl neutralizes 1 mol of NaOH, 0.001 mol of NaOH must be present in the mixture.

$$1 \text{ mol NaOH} = 40.0 \text{ g}$$
$$0.001 \text{ mol} \times 40.0 \text{ g/mol} = 0.04 \text{ g NaOH}$$

Therefore 0.04 g of NaOH was in the 0.1-g sample of the solid mixture. The percent is 0.04 g/0.1 g × 100 = 40%.

In the explanations given to this point, the reactions that took place were between monoprotic acids (single hydrogen ions) and monobasic bases (one hydroxide ion per base). This means that each mole of acid had 1 mol of hydrogen ions available, and each mole of base had 1 mol of hydroxide ions available, to interact in the following reaction until the endpoint was reached

$$H^+(aq) + OH^-(aq) \rightarrow 2H_2O$$

This is not always the case, however, and it is important to know how to deal with acids and bases that have more than one hydrogen ion and more than one hydroxide ion per formula. The following is an example of such a problem.

If 20.0 mL of an aqueous solution of calcium hydroxide, $Ca(OH)_2$, is used in a titration and an appropriate indicator is added to show the neutralization point (endpoint), the few drops of indicator that are added can be ignored in the volume considerations. Therefore, if 25.0 mL of standard 0.050 M HCl is required to reach the endpoint, what was the original concentration of the $Ca(OH)_2$ solution?

The balanced equation for the reaction gives the relationship of the moles of acid reacting to the moles of base:

$$\underbrace{2HCl}_{2 \text{ mol}} + \underbrace{Ca(OH)_2}_{1 \text{ mol}} \rightarrow CaCl_2 + 2H_2O$$

The mole relationship here is that the number of moles of acid is twice the number of moles of base:

Number of moles of acid = 2 times number of moles of base
↑ mole factor

Since the molar concentration of the acid times the volume of the acid gives the number of moles of acid:

$$M_a \times V_a = \text{moles of acid}$$

and the molar concentration of the base times the volume of the base gives the number of moles of base:

$$M_b \times V_b = \text{moles of base}$$

then, substituting these products into the mole relationship, we have

$$M_a V_a = 2M_b V_b$$

Solving for M_b gives

$$M_b = \frac{M_a V_a}{2V_b}$$

Substituting values, we obtain

$$M_b = \frac{0.050\,\text{mol/L} \times 0.0250\,L}{2 \times 0.0200\,L}$$

$$= 0.0312\,\text{mol/L or } 0.0312\,M$$

Buffer Solutions

Buffer solutions are equilibrium systems that resist changes in acidity and maintain a constant pH when acids or bases are added to them. A typical laboratory buffer can be prepared by mixing equal molar quantities of a weak acid such as $HC_2H_3O_2$ and its salt, $NaC_2H_3O_2$. When a small amount of a strong base such as NaOH is added to the buffer, the acetic acid reacts (and consumes) most of the excess OH^- ion. The OH^- ion reacts with the H^+ ion from the acetic acid, thus reducing the H^+ ion concentration in this equilibrium:

$$HC_2H_3O_2 \rightleftharpoons H^+ + C_2H_3O_2^-$$

This reduction of H^+ causes a shift to the right, forming additional $C_2H_3O_2^-$ ions and H^+ ions. For practical purposes, each mole of OH^- added consumes 1 mol of $HC_2H_3O_2$ and produces 1 mol of $C_2H_3O_2^-$ ions.

When a strong acid such as HCl is added to the buffer, the H^+ ions react with the $C_2H_3O_2^-$ ions of the salt and form more undissociated $HC_2H_3O_2$. This does not alter the H^+ ion concentration. Proportional increases and decreases in the concentrations of $C_2H_3O_2^-$ and $HC_2H_3O_2$ do not significantly affect the acidity of the solution.

Salts

A salt is an ionic compound containing positive ions other than hydrogen ions and negative ions other than hydroxide ions. The usual method of preparing a particular salt is by neutralizing the appropriate acid and base to form the salt and water.

The methods for preparing salts are

Neutralization reaction. An acid and base neutralize each other to form the appropriate salt and water. For example:

$$2HCl + Ca(OH)_2 \rightarrow CaCl_2 + 2H_2O$$
$$\text{Acid} + \text{base} \quad \rightarrow \text{Salt} + \text{water}$$

Single-replacement reaction. An active metal replaces hydrogen in an acid. For example,

$$Mg(s) + H_2SO_4(aq) \rightarrow MgSO_4(aq) + H_2(g)$$

Direct combination of elements.   An example of this method is the combination of iron and sulfur. In this reaction small pieces of iron are heated with powdered sulfur.

$$Fe + S \rightarrow Fe\ S$$
iron(II) sulfide

Double-replacement reaction.   When solutions of two soluble salts are mixed to form an insoluble salt compound.

$$AgNO_3(aq) + NaCl(aq) \rightarrow NaNO_3(aq) + AgCl(s)$$

Reaction of a metallic oxide with a nonmetallic oxide.

$$MgO + SiO_2 \rightarrow MgSiO_3$$

The naming of salts is discussed on page 79.

Amphoteric Substances

Some substances, such as the HCO_3^- ion, the HSO_4^- ion, the H_2O molecule, and the NH_3 molecule can act as either a proton donor (acid) or a proton receiver (base) depending upon which substances they come in contact with. These substances are said to be *amphoteric*. Amphoteric substances donate protons in the presence of strong bases and accept protons in the presence of strong acids.

This can be illustrated by the reactions of the bisulfate ion, HSO_4^-:

With a strong acid, HSO_4^- accepts a proton:

$$HSO_4^- + H^+ \rightarrow H_2SO_4$$

With a strong base, the HSO_4^- donates a proton:

$$HSO_4^- + OH^- \rightarrow H_2O + SO_4^{2-}$$

Acid Rain—An Environmental Concern

Acid rain is currently a subject of great concern in many countries around the world because of the widespread environmental damage it reportedly causes. It forms when the oxides of sulfur and nitrogen combine with atmospheric moisture to yield sulfuric and nitric acids—both known to be highly corrosive especially to metals. Once formed in the atmosphere, they can be carried long distances from their source before being deposited by rain. The pollution may also take the form of snow or fog or be precipitated in dry form. This dry form is just as damaging to the environment as the liquid form.

The problem of acid rain can be traced back to the beginning of the industrial revolution, and it has been growing ever since. The term "acid rain" has been in use for more than a century and is derived from atmospheric studies made in the region of Manchester, England. Although the severity of its effects has long been recognized in local settings, as indicated by spells of acid smog in heavily industrialized areas, its widespread destructiveness has been realized only in recent decades. One large area that has been studied extensively is northern Europe, where acid rain has eroded structures, injured crops and forests, and threatened or depleted life in freshwater lakes. In 1983 published reports claimed that 34% of the forested areas of West Germany had been damaged by acid rain. Many of the older monuments and buildings in Europe that have exposed metal have had to be refurbished. In 1989 the famous bronze horses on St. Mark Cathedral in Venice had to be removed for repair for this very reason.

Industrial emissions have been blamed as the major cause of acid rain. Industries have challenged this notion, because the chemical reactions involved in the production of acid rain in the atmosphere are complex and as yet little understood, and have stressed the need for further studies. Because of the high cost of pollution reduction, governments have tended to support this viewpoint. American studies released in the early 1980s, however, strongly implicated industries as the main source of acid rain in the eastern United States and Canada. In 1988, as part of the Long-Range Transboundary Air Pollution Agreement sponsored by the United Nations, the United States, along with 24 other countries, ratified a protocol freezing the rate of nitrogen oxide emissions at 1987 levels. The 1990 amendments to the Clean Air Act of 1967 put in place regulations to reduce the release of sulfur dioxide from power plants to 10 million tons per year by 2000.

Chapter 11 Review

1. The difference between HCl and $HC_2H_3O_2$ as acids is
   (A) the first has less hydrogen in solution
   (B) the second has more ionized hydrogen
   (C) the first is highly ionized
   (D) the second is highly ionized

2. The hydronium ion is represented as
   (A) H_2O^+
   (B) H_3O^+
   (C) HOH^+
   (D) H^-

3. H_2SO_4 is a strong acid because it is
   (A) slightly ionized
   (B) unstable
   (C) an organic compound
   (D) highly ionized

4. In the reaction of an acid with a base the common ionic reaction involves ions of
   (A) hydrogen and hydroxide
   (B) sodium and chloride
   (C) hydrogen and hydronium
   (D) hydroxide and nitrate

5. The pH of an acid solution is
   (A) 3
   (B) 7
   (C) 9
   (D) 10

6. The pH of a solution with a hydrogen ion concentration of 1×10^{-3} is
   (A) +3
   (B) –3
   (C) ±3
   (D) +11

7. According to the Brønsted Theory, an acid is
   (A) a proton donor
   (B) a proton acceptor
   (C) an electron donor
   (D) an electron acceptor

8. A buffer solution
   (A) changes pH rapidly with the addition of an acid
   (B) does not change pH at all
   (C) resists changes in pH
   (D) changes pH only with the addition of a strong base

9. The point at which a titration is complete is called the
   (A) endpoint
   (B) equilibrium point
   (C) calibrated point
   (D) chemical point

10. If 10 mL of 1 M HCl was required to titrate a 20-mL NaOH solution of unknown concentration to its endpoint, what was the concentration of the NaOH?
    (A) 0.5 M
    (B) 1.5 M
    (C) 2 M
    (D) 2.5 M

Answers

1. **C**	3. **D**	5. **A**	7. **A**	9. **A**
2. **B**	4. **A**	6. **A**	8. **C**	10. **A**

The formula:

$$M_{acid} \times V_{acid} = M_{base} \times V_{base}$$
$$1\,M \times 10 \text{ mL} = xM \times 20 \text{ mL}$$
$$\underline{x = 0.5\,M}$$

Terms You Should Know

acid
acid rain
amphoteric
base
buffer solution
conjugate acid
conjugate base
endpoint
equivalence point

indicator
litmus
neutralization
pH
salt
titrate
volumetric analysis

Arrhenius Theory
Brønsted Theory
Lewis Theory

Chapter 12

OXIDATION-REDUCTION AND ELECTROCHEMISTRY

Ionization

In the early 1830s, Michael Faraday discovered that water solutions of certain substances conduct an electric current. He called these substances *electrolytes*. Our definition today of an electrolyte is much the same. It is a substance that dissolves in water to form a solution that will conduct an electric current.

The usual apparatus for testing this conductivity is a light bulb placed in series with two prongs immersed in the solution being tested.

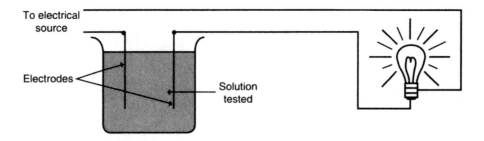

Figure 35. Conductivity of Electrolytes.

Using this type of apparatus, we can classify solutions as good, moderate, or poor electrolytes. If they do not conduct at all, they are called *nonelectrolytes*. The following table gives a classification of some common substances.

WATER SOLUTIONS TESTED FOR CONDUCTIVITY

Good	Poor	Nonelectrolyte
Sodium chloride	Acetic acid	Sugar
Hydrochloric acid	Ammonium hydroxide	Benzene

The reason that these substances conduct with varying degrees is related to the number of ions in solution.

Ionic lattice substances like sodium chloride are dissociated by water molecules, and so the individual positive and negative ions are dispersed throughout the solution. In the case of a covalent bonded substance, the degree of polarity determines the extent to which it will be ionized. The water molecules, which are polar themselves, can help weaken and finally break the polar covalent bonds by clustering around the substance. When the ions are formed in this manner, the process is called *ionization*; when the ionic lattice comes apart, it is called *dissociation*. Substances that are nonelectrolytes are usually bonded, and so the molecule is a nonpolar molecule. The polar water molecule cannot orient itself around the molecule and cause its ionization.

Let us see how the current is carried through the apparatus in Figure 35. The electricity causes one electrode to become positively charged, and one negatively charged. These are called the *anode* and *cathode*, respectively. If the solution contains ions, they will be attracted to the electrode with the charge opposite their own. This means that the positive ions migrate to the cathode, and so they are referred to as *cations*. The negative ions migrate toward the positive anode, and so they are called *anions*. When these ions arrive at the respective electrodes, the negative ions give up electrons, the positive ions accept electrons, and we have a completed path for the electric current. The more highly ionized the substance, the more current flows and the brighter the light bulb glows. It was the Swedish chemist Arrhenius who proposed a theory to explain the behavior of electrolytes in aqueous solutions.

Oxidation-Reduction and Electrochemistry

Electrochemistry

The branch of chemistry that deals with electricity-related applications of oxidation-reduction reactions is called *electrochemistry*. *Oxidation-reduction* reactions involve a transfer of electrons from the substance oxidized to the substance reduced. If the two substances are in contact with one another, a transfer of heat energy accompanies the electron transfer. This can be shown when a zinc strip is in contact with a copper(II) sulfate solution in a beaker. The zinc strip loses electrons to the copper(II) ions in solution. Copper(II) ions accept the electrons and fall out of solution as copper atoms. As electrons are transferred between zinc atoms and copper(II) ions, energy is released in the form of heat, as indicated by a rise in temperature.

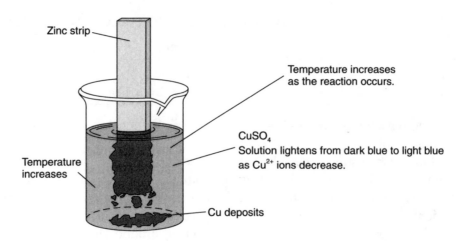

Zinc strip

Temperature increases as the reaction occurs.

CuSO$_4$
Solution lightens from dark blue to light blue as Cu^{2+} ions decrease.

Temperature increases

Cu deposits

Voltaic Cells

Another example of an oxidation-reduction reaction is one in which the substance that is oxidized during the reaction is separated from the substance that is reduced during the reaction. A transfer of electrical energy instead of heat accompanies the electron transfer. One means of separating oxidation and reduction half-reactions is with a porous barrier. This barrier prevents the metal atoms of one half-reaction from mixing with the ions of the other half-reaction. Ions in the two solutions can move through the porous barrier, and electrons can be transferred from one side to the other through an external connecting wire. Electric current moves in a closed loop path, or circuit, and so the movement of ions in the solution balances the movement of electrons through the wire. This is shown in the figure below.

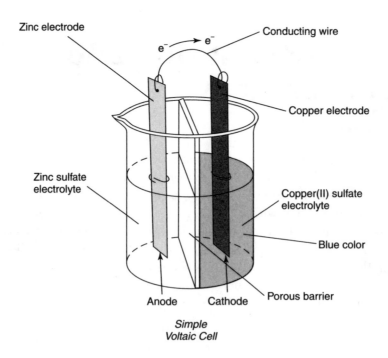

Zinc electrode

e$^-$ e$^-$

Conducting wire

Copper electrode

Zinc sulfate electrolyte

Copper(II) sulfate electrolyte

Blue color

Anode   Cathode

Porous barrier

*Simple
Voltaic Cell*

The Zn strip is in an aqueous solution of $ZnSO_4$. The Cu strip is in an aqueous solution of $CuSO_4$. Both solutions conduct electricity and are classified as electrolytes. An *electrode* is a conductor used to establish electrical contact with a nonmetallic part of a circuit, such as an electrolyte. In the figure, Zn and Cu are electrodes. A single electrode immersed in a solution of its ions is a *half-cell*. The Zn strip in aqueous $ZnSO_4$ is an *anode*, the electrode where oxidation takes place. The Cu strip in $CuSO_4$ is a *cathode*, the electrode where reduction takes place. The copper half-cell can be written as Cu^{2+}/Cu, and the zinc half-cell can be written as Zn^{2+}/Zn. The two half-cells together make an electrochemical cell. An *electrochemical cell* is a system of electrodes and electrolytes in which either chemical reactions produce electrical energy or an electric current produces chemical change. The following notation represents an electrochemical cell: cathode | anode. For the example, the cell made up of zinc and copper can be written as Cu | Zn. There are two types of electrochemical cells: voltaic (also called galvanic) and electrolytic.

If the redox reaction in an electrochemical cell occurs naturally and produces electrical energy as shown in the drawing immediately above, the cell is a *voltaic cell*. Cations in the solution are reduced when they gain electrons at the surface of the cathode to become metal atoms. This half-reaction for the voltaic cell shown above is

$$Cu^{2+}(aq) + 2e^- \rightarrow Cu(s)$$

The half-reaction occurring at the anode is

$$Zn(s) \rightarrow Zn^{2+}(aq) + 2e^-$$

Electrons given up at the anode pass along the external connecting wire to the cathode.

The movement of ions in the solution must balance the movement of electrons through the wire. Anions move toward the anode to replace the negatively charged electrons that are moving away. Cations move toward the cathode as positive charge is lost through reduction. Thus, sulfate ions in the $CuSO_4$ solution can move through the barrier into the $ZnSO_4$ solution. Likewise, Zn^{2+} ions pass through the barrier into the $CuSO_4$ solution.

Electrode Potentials

In the discussion of acids and metals reacting to produce hydrogen (page 96), the activities of metals compared to the activity of hydrogen are shown in the form of a chart. A more inclusive presentation of this activity of metals, called the *electromotive series*, is shown in Table 10.

From this chart note that zinc is above copper and, therefore, more active. This means zinc can displace copper ions in a solution of copper sulfate:

$$Zn^0 + Cu^{2+} + SO_4^{2-} \rightarrow Cu^0 + Zn^{2+} + SO_4^{2-}$$

The zero indicates the zinc atom is neutral. ⌐———→ The zinc ion carries a 2+ charge.

The Zn atom must have lost 2 electrons to become Zn^{2+} ions.

$$Zn^0 \rightarrow Zn^{2+} + 2e^- \text{ (electrons)}$$

At the same time, the Cu^{2+} must have gained 2 electrons to become the Cu^0 atom.

$$Cu^{2+} + 2e^- \text{ (electrons)} \rightarrow Cu^0$$

TABLE 10

ELECTROMOTIVE SERIES

Element			Standard Half-Cell Voltages	
			Oxidation Electron Reaction	Electrode Potential,* E^0_{ox}
Potassium			$K \rightarrow K^+ + e^-$	+2.93 V
Calcium			$Ca \rightarrow Ca^{2+} + 2e^-$	+2.87 V
Sodium	Increasing	Increasing	$Na \rightarrow Na^+ + e^-$	+2.71 V
Magnesium	tendency	tendency	$Mg \rightarrow Mg^{2+} + 2e^-$	+2.37 V
Aluminum	for	for	$Al \rightarrow Al^{3+} + 3e^-$	+1.67 V
Zinc	atoms	ions	$Zn \rightarrow Zn^{2+} + 2e^-$	+0.76 V
Iron	to	to gain	$Fe \rightarrow Fe^{2+} + 2e^-$	+0.44 V
Tin	lose	electrons	$Sn \rightarrow Sn^{2+} + 2e^-$	+0.14 V
Lead	electrons	and	$Pb \rightarrow Pb^{2+} + 2e^-$	+0.13 V
Hydrogen	and	form	$H_2 \rightarrow 2H^+ + 2e^-$	0.00 V
Copper	form	atoms	$Cu \rightarrow Cu^{2+} + 2e^-$	−0.34 V
Mercury	positive	of the	$2Hg \rightarrow Hg_2^{2+} + 2e^-$	−0.79 V
Silver	ions	metal	$Ag \rightarrow Ag^+ + e^-$	−0.80 V
Mercury			$Hg \rightarrow Hg^{2+} + 2e^-$	−0.85 V
Gold			$Au \rightarrow Au^{3+} + 3e^-$	−1.50 V

*A measure in volts of the tendency of atoms to gain or lose electrons.

These two equations are called *half-reactions*. The loss of electrons by the zinc is called oxidation, and the gain of electrons by the copper ion, reduction. It is important to remember that the gain of electrons is *reduction* and the loss of electrons is *oxidation*. A way to remember this is "LEO the lion says GER." LEO translates into *L*oss of *E*lectrons is *O*xidation, and GER into *G*ain of *E*lectrons is *R*eduction. Because giving up electrons oxidizes zinc in this reaction, it makes it possible for electrons to be gained by the copper. Zinc acts as a *reducing agent*. Because gaining electrons reduces the copper in this reaction, it makes it possible for electrons to be lost by the zinc. Copper acts as an oxidizing agent. Remember, the substance oxidized is called the reducing agent and the substance reduced is called the oxidizing agent.

The metal elements that lose electrons easily and become positive ions are placed high in the electromotive series. The metal elements that lose electrons with more difficulty are placed lower on the chart.

The energy required to remove electrons from metallic atoms can be assigned numerical values called *electrode potentials*. The energy required for the reduction of nonmetals is shown in Table 11.

These voltages depend on the nature of the reaction, the concentrations of reactants and products, and the temperature. Throughout this discussion, we use standard concentrations; that is, all ions or molecules in aqueous solution are at a concentration of 1 *M*. Furthermore, all gases taking part in the reactions are at 1 atm pressure and the temperature is 25°C. The voltage measured under these conditions is called *standard voltage*.

TABLE 11

ACTIVITY OF NONMETALS

Element			Reduction Electron Reaction	E^0_{red}
Fluorine	Increasing tendency	Increasing tendency	$F_2 + 2e^- \rightarrow 2F^-$	+2.87 V
Chlorine	for atoms to gain	for ions to lose	$Cl_2 + 2e^- \rightarrow 2Cl^-$	+1.36 V
Bromine	electrons and form	electrons and form	$Br_2 + 2e^- \rightarrow 2Br^-$	+1.09 V
Iodine	negative ions	atoms	$I_2 + 2e^- \rightarrow 2I^-$	+0.54 V

The currently accepted convention is to give the potentials of half-reactions as reduction processes. For example,

$$2H^+ + 2e^- \rightarrow H_2$$

The E^0 values corresponding to these half-reactions are called *standard reduction potentials*. The most common ones are given in Table Ⓚ in the Tables for Reference section at the back of the book.

Some oxidation electrode potentials are shown in the last column of Table 10. Notice that hydrogen is used as the standard with an electrode potential of zero. These values help to predict what reactions will occur and how readily they will occur. These reactions are called *voltaic cell reactions* or galvanic cell reactions.

The following examples will clarify the use of electrode potentials.

If magnesium reacts with chlorine, we can write the equation

$$Mg + Cl_2 \rightarrow MgCl_2$$

The two half-reactions with the electrode potentials are

Oxidation half-reaction:	$Mg \rightarrow Mg^{2+} + 2e^-$	$E^0 = +2.37$ V
Reduction half-reaction:	$Cl_2 + 2e^- \rightarrow 2Cl^-$	$E^0 = +1.36$ V
Net reaction:	$Mg + Cl_2 \rightarrow Mg^{2+} + 2Cl^-$	$E^0 = +3.73$ V

In the net reaction, E^0 is a positive number. This indicates that the reaction occurs spontaneously.

You should also note that the total number of electrons lost in oxidation is equal to the total number of electrons gained and so the net reaction (arrived at by adding the two reactions) does not contain any electrons.

Another example is sodium reacting with chlorine:

$$2Na + Cl_2 \rightarrow 2NaCl$$

The two half-reactions are

Oxidation half-reaction:	$2Na \rightarrow 2Na^+ + 2e^-$	$E^0 = +2.71$ V
Reduction half-reaction:	$Cl_2 + 2e^- \rightarrow 2Cl^-$	$E^0 = +1.36$ V
Net reaction:	$2Na + Cl_2 \rightarrow 2Na^+ + 2Cl^-$	$E^0 = +4.07$ V

Note: The standard potentials (E^0) are *not* multiplied by the coefficients in calculating the E^0 for the reaction.

Again, the E^0 for this reaction is positive, and the reaction is spontaneous.

The next example shows a negative E^0.

Copper metal placed in an acid solution is shown as follows:

Oxidation:	$Cu \rightarrow Cu^{2+} + 2e^-$	$E^0 = -0.34$ V
Reduction:	$2H^+ + 2e^- \rightarrow H_2$	$E^0 = 0.00$ V
Net reaction:	$Cu + 2H^+ \rightarrow Cu^{2+} + H_2$	$E^0 = -0.34$ V

Since E^0 is negative, we know the reaction will not take place.

Notice in Table Ⓚ in the Equations and Tables for Reference section that the reduction reactions with their E^0 values are shown for metals. If you use the oxidation reactions of these metals, the equation must be reversed and the sign of E^0 changed. An example of this is placing a piece of copper in a solution of silver ions.

Oxidation (reversed):	$Cu \rightarrow Cu^{2+} + 2e^-$	$E^0 = -0.34$ V
Reduction	$2Ag^+ + 2e^- \rightarrow 2Ag^0$	$E^0 = +0.80$ V
	$Cu + 2Ag^+ \rightarrow Cu^{2+} + 2Ag^0$	$E^0 = +0.46$ V

This reaction occurs spontaneously since E^0 is positive.

Electrolytic Cells

The second type of electrochemical reaction is where redox reactions that do not occur spontaneously can be forced to take place by supplying energy with an external current. These reactions are called *electrolytic reactions*. Some examples of this type of reaction are electroplating, electrolysis of salt solution, electrolysis of water, and electrolysis of molten salts. An example of the electrolytic setup is shown in Figure 36.

If this solution contains Cu^{2+} ions and Cl^- ions, the half-reactions will be

Anode half-reaction:
Oxidation	$2Cl^- \rightarrow Cl_2(g) + 2e^-$	$E^0 = -1.36$ V

Cathode half-reaction:
Reduction:	$Cu^{2+} + 2e^- \rightarrow Cu^0$	$E^0 = +0.34$ V
Net reaction:	$Cu^{2+} + 2Cl^- \rightarrow Cu^0 + Cl_2(g)$	$E^0 = -1.02$ V

Notice that the E^0 for this reaction is negative, and so an outside source of energy must be used to make it occur.

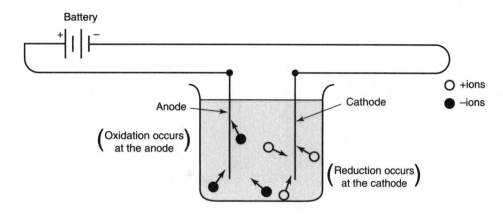

Figure 36. Electrochemical Reactions.

In electroplating, where electrolysis is used to coat a material with a layer of metal, the object to be plated is made the cathode in the reaction. A bar of the plating metal is made the anode, and the solution contains ions of the plating metal.

For example, silver nitrate, $AgNO_3$, can be used to silverplate because when dissolved in H_2O it forms a solution containing silver ions, Ag^+. Assume a metal fork is made the cathode, and a bar of silver is made the anode. When the current is switched on, the positive silver ions in the solution are attracted to the fork. When the silver ions make contact, they are reduced and change from ions to atoms of silver. These atoms gradually form a metallic coating on the fork. At the anode, oxidation occurs, and the anode itself is oxidized. These two half-reactions are

$$\text{Cathode reaction:} \qquad Ag^+ + e^- \rightarrow Ag$$
$$\text{Anode reaction:} \qquad Ag \rightarrow Ag^+ + e^-$$

The silver ions formed at the anode replace those that are plated onto the fork during the cathode reaction.

Another example is the electrolysis of a water solution of sodium chloride. This water solution contains chloride ions, which are attracted to the anode and set free as chlorine molecules. The cathode reaction is somewhat more complicated. Although the sodium ions are attracted to the cathode, they are not set free as atoms. Remember that water can ionize to some extent, and the electromotive series shows that the hydrogen ion is reduced more easily than the sodium ion. Therefore, hydrogen, *not* sodium, is set free at the cathode. The reaction can be summarized as

$$\text{Cathode reaction:} \qquad 2H_2O + 2e^- \rightarrow H_2(g) + 2OH^-$$
$$\text{Anode reaction:} \qquad 2Cl^- \rightarrow 2e^- + Cl_2$$
$$\overline{}$$
$$\text{Net reaction:} \qquad 2H_2O + 2Cl^- \rightarrow H_2(g) + Cl_2(g) + 2OH^-$$

Another example of electrolysis is the decomposition of water by using an apparatus like the one shown in Figure 37.

The solution in this apparatus contains distilled water and a small amount of H_2SO_4. The reason for adding H_2SO_4 is to make the solution an electrolyte since distilled water alone will not conduct an electric current. The solution, therefore, contains ions of H^+, HSO_4^-, and SO_4^{2-}.

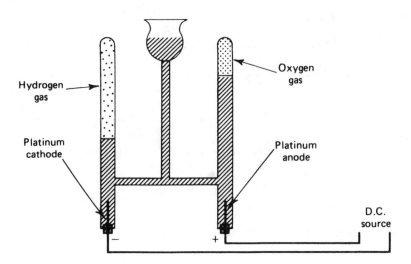

Figure 37. Electrolysis of Water.

The hydrogen ions (H⁺) migrate to the cathode where they are reduced to hydrogen atoms and form hydrogen molecules (H₂) in the form of a gas. The SO_4^{2-} and HSO_4^- migrate to the anode but are not oxidized since the oxidation of water occurs more readily. These ions then are merely spectator ions. The oxidized water reacts as shown in the half-reaction below.

Cathode reaction:	
Reduction:	$4H_2O + 4e^- \rightarrow 2H_2(g) + 4OH^-$
Anode reaction:	
Oxidation:	$2H_2O \rightarrow O_2(g) + 4H^+ + 4e^-$
Net reaction:	$2H_2O \rightarrow 2H_2(g) + O_2(g)$

Notice that the equation shows 2 vol of hydrogen gas are released while only 1 vol of oxygen gas is liberated. This agrees with the discussion of the composition of water in Chapter 7.

There are two important differences between voltaic cells and electrolytic cells.

1. The anode and cathode of an electrolytic cell are connected to a battery or other direct-current source, whereas a voltaic cell serves as a source of electrical energy.
2. Electrolytic cells are those in which electrical energy from an external source causes nonspontaneous redox reactions to occur. Voltaic cells are those in which spontaneous redox reactions produce electricity. In an electrolytic cell, electrical energy is converted to chemical energy; in a voltaic cell, chemical energy is converted to electrical energy.

Applications of Electrochemical Cells (Commercial Voltaic Cells)

One of the most common voltaic cells is the ordinary "dry cell" used in flashlights. Its makeup is shown in the drawing below, along with anode, cathode, and paste reactions.

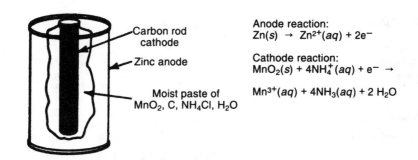

Anode reaction:
$Zn(s) \rightarrow Zn^{2+}(aq) + 2e^-$

Cathode reaction:
$MnO_2(s) + 4NH_4^+(aq) + e^- \rightarrow$

$Mn^{3+}(aq) + 4NH_3(aq) + 2 H_2O$

The automobile lead storage battery is also a voltaic cell. When it discharges, the reactions are as shown in the drawing below.

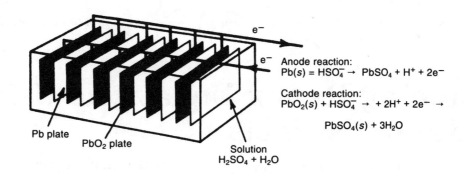

Anode reaction:
$Pb(s) = HSO_4^- \rightarrow PbSO_4 + H^+ + 2e^-$

Cathode reaction:
$PbO_2(s) + HSO_4^- \rightarrow + 2H^+ + 2e^- \rightarrow$

$PbSO_4(s) + 3H_2O$

Quantitative Aspects of Electrolysis
Relationship between Quantity of Electricity and Amount of Products

The amounts of products liberated at the electrodes of an electrolytic cell are related to the quantity of electricity passed through the cell and the electrode reactions.

In electrolysis, 1 mol of electrons is called a *faraday* of electric charge. In the reaction of electrolysis of molten NaCl, 1 faraday will liberate 1 mol of sodium atoms.

$$\text{Cathode reactions: } Na^+ + e^- \rightarrow Na(s)$$

At the same time , 1 mol of Cl^- ions at the anode will form 1 mol of chlorine atoms and thus 0.5 mol of chlorine molecules.

If the metallic ion had been Ca^{2+}, 1 faraday would have released only 0.5 mol of calcium atoms since each calcium ion requires 2 electrons, as shown here:

$$Ca^{2+} + 2e^- \rightarrow Ca(s)$$

Example

How many moles of electrons are required to reduce 2.93 g of nickel ions from melted $NiCl_2$?

Solution:
The reaction is

$$Ni^{2+} + 2e^- \rightarrow Ni(s)$$

The 2.93 g represents

$$2.93 \ g \times \frac{1 \ mol \ Ni}{58.7 \ g} = 0.05 \ mol \ Ni$$

Since 2 mol of electrons are needed to produce 1 mol of Ni(s),

$$0.05 \ mol \ Ni \times \frac{2 \ mol \ of \ electrons}{1 \ mol \ Ni} = 0.1 \ mol \ of \ electrons$$

Balancing Redox Equations Using Oxidation Numbers

In general, many chemical reactions are so simple that once the reactants and products are known, the equations can be readily written. Many redox reactions, however, are of such complexity that the process of writing the equations by trial and error is time-consuming. In these cases the operation can be simplified by limiting the change involved to the actual electron shift, and balancing that—an operation that usually can be done without difficulty.

That part of the process being accomplished, the remainder of the equation can easily be adjusted to it. There are several methods by which this can be done, and a number of techniques that can be employed for each method. We will show two methods: the electron shift method and the ion-electron method.

The Electron Shift Method

Example 1

$$\overset{+1}{H}\overset{-1}{Cl} \quad + \quad \overset{+4}{Mn}\overset{-2}{O_2} \quad \rightarrow \quad \overset{+1}{H_2}\overset{-2}{O} \quad + \quad \overset{+2}{Mn}\overset{-1}{Cl_2} \quad + \quad \overset{0}{Cl_2}$$

Solution:
1. The molecular formulas are written in a statement of the reaction, and the oxidation states are assigned to the elements. These appear in the above equation.
2. Inspection shows that the oxidation state of the manganese atoms has changed from +4 to +2 and that the oxidation state of the chlorine atoms which emerge from the reaction as free elements has changed from −1 to 0:

$$1(Mn^{4+} + 2e^+ \rightarrow Mn^{2+})$$
$$2(Cl^{1-} \qquad \rightarrow Cl^0 + 1e^-)$$

3. The reason for multiplying the two half-reactions by 1 and 2 is to balance the electrons gained with the electrons lost.
4. It appears that $1Mn^{4+} + 2Cl^{1-} \rightarrow 1Mn^{2+} + 2Cl^0$
5. Placing these coefficients in the equation gives

$$2HCl + 1MnO_2 \rightarrow *?H_2O + 1MnCl_2 + Cl_2$$

6. From the numbers thus established, the remaining coefficients can be easily deduced. Two *more* molecules of HCl are required to furnish the chlorine for $MnCl_2$, and the 2 atoms of oxygen in MnO_2 form $2H_2O$.
7. *The final equation is* $4HCl + MnO_2 \rightarrow 2H_2O + MnCl_2 + Cl_2$.

Now consider a more complicated reaction. Only the elements that show a change in oxidation state are indicated.

Example 2

$$\overset{-1}{HCl} + \overset{+7}{KMnO_4} \rightarrow H_2O + KCl + \overset{+2}{MnCl_2} + \overset{0}{Cl_2}$$

Solution:
The electron change is: $5\ (Cl^{-1} \qquad = Cl^0 + 1e^-)$
$\qquad\qquad\qquad\qquad\ \underline{1\ (Mn^{7+} + 5e^- \quad = Mn^{2+})}$
Therefore, $\qquad\qquad 5HCl + 1KMnO_4 = \frac{5}{2}Cl_2 + 1MnCl_2$

5 atoms of chlorine
equal $\frac{5}{2}Cl_2$ molecules

Multipling by 2 to avoid fractions gives

$$10HCl + 2KMnO_4 \rightarrow 5Cl_2 + 2MnCl_2$$

Substituting in the original equation,

$$10HCl + 2KMnO_4 \rightarrow *?H_2O + *?KCl + 2MnCl_2 + 5Cl_2$$

For the remainder of the coefficients, $2KMnO_4$ produces $2KCl$ and $8H_2O$, while $2KCl + 2MnCl_2$ call for six additional molecules of HCl. So finally the expression becomes

$$16HCl + 2KMnO_4 \rightarrow 8H_2O + 2KCl + 2MnCl_2 + 5Cl_2$$

Example 3

$$As_2S_5 + KClO_3 + H_2O \rightarrow H_3AsO_4 + H_2SO_4 + KCl$$

Solution:
The equation with only the oxidation numbers of the elements that show a change in oxidation number is

*? indicates that this coefficient is not yet known.

$$\overset{-2 \quad +5}{As_2S_5 + KClO_3 + H_2O} \rightarrow \overset{+6 \quad -1}{H_3AsO_4 + H_2SO_4 + KCl}$$

The electron change is:

$$4 \times (Cl^{5+} + 6e^- \rightarrow Cl^-)$$
$$3 \times (S^{2-} \qquad \rightarrow S^{6+} + 8e)$$

or

$$4Cl^{5+} + 24e^- \rightarrow 4Cl^-$$
$$\underline{3S^{2-} \qquad \rightarrow 3S^{6+} + 24e^-}$$
$$4Cl^- + 3S^{2-} \rightarrow 4Cl^- + 3S^{6+}$$

When you try to fit these numbers to the equation, you will find that, because the formula As_2S_5 has sulfur occurring 5 atoms per molecule, only three sulfurs can't be used. You must multiply the simplest equation by an appropriate number to get a multiple of 5.

$$5 \times (3S^{2-} + 4Cl^- \rightarrow 4Cl^- + 3S^{6+})$$

or

$$15S^{2-} + 20Cl^- \rightarrow 20Cl^- + 15S^{6+}$$

Transfer these to the equation:

$$\underline{3}As_2S_5 + \underline{20}KClO_3 + H_2O \rightarrow H_3AsO_4 + \underline{15}H_2SO_4 + \underline{20}KCl$$

Now set the coefficients of H_2O and H_3AsO_4 by inspection:

$$\underline{3}As_2S_5 + \underline{20}KClO_3 + \underline{24}H_2O \rightarrow \underline{6}H_3AsO_4 + \underline{15}H_2SO_4 + \underline{20}KCl$$

The Ion-Electron Method

The second method is more complex but seems to represent the true mechanism of such reactions more closely.

In this method, only units that actually have an individual existence (atoms, molecules, or ions) in the particular reaction being studied are taken into consideration. The principal oxidizing agent and the principal reducing agent are chosen from these (by a method to be indicated later), and the electron loss or gain of these two principal actors is then determined by taking into consideration the fact that since electrons can be neither created nor destroyed, electrons lost by one of these actors must be gained by the other. This is accomplished by using two separate partial equations representing the changes undergone by each of the two principal actors.

Probably the best way to show how the method operates is to follow in detail the steps taken in balancing an example of an actual oxidation-reduction reaction.

Example 1

Assume that the equation $K_2CrO_4 + HCl \rightarrow KCl + CrCl_3 + H_2O + Cl_2$ is to be balanced.

Solution:
1. Determine which of the substances present are involved in the oxidation-reduction.

This is done by listing all substances present on each side of the arrow and crossing out those that appear on both sides of the arrow without any portion being changed in any way.

$$K^+, CrO_4^{2-}, H^+, Cl^- \rightarrow K^+ Cl^-, Cr^{3+}, H_2O, Cl_2$$

Note that Cl^- is not crossed out although it appears on both sides because some of the Cl^- from the left side appears in a changed form, namely, Cl_2, on the right.

The two substances on the left side that are *not* crossed out are the ones involved in the oxidation-reduction.

If, as *in this case* there are *more* than two, disregard H^+, OH^-, or H_2O; this will leave the two principal actors.

2. Indicate in two as yet unbalanced partial equations the fate of each of the two active agents:

$$CrO_4^{2-} \rightarrow Cr^{3+}$$
$$Cl^- \rightarrow Cl_2$$

3. Balance these equations chemically, inserting any substance necessary.

$$CrO_4^{2-} + 8H^+ \rightarrow Cr^{3+} + 4H_2O$$
$$2Cl^- \rightarrow Cl_2$$

In the upper partial equation, H^+ had to be added in order to remove the oxygen from the CrO_4^{2-} ion.

H^+ is always used for this purpose in acid solutions. If the solution is basic, H_2O must be used for this purpose:

$$CrO_4^{2-} + 4H_2O \rightarrow Cr^{3+} + 8OH^-$$

4. Balance these equations electrically by adding electrons *on either side* so that the total electric charge is the same on the left and right sides:

$$CrO_4^{2-} + 8H^+ + 3e^- \rightarrow Cr^{3+} + 4H_2O$$
$$2Cl^- \rightarrow Cl_2 + 2e^-$$

5. Add these partial equations.

But before we add we must realize that electrons can neither be created nor destroyed. Electrons are gained in one of these equations and lost in the other. THOSE GAINED IN ONE MUST COME FROM THE OTHER. Therefore we must multiply each of these equations through by numbers so chosen that the number of electrons gained in one equation will be the same as the number lost in the other:

$$2[CrO_4^{2-} + 8H^+ 3e^- \rightarrow Cr^{3+} + 4H_2O]$$
$$3[2Cl^- \rightarrow Cl_2 + 2e^-]$$

THIS IS THE KEY TO THE METHOD!
Adding the multiplied equations, we have

$$2CrO_4^{2-} + 16H^+ + 6Cl^- \rightarrow 2Cr^{3+} + 8H_2O + 3Cl_2$$

This sum tells us all that really happens in the oxidation-reduction, but if a conventionally balanced equation is desired for problem purposes or for some other reason, it can be obtained by carrying the coefficients into the original skeleton equation, thus

$$2K_2CrO_4 + 16HCl \rightarrow 4KCl + 2CrCl_3 + 8H_2O + 3Cl_2$$
$$\uparrow$$

(4 inserted by inspection)

Example 2

$$2KMnO_4 + H_2O + KI \rightarrow 2MnO_2 + 2KOH + KIO_3$$

Solution:

$$
\begin{array}{ll}
2[MnO_4^- + 2H_2O + 3e^- & \rightarrow MnO_2 + 4OH^-] \\
\underline{I^- + 6OH^- & \rightarrow IO_3^- + 3H_2O + 6e^-} \\
2MnO_4^- + I^- + H_2O & \rightarrow 2MnO_2 + IO_3^- + 2OH^-
\end{array}
$$

Adding:

Note that since this takes place in the basic solution, H_2O was used to remove oxygen in the upper partial.

Note also that in the second partial it was necessary to *add* oxygen and that this was done by means of OH^- ion.

If the solution had been *acid*, then H_2O would have been used for this purpose:

$$I^- + 3H_2O \rightarrow IO_3^- + 6H^+ + 6e^-$$

In general, you will meet three types of partial equations:

1. Where only electrons are needed to balance, as in the second partial of the first example above.
2. Where oxygen must be removed from an ion.
    In acid solution, H^+ is used for this purpose as in the first partial of the first example above.
    In basic solution, H_2O is used for this as in the first partial of the second example above.
3. Where oxygen must be added to an ion.
    In basic solution, OH^- is used for this as in the second partial of the second example above.
    In acid solution, H_2O is used for this as in

$$SO_3^{2-} + H_2O \rightarrow SO_4^{2-} + 2H^+ + 2e^-$$

Example 3

$$KMnO_4 + H_2SO_3 \rightarrow H_2SO_4 + MnSO_4 + K_2SO_4 + H_2O$$

Solution:

Since this takes place in acid solution, you will have to use H_2O to add oxygen to SO_3^{2-}.

$$KMnO_4 + H_2SO_3 \rightarrow H_2SO_4 + MnSO_4 + K_2SO_4 + H_2O$$

The half-reactions are

$$MnO_4^- \rightarrow Mn^{2+}$$
$$SO_3^{2-} \rightarrow SO_4^{2-}$$

Balancing these half-reactions chemically using H_2O and H^+ gives

$$MnO_4^- + 8H^+ \rightarrow Mn^{2+} + 4H_2O$$
$$SO_3^{2-} + H_2O \rightarrow SO_4^{2-} + 2H^+$$

Balancing them electrically and equalizing the electrons lost and gained:

$$2(MnO_4^- + 8H^+ + 5e^- \rightarrow Mn^{2+} + 4H_2O)$$
$$5(SO_3^{2-} + H_2O \rightarrow SO_4^{2-} + 2H^+ + 2e^-)$$

Adding the multiplied half-reactions:

$$2MnO_4^- + 5SO_3^{2-} + 6H^+ \rightarrow 2Mn^{2+} + 5SO_4^{2-} + 3H_2O$$

Carrying these coefficients into the equation gives

$$2KMnO_4 + 5H_2SO_3 \rightarrow H_2SO_4 + 2MnSO_4 + K_2SO_4 + 3H_2O$$

This sets this other coefficient as 1 ↑

Now this far you have used $3SO_4^{2-}$ in $2MnSO_4 + K_2SO_4$. Since you must have $5SO_4^{2-}$, you need 2 more, which sets the coefficient of H_2SO_4 at 2.

So

$$\underline{2}KMnO_4 + \underline{5}H_2SO_3 \rightarrow \underline{2}H_2SO_4 + \underline{2}MnSO_4 + K_2SO_4 + \underline{3}H_2O$$

Example 4

$$Zn + HNO_3 \rightarrow Zn(NO_3)_2 + NH_4NO_3 + H_2O$$

Solution:
The half-reactions balanced chemically:

$$Zn \rightarrow Zn^{2+}$$
$$NO_3^- + 10H^+ \rightarrow NH_4^+ + 3H_2O$$

Balancing electrically and adding:

$$4(Zn \rightarrow Zn^{2+} + 2e^-)$$
$$\frac{NO_3^- + 10H^+ + 8e^- \rightarrow NH_4^+ + 3H_2O}{4Zn + NO_3^- + 10H^+ \rightarrow 4Zn^{2+} + NH_4^+ + 3H_2O}$$

In the equation:

$$\underline{4}Zn + \underline{10}HNO_3 \rightarrow \underline{4}Zn(NO_3)_2 + NH_4NO_3 + \underline{3}H_2O$$

Example 5

$$Cu + HNO_3 \rightarrow Cu(NO_3)_2 + H_2O + NO$$

Solution:
The half-reactions balanced chemically:

$$Cu^0 \rightarrow Cu^{2+}$$
$$NO_3^- + 4H^+ \rightarrow NO + 2H_2O$$

Balancing electrically and adding:

$$3(Cu^0 \rightarrow Cu^{2+} + 2e^-)$$
$$\underline{2(NO_3^- + 4H^+ + 3e^- \rightarrow NO + 2H_2O)}$$
$$3Cu + 8H^+ + 2NO_3^{3-} \rightarrow 3Cu^{2+} + 2NO + 4H_2O$$

In the equation:

$$\underline{3}Cu + \underline{8}HNO_3 \rightarrow \underline{3}Cu(NO_3)_2 + \underline{4}H_2O + \underline{2}NO$$

Example 6

$$KMnO_4 + HCl \rightarrow KCl + MnCl_2 + H_2O + Cl_2$$

Solution:
The half-reactions balanced chemically:

$$MnO_4^- + 8H^+ \rightarrow Mn^{2+} + 4H_2O$$
$$2Cl^- \rightarrow Cl_2$$

Balancing electrically and adding:

$$2(MnO_4^- + 8H^+ + 5e^- \rightarrow Mn^{2+} + 4H_2O)$$
$$\underline{5(2Cl^- \rightarrow Cl_2^0 + 2e^-)}$$
$$2MnO_4^- + 16H^+ + 10Cl^- \rightarrow 2Mn^{2+} + 8H_2O + 5Cl_2$$

In the equation:

$$\underline{2}KMnO_4 + \underline{16}HCl^+ \rightarrow \underline{2}KCl + \underline{2}MnCl_2 + \underline{8}H_2O + \underline{5}Cl_2$$
$$\text{sets this as 2} \quad \uparrow$$
$$\text{by inspection}$$

Chapter 12 Review

1. Which of the following when dissolved in water and placed in the conductivity apparatus will cause the light to glow?
   (A) table salt
   (B) ethyl alcohol
   (C) sugar
   (D) glycerine

2. The reason that a current can flow is because
   (A) ions combine to form molecules
   (B) molecules migrate to the charge plates
   (C) ions migrate to the charge plates
   (D) sparks cross the gap

3. The extent of ionization depends on the
   (A) nature of the solvent
   (B) nature of the solute
   (C) concentration of the solution
   (D) temperature of the solution
   (E) all the above

4. Which of the following is TRUE?
   (A) The number of positive ions in solution equals the number of negative ions.
   (B) The positive ions are called anions.
   (C) The positive ions are called cathodes.
   (D) The total positive charge equals the total negative charge.
   (E) None of the above

5. The hydronium ion is represented as
   (A) H_2O^+
   (B) H_3O^+
   (C) HOH^+
   (D) H^-

6. In the electrolysis of copper chloride, the substance liberated at the anode is
   (A) copper
   (B) chlorine
   (C) hydrogen
   (D) copper chloride

7. Ions are particles that exist
   (A) only in water solutions
   (B) in some crystals
   (C) in polar covalent compounds
   (D) in covalent compounds that are not polar

8. Ionic compounds conduct an electric current when they are
   (A) solidified
   (B) melted
   (C) frozen
   (D) dehydrated

9. The cathode in a direct-current circuit is an electrode that is
   (A) always negative
   (B) always positive
   (C) always neutral
   (D) sometimes negative and sometimes positive

10. Electrolysis of a dilute solution of sodium chloride results in the cathode product
    (A) sodium
    (B) chlorine
    (C) hydrogen
    (D) oxygen

11. Electrode potentials are:

    $Zn^0 = Zn^{2+} + 2e^-$ $E^0 = +0.76$ V
    $Au^0 = Au^{3+} + 3e^-$ $E^0 = -1.42$ V

    If a gold foil is placed in a solution containing Zn ions, the reaction potential will be
    (A) -1.34 V
    (B) -2.18 V
    (C) -0.66 V
    (D) $+2.18$ V
    (E) $+1.34$ V

12. If the reaction potential is positive, this indicates that
    (A) the reaction will not occur
    (B) the reaction will occur and give off energy
    (C) the reaction will occur if heat or energy is added
    (D) the reaction will power an outside alternating electric current

Answers

1. **A**	4. **D**	7. **B**	9. **A**	11. **B**
2. **C**	5. **B**	8. **B**	10. **C**	12. **B**
3. **E**	6. **B**			

Terms You Should Know

anions
anode
cathode
cations
dissociation
electrochemical cell
electrode potential
electrolyte
electrolytic cell

electrolytic reaction
electrolysis
electromotive series
electroplating
half-cell
ionization
Nernst equation
oxidation
oxidizing agent
reducing agent
reduction
standard voltage
voltaic cell

PART VI: REPRESENTATIVE GROUPS AND FAMILIES

Chapter 13

SOME REPRESENTATIVE GROUPS AND FAMILIES

In the following section, a brief description is given of some of the important and representative groups of elements usually referred to in most first-year chemistry courses. The presentation of each group will follow along this outline: the important element or elements, their occurrence, their preparation both in laboratory and in industrial processes, their properties, and some of their important uses.

Sulfur Family

Since we discussed oxygen in Chapter 5, the most important element left to discuss is sulfur.

Sulfur is found free in the volcanic regions of Japan, Mexico, and Sicily. It was removed from the rock mixtures by heating in retorts or furnaces.

In 1867 sulfur was discovered in underground deposits in Louisiana and Texas. The overlying layer of earth was "quicksand" which prohibited the usual mining operations. An American chemist, Herman Frasch, devised an ingenious plan for its removal which bears his name, the Frasch process. It consists of drilling a hole and inserting three concentric pipes. Through the outer pipe superheated water (at 160°C) is forced into the sulfur bed, melting the sulfur. Compressed air, which is forced down the innermost pipe, causes the liquid sulfur to rise up through the middle pipe to the surface. At the surface the liquid sulfur is pumped into vast molds in which it solidifies. See Figure 38. The sulfur obtained in this process is usually more than 99.5% pure.

Sulfur has three allotropic forms. Their differences are summarized in Table 12.

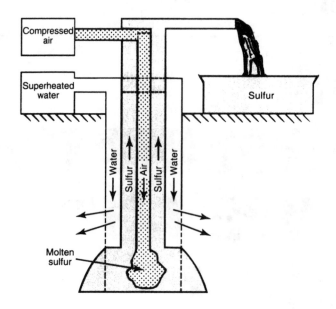

Figure 38. Frasch Process.

TABLE 12

ALLOTROPIC FORMS OF SULFUR

	Allotropic Form		
Characteristic	**Rhombic**	**Monoclinic**	**Amorphous**
Shape	Rhombic or octahedral crystals	Needle-shaped monoclinic crystals	Noncrystalline
Color	Pale-yellow, opaque, brittle	Yellow, waxy, translucent, brittle	Dark, tough, elastic
Preparation	Crystallizes from CS_2 solution of sulfur	Crystallizes from molten sulfur as it cools	Obtained by quick-cooling very hot liquid sulfur
Stability	Stable below 96°C	Stable between 96° and 119°C	Unstable; slowly changes to rhombic

The solid sulfur is composed of S_8 molecules as shown below:

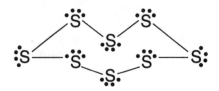

Since each covalent bond is sharing of a pair of electrons, there is a potential for extending this structure: $\left(\overset{s}{\underset{}{\cdot}}\overset{s}{S}\overset{s}{\underset{}{\cdot}} \right)$ which is found in the S_8 molecule. As heat is applied, the kinetic molecular motion increases and eventually breaks this ring structure, leading to a chain that becomes entangled with other chains and explains the increased viscosity of the liquid sulfur at high temperatures. Because the broken chain has unshared electrons on its ends, it is able to combine with other chains to form even longer chains and thus increase the viscosity further. The unpaired electrons are also associated with the change in color that occurs.

Sulfur is used in making sprays to control plant diseases and harmful insects. It is used in rubber to provide added hardness or to "vulcanize" it. Sulfur enters into the preparation of medicines, matches, gunpowder, and fireworks. Its most important use is making sulfuric acid, which is often referred to as the "king of chemicals" because of its widespread industrial use.

Sulfuric Acid

There are two basic methods for preparing *sulfuric acid*: the lead-chamber process and the contact process. In both, essentially the same reactions occur, but the methods vary in the catalyst used in the second step.

Step 1: Sulfur is burned to form SO_2.

$$S + O_2 \rightarrow SO_2$$

Step 2: SO_2 is oxidized with the aid of a catalyst to form SO_3. The contact process uses insulated steel towers containing layers of V_2O_5 pellets as the catalyst. The lead-chamber process employs a series of lead chambers in which SO_2 is oxidized by means of NO_2, supplied for that purpose. The basic reaction in both cases is

$$2SO_2 + O_2 \rightarrow 2SO_3$$

Step 3: H_2O is added to SO_3 to form H_2SO_4.

$$SO_3 + H_2O \rightarrow H_2SO_4$$

In the contact process water can be added as fast as the SO_3 is absorbed to give 98 or 99% H_2SO_4. If less water is added, "oleum" or "fuming" sulfuric acid is formed. The older lead-chamber process generally makes a more dilute solution of H_2SO_4.

Important Properties of Sulfuric Acid

Sulfuric acid ionizes in two steps:

$$H_2SO_4 + H_2O \rightleftharpoons H_3O^+ + HSO_4^- \quad (Ka_1 \text{ is very large.})$$
$$HSO_4^- + H_2O \rightleftharpoons H_3O + SO_4^{2-} \quad (Ka_2 \text{ is very small.})$$

to form a strong acid solution. The ionization is more extensive in a dilute solution. Most hydronium ions are formed in the first step as indicated by the longer forward-reaction arrow. Salts formed with the HSO_4^- (bisulfate ion) are called *acids salts*; the SO_4^{2-} (sulfate ion) forms *normal salts*. The test for sulfate ion consists of adding a solution of HCl and

$BaCl_2$ or $Ba(NO_3)_2$ to the solution to be tested. If sulfate is present, a white precipitate of $BaSO_4$ will form.

Sulfuric acid reacts like other acids, as shown below:

With active metals: $Zn + H_2SO_4 \rightarrow ZnSO_4 + H_2(g)$ (for dilute H_2SO_4)
With bases: $\qquad 2NaOH + H_2SO_4 \rightarrow Na_2SO_4 + 2H_2O$
With metal oxides: $MgO + H_2SO_4 \rightarrow MgSO_4 + H_2O$
With carbonates: $\quad CaCO_3 + H_2SO_4 \rightarrow CaSO_4 + H_2O + CO_2(g)$

Sulfuric acid has other particular characteristics as shown below:

As an oxidizing agent:
$Cu + 2H_2SO_4 \text{ (concentrated)} \rightarrow CuSO_4 + SO_2(g) + 2H_2O$
As a dehydrating agent with carbohydrates:
$C_{12}H_{22}O_{11}(\text{sugar}) \xrightarrow[H_2SO_4]{\text{conc.}} 12C + 11H_2O$

Important Uses of H_2SO_4

1. Making other acids (because of its high boiling point, 338°C):

$$NaCl + H_2SO_4 \xrightarrow{\text{heat}} NaHSO_4 + HCl(g)$$

(With more heat and an excess of salt, the second hydrogen ion can also be removed to form the salt Na_2SO_4.)
2. Washing objectionable colors from gasoline made by the "cracking" process.
3. Dehydrating agent in making explosives, dyes, and drugs.
4. Freeing metals of iron or steel of scale and rust. This is called "pickling."
5. Electrolyte in ordinary lead storage batteries.
6. Used in precipitating bath in making rayon.

This is a list of just the more important uses of sulfuric acid. It is by no means complete.

Other Important Compounds of Sulfur

Hydrogen sulfide is a colorless gas having an odor of rotten eggs. It is fairly soluble in water and poisonous in rather small concentrations. It can be prepared by reacting ferrous sulfide with an acid, such as dilute HCl:

$$FeS + 2HCl \rightarrow FeCl_2 + H_2S(g)$$

It does burn in excess oxygen to form compounds of water and sulfur dioxide. If insufficient oxygen is available, some free sulfur will form. It is only a weak acid in a water solution. Hydrogen sulfide is used widely in qualitative laboratory tests since many metallic sulfides precipitate with recognizable colors. These sulfides are sometimes used as paint pigments. Some common sulfides and their colors are

ZnS—White
CdS—Bright yellow

As$_2$S$_3$—Lemon yellow
Sb$_2$S$_3$—Orange
CuS—Black
HgS—Black
PbS—Brown-black

Another important compound of sulfur is *sulfur dioxide*. It is a colorless gas with a suffocating odor. It can be prepared by burning sulfur in air:

$$S + O_2 \rightarrow SO_2$$

or decomposing sulfites:

$$Na_2SO_3 + H_2SO_4 \rightarrow Na_2SO_4 + H_2O + SO_2(g)$$

Sulfur dioxide is the acid anhydride of sulfurous acid:

$$SO_2 + H_2O \rightarrow H_2SO_3$$

It is used as a bleach on moist straw, silk, wool, and paper because of its ability to reduce the coloring agent. This process is reversed by oxidation in sunlight as shown by the yellowing of straw and paper. Potassium permanganate can also be bleached by a water solution of SO$_2$ in the form of H$_2$SO$_3$:

$$5H_2SO_3 + 2KMnO_4 \rightarrow K_2SO_4 + 2MnSO_4 + 2H_2SO_4 + 3H_2O$$

The structure of sulfur dioxide is a good example of a *resonance* structure. Its molecule is depicted in Figure 39.

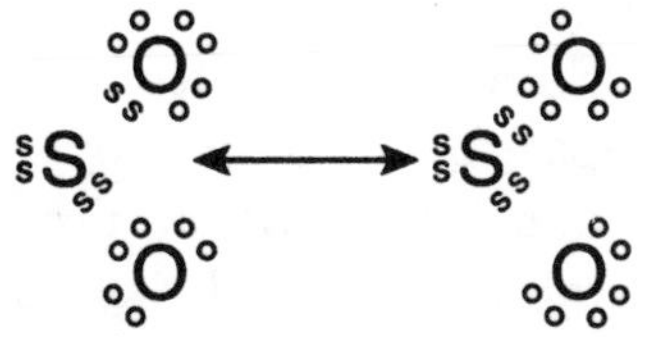

Figure 39. Sulfur Dioxide Molecule.

Notice that the covalent bonds between sulfur and oxygen are shown in one figure as single bonds and in the other as double bonds. This signifies that the bonds between the sulfur and oxygens have been shown by experimentation to be neither single nor double bonds but "hybrids" of the two. Sulfur trioxide also is a resonance structure.

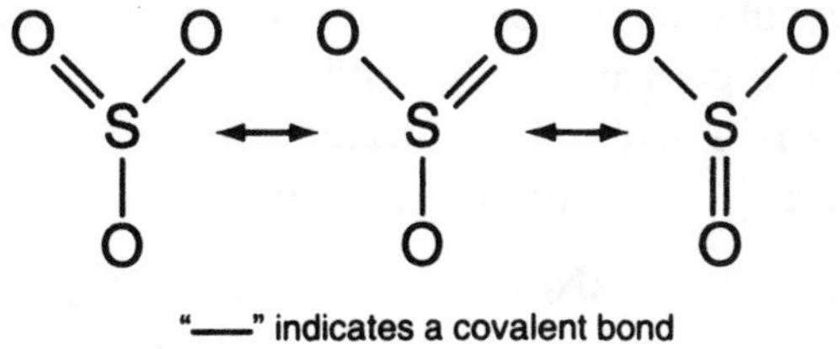

"——" indicates a covalent bond

Halogen Family

The common members of the halogen family are shown in Table 13 with some important facts concerning them.

TABLE 13

HALOGEN FAMILY

Item	Fluorine	Chlorine	Bromine	Iodine
Molecular formula	F_2	Cl_2	Br_2	I_2
Atomic number	9	17	35	53
Activity	Most active	← to →		Least active
Outer energy level structure	7 electrons	7 electrons	7 electrons	7 electrons
State and color at room temperature	Gas—pale yellow	Gas—green	Liquid— dark red	Solid— purplish black crystals

Because all the halogens lack one electron in their outer principal energy level, they usually are acceptors of electrons (oxidizing agents). Fluorine is the most active nonmetal in the periodic chart. The halogens can be prepared as follows:

1. Fluorine can be prepared by electrolysis of a mixture of KHF_2 and liquid anhydrous hydrogen fluoride. Electrolysis is necessary because the extreme electronegativity of fluorine makes it impossible to prepare by simple displacement.
2. Chlorine, bromine, and iodine can be prepared by reacting a halide salt of each, like NaCl, NaBr, or NaI, with $H_2SO_4 + MnO_2$. The ionic equations are

$$2Cl^- + 2H_2SO_4 + MnO_2 \rightarrow SO_4^{2-} + Mn^{2+}SO_4^{2-} + 2H_2O + Cl_2(g)$$
$$2Br^- + 2H_2SO_4 + MnO_2 \rightarrow SO_4^{2-} + Mn^{2+}SO_4^{2-} + 2H_2O + Br_2(g)$$
$$2I^- + 2H_2SO_4 + MnO_2 \rightarrow SO_4^{2-} + Mn^{2+}SO_4^{2-} + 2H_2O + I_2(g)$$

A laboratory preparation of Cl_2 is

$$4HCl + MnO_2 \rightarrow MnCl_2 + 2H_2O + Cl_2(g)$$

The resulting chlorine is collected by the upward displacement of air since Cl_2 is heavier than air.

The commercial preparation of chlorine involves the electrolysis of a salt solution. When a direct current is passed through the solution, chlorine gas is released at the anode and hydrogen gas is released at the cathode. One type of commercial electrolysis apparatus is called a Hooker cell. The overall reaction is

$$2NaCl + 2H_2O \rightarrow H_2(g) + Cl_2(g) + 2NaOH$$

Testing for Halides

This ability of chlorine to replace I^- and Br^- ions is also used in testing for iodides and bromides. The test involves putting the salt into a water solution and then adding fresh chlorine water and some carbon tetrachloride. The resulting mixture is shaken vigorously and observed for a purple or red color in the carbon tetrachloride, which will sink to the bottom of the container. Purple indicates a positive test for iodine; deep red shows the presence of bromine. The reactions that release these halogens are as follows:

$$2I^- + Cl_2 \rightarrow 2Cl^- + I_2 \text{ (purple in CCl}_4)$$
$$2Br^- + Cl_2 \rightarrow 2Cl^- + Br_2 \text{ (deep red in CCl}_4)$$

Test for fluoride   Add sulfuric acid and test if the gas released, which will be hydrogen fluoride (HF), will etch glass. (Use **caution** not to get on hands.)

Test for chloride   Add a solution of silver salt (usually silver nitrate) and look for a white precipitate that is soluble in ammonium hydroxide but insoluble in nitric acid. The white precipitate darkens to gray when exposed to light.

Some Important Halides and Their Uses

Hydrochloric acid   Common acid prepared in the laboratory by reacting sodium chloride with concentrated sulfuric acid. It is used in many important industrial processes.

Silver bromide and silver iodide   Halides used on photographic film. Light intensity is recorded by developing as black metallic silver those portions of the film upon which the light fell during exposure.

Hydrofluoric acid   Acid used to etch glass by reacting with the SiO_2 to release silicon fluoride gas. Also used to frost light bulbs.

Uses of Halogens

Fluorides are added to drinking water and to toothpaste to reduce tooth decay. The halogen chlorine is used to purify water supplies, to bleach, and to prepare pure hydrochloric acid, along with many other industrial applications. Its reaction as a bleach is

$$Cl_2 + H_2O \rightarrow HCl + HClO \text{ (hypochlorous acid)}$$

It is the hypochlorous acid that does the bleaching by oxidation.

$$\text{Colored material} + HClO \rightarrow \text{decolorized material} + HCl$$

Chlorine also has the ability to withdraw hydrogen from hydrocarbons. An example:

$$C_{10}H_{16} + 8Cl_2 \rightarrow 10C + 16HCl$$

Warm turpentine   Carbon released
as dense, black
smoke

Bromine is used as ethylene dibromide in antiknock gasolines to remove the lead deposits that form from the other additive, tetraethyl lead. It is also used as a bromide in many nerve sedatives.

Iodine is most widely known for its use as an antiseptic in an alcohol solution of iodine called tincture of iodine. Iodine is also used in small amounts to treat goiter conditions.

Nitrogen Family

The most common member of this family is nitrogen itself. It is a colorless, odorless, tasteless, and rather inactive gas that makes up about four fifths of the air in our atmosphere. The inactivity of N_2 gas can be explained by the fact that the two atoms of nitrogen are bonded by three covalent bonds ($:N::N:$) that require a great deal of energy to break. Since nitrogen must be "pushed" into combining with other elements, many of its compounds tend to decompose violently with a release of the energy that went into forming them.

Nitrogen can be prepared in the laboratory by the equation

$$NaNO_2 + NH_4Cl \rightarrow NaCl + 2H_2O + N_2(g)$$

It is most often prepared industrially by the fractional distillation of liquid air. This product, of course, contains some of the inert gases but is pure enough for most uses.

Nature "fixes" nitrogen, or makes nitrogen combine, by a nitrogen-fixing bacteria found in the roots of beans, peas, clover, and other leguminous plants. Discharges of lightning also cause some nitrogen fixation with oxygen to form nitrogen oxides.

Nitric Acid

An important compound of nitrogen is nitric acid. This acid is useful in making dyes, celluloid film, and many of the lacquers used on cars.

Its physical properties are as follows: It is a colorless liquid (when pure), it is $1\frac{1}{2}$ times as dense as water, it has a boiling point of 86°C, the commercial form is about 68% pure, and it is miscible with water in all proportions.

Its outstanding chemical properties are: the dilute acid shows the usual properties of an acid except it rarely produces hydrogen when it reacts with metals, and it is quite unstable and decomposes as follows:

$$4HNO_3 \rightarrow 2H_2O + 4NO_2(g) + O_2(g)$$

Because of this ease of decomposition, nitric acid is a good oxidizing agent. When it reacts with metals, the nitrogen product formed will depend on the conditions of the reaction, especially the concentration of the acid, the activity of the metal, and the temperature. If the nitric acid is concentrated and the metal is copper, the principal reduction product will be nitrogen dioxide (NO_2), a heavy, red-brown gas with a pungent odor.

$$Cu + 4HNO_3 \rightarrow Cu(NO_3)_2 + 2NO_2(g) + 2H_2O$$

With dilute nitric acid, this reaction is

$$3Cu + 8HNO_3 \rightarrow 3Cu(NO_3)_2 + 4H_2O + 2NO(g)$$

The product, called nitric oxide (NO), is colorless and is immediately oxidized in air to NO_2 gas.

With still more dilute nitric acid, considerable quantities of nitrous oxide (N_2O) are formed; with an active metal like zinc, the product may be the ammonium ion (NH_4^+).

When nitric acid is mixed with hydrochloric acid, the mixture is called *aqua regia* because of its ability to dissolve gold. The great activity of aqua regia is due to the formation of the nitrosyl chloride (NOCl) and chlorine (Cl_2). The Cl_2 reacts with gold to form the soluble compound of gold.

Nitric acid reacts with protein to give a yellow color and also reacts with other organic compounds to form unstable compounds, such as nitrobenzene, trinitrotoluene (TNT), and picric acid. The last two are powerful explosives. Another explosive is made by reacting glycerine with nitric acid to form glyceryl trinitrate ("nitroglycerine").

The preparation of nitric acid in the laboratory involves reacting a nitrate salt with sulfuric acid in a retort.

The industrial preparation, called the Ostwald process, is the greatest source of nitric acid. Ammonia and air are passed through a heated platinum gauze, which acts as a catalyst, to form nitric oxide (NO).

$$4NH_3 + 5O_2 \rightarrow 4NO(g) + 6H_2O$$

The nitric oxide is further oxidized to NO_2.

$$2NO + O_2 \rightarrow 2NO_2(g)$$

Then $3NO_2 + H_2O \rightarrow 2HNO_3 + NO(g)$ forms the nitric acid

Other Important Compounds of Nitrogen

Ammonia is one of the oldest known compounds of nitrogen. In times past, it was prepared by distilling leather scraps, hoofs, and horns. It can be prepared in the laboratory by heating an ammonium salt with a strong base. An example is

$$Ca(OH_2) + (NH_4)_2SO_4 \rightarrow CaSO_4 + 2H_2O + 2NH_3(g)$$

Industrial methods include the Haber process, which combiines nitrogen and hydrogen (discussed further on page 186)

$$3H_2 + N_2 \rightarrow 2NH_3$$

and the destructive distillation of coal, which gives off NH_3 as a by-product.

The physical properties of ammonia are: it is a colorless, pungent gas which is extremely soluble in water; it is lighter than air and has a high critical temperature (the critical temperature of a gas is the temperature above which it cannot be liquefied by pressure alone).

Because ammonia can be liquefied easily and has a high heat of vaporization, one of its uses is as a refrigerant. Also, a water solution of ammonia is used as a household cleaner to cut grease. Another use of ammonia is as a fertilizer.

Some other nitrogen compounds are summarized in Table 14.

TABLE 14

SOME NITROGEN COMPOUNDS

Name	Formula	Notes of Importance
Nitrate ion	NO_3^-	1. Nitrates of metals can be decomposed with heat to release oxygen. 2. Nitrate test—add freshly prepared ferrous sulfate solution; hold test tube at an angle and slowly add concentrated sulfuric acid down the side; if a brown ring forms just above the sulfuric acid level, it indicates a positive test for nitrate ion. (Test can be repeated using acetic instead of sulfuric acid; if brown ring again forms, it indicates a nitrite.)
Nitrous oxide	N_2O	1. Colorless; $1\frac{1}{2}$ times heavier than air; slight sweet smell and odor; supports combustion much like oxygen. 2. Sometimes used as an anesthetic called "laughing gas."
Nitric oxide	NO	1. Colorless, slightly heavier than air, does not support combustion, is easily oxidized to NO_2 when exposed to air.
Nitrogen dioxide	NO_2	1. Reddish-brown gas, 1.6 times heavier than air, very soluble in water, and extremely poisonous; has unpleasant odor.
Nitrogen tetroxide	N_2O_4	1. When NO_2 is cooled, its brown color fades to a pale yellow (N_2O_4). Two molecules of NO_2 combine to form this new compound.

Other Members of the Nitrogen Family

As you proceed down this family (see Table 15), the elements change from gases to volatile solids and then to less volatile solids. At the same time, acid-forming properties are decreasing while base-forming properties are increasing. These changes can be attributed to the increased atomic radii and the decreased ability to attract and hold the outer energy level electrons.

TABLE 15

THE NITROGEN FAMILY

Name	Symbol	Atomic Number	Properties and Uses
Nitrogen	N	7	(Discussed previously)
Phosphorus	P	15	Two allotropic forms: white (kindling temperature near room temperature, thus kept under water) and red. When burned in oxygen, forms P_2O_5, the acid anhydride of phosphoric acid. When P_2O_5 reacts with water, it first forms metaphosphoric acid, HPO_3; then more water reacts to form the orthophosphoric acid. Phosphorus is used in making matches.
Arsenic	As	33	Slightly volatile, gray solid. Most abundant compound is As_2O_3, which is used as a preservative, in medicines, and in glassmaking (highly poisonous).
Antimony	Sb	51	Lustrous, gray solid. Forms very weak acid, mostly basic solutions.
Bismuth	Bi	83	Almost wholly metallic properties. Shiny, dense metal with slight reddish tinge. Low melting point; therefore used in alloys for electric fuses and fire sprinkler systems. "Woods metal" is the most outstanding of these (melting point 70°C).

Metals

Properties of Metals

Some physical properties of metals are: they have metallic luster, they can conduct heat and electricity, they can be pounded into sheets (malleable), they can be drawn into wires (ductile), most have a silvery color, they have densities usually between 7 and 14 g/cm^3 and none is soluble in any ordinary solvent without a chemical change.

The general chemical properties of metals are: they are electropositive, and the more active metallic oxides form bases, although some metals form amphoteric hydroxides which can react as both acid and base.

Metals can be removed from their ores by various means depending on their degree of activity (see Table 16).

TABLE 16

MEANS OF EXTRACTING METALS

Occurrence	Activity Decreasing	Metal	Ease of Reduction	Reduced by
In combined form only		Potassium Calcium Sodium Magnesium Aluminum	Very difficult	Electrolysis
		Manganese Zinc Chromium	Difficult	Carbon with diminishing difficulty
Usually combined		Iron Cadmium Cobalt Nickel		
		Tin Lead		
Free and combined		HYDROGEN Antimony Bismuth Arsenic Copper	Easy	Carbon
		Mercury Silver	Very easy	By heating oxides
Free		Platinum Gold	Easiest	All compounds decomposed by heat

In some cases the ore is concentrated before it is reduced. This might be done by a merely physical means, such as flotation, in which the ore is crushed, mixed with a water solution containing a suitable oil (e.g., pine oil), and then vigorously agitated so that a frothy mass of oil rises to the top with the adhering metal-bearing portion of the ore. The froth is removed from the top of the machine and filtered to recover the metal ore.

A chemical process is used to convert sulfide or carbonate ores to oxides. It is called *roasting* and consists of heating the compounds in air. For example,

$$2ZnS + 3O_2 \rightarrow 2ZnO + 2SO_2(g)$$

and

$$PbCO_3 \rightarrow PbO + CO_2(g)$$

These oxides can then be treated with appropriate reducing agents to free the metal.

The following table summarizes the properties of the first two families of metals found in Group IA and Group IIA or 2 of the periodic chart. The table shows the distinct similarities in the properties and bonding of these metals. These similarities are directly related to the similarities that exist in the outer orbital electron configuration.

Element	Outer Orbital	Bonding and Charge	Physical Properties
Alkali Metals			
Lithium (Li)	$2s^1$	Ionic, +1	Silvery white
Sodium (Na)	$3s^1$	Ionic, +1	Silvery white
Potassium (K)	$4s^1$	Ionic, +1	Silvery white
Rubidium (Rb)	$5s^1$	Ionic, +1	Silvery white
Cesium (Cs)	$6s^1$	Ionic, +1	Silvery white
Francium (Fr)	$7s^1$	Ionic, +1	Silvery white
Alkaline-Earth Metals			
Beryllium (Be)	$2s^2$	Ionic, +2	Gray-white
Magnesium (Mg)	$3s^2$	Ionic, +2	Silvery white
Calcium (Ca)	$4s^2$	Ionic, +2	Silvery white
Strontium (Sr)	$5s^2$	Ionic, +2	Silvery white
Barium (Ba)	$6s^2$	Ionic, +2	Yellow white; lumpy
Radium (Ra)	$7s^2$	Ionic, +2	White; radioactive

Some Important Reduction Methods

The Hall process is the electrolytic method of preparing *aluminum*. It uses molten cryolite (Na_3AlF_6) as the solvent for bauxite ore ($Al_2O_3 \cdot 2H_2O$). This mixture is electrolyzed in a rectangular iron tank lined with carbon which acts as the cathode. The anodes are heavy carbon rods suspended in the tank. The aluminum released at the cathode is tapped periodically and allowed to flow out of the tank so that the process can continue.

The outstanding properties of aluminum are its light weight with high strength, its ability to form an aluminum oxide coating that protects this metal from further oxidation, and its ability to conduct an electric current.

The Dow process is an electrolytic method of preparing *magnesium* from seawater. The magnesium salts are precipitated with $Ca(OH)_2$ solution to form $Mg(OH)_2$. The hydroxide is then converted to a chloride by adding hydrochloric acid to the hydroxide. This chloride is mixed with potassium chloride, melted, and electrolyzed to release the magnesium at the cathode and chlorine at the anode.

Magnesium has the important property of being a light, rigid, and inexpensive metal. It, too, can form a protective adherent oxide coating.

Another major metal prepared by electrolysis is *copper*. After the ore is concentrated by flotation and roasted to change sulfides to oxides, it is heated in a converter to obtain impure copper called "blister" copper. This copper is then used as anodes in the electrolysis apparatus, in which pure copper is used to make the cathode plates. As the process progresses, the copper in the anode becomes ions in the solution which are reduced at the cathode as pure copper.

Copper has wide usage in electrical appliances and wires because of its high conductivity. It is widely employed in alloys of bronze (copper and tin) and brass (copper and zinc).

Iron ore is refined by reduction in a blast furnace, that is, a large, cylinder-shaped furnace charged with iron ore (usually hematite, Fe_2O_3), limestone, and coke. A hot air blast, often enriched with oxygen, is blown into the lower part of the furnace through a series of pipes called tuyeres. The chemical reactions that occur can be summarized as follows:

Burning coke: $2C + O_2 \rightarrow 2CO(g)$
       $C + O_2 \rightarrow CO_2(g)$
Reduction of CO_2: $CO_2 + C \rightarrow 2CO(g)$
Reduction of ore: $Fe_2O_3 + 3CO \rightarrow 2Fe + 3CO_2(g)$
       $Fe_2O_3 + 3C \rightarrow 2Fe + 3CO(g)$
Formation of slag: $CaCO_3 \rightarrow CaO + CO_2(g)$
       $CaO + SiO_2 \rightarrow CaSiO_3$

The molten iron from the blast furnace is called "pig iron."

From pig iron, the molten metal might undergo one of three steel-making processes that burn out impurities and set the contents of carbon, manganese, sulfur, phosphorus, and silicon. Often nickel and chromium are alloyed in steel to give particular properties of hardness needed for tool parts. The three most important means of making steel involve the basic oxygen, the open-hearth, and the electric furnace. The first two methods are the most common.

The basic oxygen furnace uses a lined "pot" into which the molten pig iron is poured. Then a high-speed jet of oxygen is blown from a watercooled lance into the top of the pot. This "burns out" impurities to make a batch of steel rapidly and cheaply.

The open-hearth furnace is a large oven containing a dish-shaped area to hold the molten iron, scrap steel, and other additives with which it is charged. Alternating blasts of flame are directed across the surface of the melted metal until the proper proportions of additives are established for that "heat" so that the steel will have the particular properties needed by the customer. The tapping of one of these furnaces holding 50 to 400 tons of steel is a truly beautiful sight.

The final method of making steel involves the electric furnace. This method uses enormous amounts of electricity through graphite cathodes that are lowered into the molten iron to purify it and produce a high grade of steel.

Alloys

An *alloy* is a mixture of two or more metals. In a mixture certain properties of the metals involved are affected. Some of these are

Melting point The melting point of an alloy is lower than that of its components.

Hardness An alloy is usually harder than the metals that compose it.

Crystal structure The size of the crystalline particles in the alloy determine many of the physical properties. The size of these particles can be controlled by heat treatment. If the alloy cools slowly, the crystalline particles tend to be larger. Thus, by heating and cooling an alloy, its properties can be altered considerably.

Metalloids

In the preceding sections representative metals and nonmetals have been reviewed, along with the properties of each. Some elements, however, are difficult to classify as one or the other. One example is carbon. The diamond form of carbon is a poor conductor, yet the graphite form conducts fairly well. Neither form looks metallic, and so carbon is classified as a nonmetal.

Silicon looks like a metal. However, its conductivity properties are closer to those of carbon.

Since some elements are neither distinctly metallic nor clearly nonmetallic, a third class, called the *metalloids*, is recognized.

The properties of metalloids are intermediate between those of metals and those of non-metals. Although most metals form ionic compounds, metalloids as a group may form ionic or covalent bonds. Under certain conditions pure metalloids conduct electricity but do so poorly and are thus termed *semiconductors*. This property makes them important in micro-circuitry.

The metalloids are located in the periodic table along the heavy dark line that starts alongside boron and drops down in steplike fashion between the elements found lower in the table (see the chart below).

	13 IIIA	14 IVA	15 VA	16 VIA	17 VIIA
	5 B	6 C	7 N	8 O	9 F
12 IIB	13 Al	14 Si	15 P	16 S	17 Cl
30 Zn	31 Ga	32 Ge	33 As	34 Se	35 Br
48 Cd	49 In	50 Sn	51 Sb	52 Te	53 I
80 Hg	81 Ti	82 Pb	83 Bi	84 Po	85 At

Location of Metalloids

Chapter 13 Review

1. The most active nonmetallic element is
   (A) chlorine
   (B) fluorine
   (C) oxygen
   (D) sulfur

2. The insertion of a burning paraffin taper into a bottle of chlorine will result in
   (A) extinguishing the flame
   (B) the production of carbon
   (C) the production of hydrogen
   (D) the production of carbon tetrachloride

3. The bromide ion is
   (A) colorless
   (B) red-orange
   (C) violet
   (D) yellow

4. When chlorine water and carbon tetrachloride are mixed with an unknown, a violet layer is produced. The unknown is a(n)
   (A) fluoride
   (B) chloride
   (C) bromide
   (D) iodide

5. The order of decreasing activity of the halogens is
   (A) Fl, Cl, I, Br
   (B) F, Cl, Br, I
   (C) Cl, F, Br, I
   (D) Cl, Br, I, F

6. A light-sensitive substance used on photographic films has the formula
   (A) $AgBr$
   (B) CaF_2
   (C) $CuCl$
   (D) $MgBr_2$

7. Aqua regia is
   (A) HCl and H_2SO_4
   (B) HCl and HNO_3
   (C) HCl and HBr
   (D) HCl and HF

8. The metal that has an orange sulfide is
   (A) Cd
   (B) As
   (C) Pb
   (D) Sb

9. Sulfur dioxide is the anhydride of
   (A) hydrosulfuric acid
   (B) sulfurous acid
   (C) sulfuric acid
   (D) hyposulfurous acid

10. The charring action of sulfuric acid is due to its being
    (A) a strong acid
    (B) an oxidizing agent
    (C) a reducing agent
    (D) a dehydrating agent

11. Sulfuric acid is used in the manufacture of many other acids because of its
    (A) high specific gravity
    (B) high boiling point
    (C) oxidizing ability
    (D) solubility

12. Nitrogen for commercial use is generally obtained from
    (A) ammonia
    (B) liquid air
    (C) sodium nitrate
    (D) nitrogen-fixing bacteria

13. Ammonia is prepared commercially by the
    (A) Ostwald process
    (B) arc process
    (C) Haber process
    (D) contact process

14. A nitrogen compound that has a color is
    (A) nitric oxide
    (B) nitrous oxide
    (C) nitrogen dioxide
    (D) ammonia

15. If a student heats a mixture of ammonium chloride and calcium hydroxide in a test tube, he or she will detect
    (A) no reaction
    (B) the odor of ammonia
    (C) the odor of rotten eggs
    (D) nitric acid fumes

16. The difference between ammonia and ammonium is
    (A) an electron
    (B) a neutron
    (C) a proton
    (D) radioactivity

17. An important ore of iron is
    (A) bauxite
    (B) galena
    (C) hematite
    (D) smithsonite

18. The material added to the charge of a blast furnace to react with the sand is
    (A) calcium carbonate
    (B) coke
    (C) silicon dioxide
    (D) slag

19. A reducing agent used in the blast furnace is
    (A) $CaCO_3$
    (B) CO
    (C) O_2
    (D) SiO_2

20. The most abundant of all metals in the earth's crust is
    (A) iron
    (B) aluminum
    (C) copper
    (D) silver

21. The solvent used in obtaining aluminum from its ore is
    (A) bauxite
    (B) cryolite
    (C) dolomite
    (D) carnalite

22. The process of extracting aluminum is called
    (A) the Dow process
    (B) the Haber process
    (C) the Hall process
    (D) electroplating

23. Aluminum is extracted from its oxide by
    (A) roasting
    (B) reduction with carbon
    (C) smelting
    (D) electrolysis

24. The Dow process extracts magnesium from
    (A) sea shells
    (B) dolomite ($CaCO_3$_ $MgCO_3$)
    (C) seawater
    (D) oyster shells

25. An alloy of bismuth is used in automatic fire sprinklers because
    (A) it is soft
    (B) it has a low melting point
    (C) it has a high melting point
    (D) it resists corrosion

Answers

1. **B**	6. **A**	11. **B**	16. **C**	21. **B**
2. **B**	7. **B**	12. **B**	17. **C**	22. **C**
3. **A**	8. **D**	13. **C**	18. **A**	23. **D**
4. **D**	9. **B**	14. **C**	19. **B**	24. **C**
5. **B**	10. **D**	15. **B**	20. **B**	25. **B**

Terms You Should Know

acid salt
allotropic forms
alloys
aqua regia
contact process
electric furnace
Haber process
lead-chamber process
leguminous plants

metalloids
nitrogen-fixing
normal salt
Ostwald process
pig iron
resonance structure
rhombic, monoclinic, amorphous
semiconductors

PART VII:
ORGANIC AND
NUCLEAR CHEMISTRY

Chapter 14

CARBON AND ORGANIC CHEMISTRY

Carbon

Forms of Carbon

The element carbon occurs in three allotropic forms: diamond, graphite, and amorphous (although some evidence shows the amorphous forms have some crystalline structure). In the mid-1980s, fullerenes were identified as a new allotropic form of carbon. They are found in soot that forms from burning carbon-containing materials with limited oxygen. Their structure consists of near-spherical cages of carbon atoms resembling geodesic domes.

The diamond form has a close-packed crystal structure that gives it its property of extreme hardness. In it each carbon is bonded to four other carbons in a tetrahedron arrangement like this:

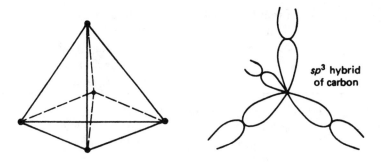

sp^3 hybrid of carbon

It has been possible to make synthetic diamonds in machines that subject carbon to extremely high pressures and temperatures. Most of these diamonds are used for industrial purposes, such as dies.

The graphite form is made up of planes of hexagonal structures that are weakly bonded to the planes above and below. This explains graphite's slippery feeling and makes it useful as a dry lubricant. Its structure can be seen below. Graphite also has the property of being an electrical conductor.

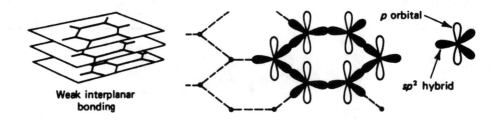

Some of the common amorphous forms of carbon are shown in Table 17 along with their properties and uses.

The process of destructive distillation is used in the preparation of both charcoal and coke. This process involves the heating of a substance in an enclosed container so that volatile materials, which would cause wood and coal to burn with a flame, can be forced out. After these substances are removed, the charcoal and coke merely glow when burned. Many of these volatile substances have proved to be useful by-products of destructive distillation. From the destructive distillation of coal we get coal gas, coal tar, and ammonia. Coal gas can be used as a fuel; coal tars are the primary source of most synthetic dyes and many other useful organic compounds; and ammonia is also a useful chemical.

TABLE 17

ALLOTROPES OF CARBON

Forms	Occurrence or Preparation	Properties	Uses
1. Crystalline			
a. Diamond	South Africa; machines using very high pressures and temperatures	Hardest substance Brilliant: reflects and refracts light	Abrasive: drills, saws, polish bearings, gems, dies
b. Graphite	From hard coal in an electric furnace	Soft, gray, greasy Electrical conductor Refractory: high melting and kindling points	Lubricant, crucibles, electrodes, lead pencils, atomic pile
2. Amorphous (noncrystalline)—no definite shape			
a. Charcoal	Destructive distillation of wood	Burns with a glow and no flame; porous Adsorbs gases Reducing agent	Fuel Gas masks (activated) Reducing agent
b. Coke	Destructive distillation of soft coal	Burns with little smoke or flame; porous gray Reducing agent	Fuel and reducing agent in metallurgy Manufacture of water, gas, SiC, CaC_2, CS_2

TABLE 17 CONTINUED

Forms	Occurrence or Preparation	Properties	Uses
c. Bone black	Destructive distillation of bones	Adsorbs coloring matter	Decolorizing sugar
d. Lampblack (carbon black)	Incomplete combustion of natural gas (or oil)	Soft, black, minute particles	Rubber tires Pigment: india and printing ink, shoe polish
e. Anthracite coal	Pennsylvania	Almost pure carbon, burns with little smoke	Fuel, production of water gas

Carbon Dioxide

Carbon dioxide is a widely distributed gas that makes up 0.04% of the air. There is a cycle that keeps this figure relatively stable. It is shown in Figure 40.

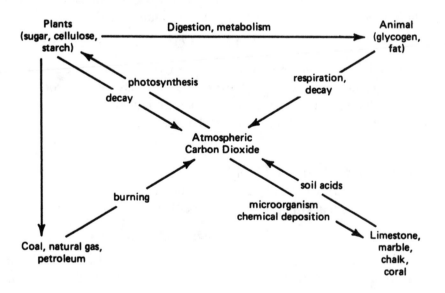

Figure 40. Carbon Dioxide Cycle.

Laboratory Preparation of CO_2

The usual laboratory preparation consists of reacting calcium carbonate (marble chips) with hydrochloric acid, although any carbonate or bicarbonate and any common acid can be used. The gas is collected by water displacement or air displacement. (See Figure 41).

The test for carbon dioxide consists of passing it through limewater ($Ca(OH)_2$). If CO_2 is present, the limewater turns cloudy because of the formation of a white precipitate of finely divided $CaCO_3$:

$$Ca(OH)_2 + CO_2 \rightarrow CaCO_3(s) + H_2O$$

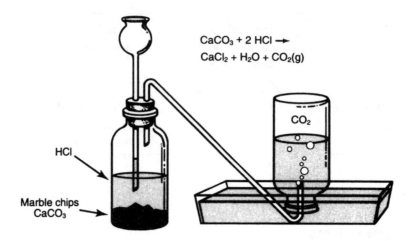

$$CaCO_3 + 2\ HCl \rightarrow CaCl_2 + H_2O + CO_2(g)$$

Figure 41. Laboratory Preparation of Carbon Dioxide.

Continued passing of CO_2 into the solution will clear the cloudy condition because the insoluble $CaCO_3$ becomes soluble calcium bicarbonate [$Ca(HCO_3)_2$].

$$CaCO_3(s) + H_2O + CO_2 \rightarrow Ca^{2+}(HCO_3)_2^-$$

This reaction can easily be reversed with increased temperature or decreased pressure. This is the way stalagmites and stalactites form on the floors and roofs of caves, respectively. The groundwater containing calcium bicarbonate is deposited on the roof and floor of the cave and decomposes into solid calcium carbonate formations.

Important Uses of CO_2

1. Since CO_2 is the acid anhydride of carbonic acid, it forms the acid when reacted with soft drinks, thus marking them "carbonated" beverages.

$$CO_2 + H_2O \rightarrow H_2CO_3$$

2. Solid carbon dioxide (–78°C), or "dry ice," is used as a refrigerant because it has the advantage of not melting into a liquid but sublimes and in the process absorbs 3 times as much heat per gram as ice.

3. Fire extinguishers make use of CO_2 because it is $1\frac{1}{2}$ times heavier than air and does not support ordinary combustion. It is used in extinguishers that release CO_2 from a steel cylinder in the form of a gas to smother the fire.

4. Leavening agents are usually made up of a dry acid-forming salt, dry baking soda ($NaHCO_3$), and a substance to keep these ingredients dry, such as starch or flour. Some of the acid-forming salts are alum [$NaAl(SO_4)_2$] in alum baking powders, potassium acid

tartrate ($KHC_4H_4O_6$) in cream of tartar baking powders, and sodium hydrogen phosphate (NaH_2PO_4) or calcium hydrogen phosphate [$Ca(H_2PO_4)_2$] in phosphate baking powders. In the moist batter, the acid reacts with the baking soda to release CO_2 as the leavening agent. This gas is trapped and raises the batter, which hardens in this raised condition.

Baking soda, when used independently, must have some source of acid, such as buttermilk, molasses, or sour milk, to cause the release of CO_2. The simple ionic equation is

$$H^+ + HCO_3^- \rightarrow H_2O + CO_2(g)$$

5. Plants use up CO_2 in the photosynthesis process. In this process, chlorophyll (the catalyst) and sunlight (the energy source) must be present. The reactants and products of this reaction are

$$6CO_2 + 6H_2O \rightarrow C_6H_{12}O_6 + 6O_2(g)$$
$$\text{simple}$$
$$\text{sugar}$$

or

$$6CO_2 + 5H_2O \rightarrow C_6H_{10}O_5 + 6O_2(g)$$
$$\text{cellulose}$$

6. The Solvay process for making $NaHCO_3$ and Na_2CO_3 uses CO_2. The CO_2 is combined NH_3 and water, and the product is reacted with $NaCl$. Sodium bicarbonate ($NaHCO_3$) precipitates out of solution since its solubility in the $NaCl$ solution is exceeded. The overall equation can be shown as

$$NaCl + H_2O + NH_3 + CO_3 \rightarrow NaHCO_3(s) + NH_4Cl$$

Sodium carbonate is made by heating the $NaHCO_3$:

$$2Na^+HCO_3^- \rightarrow Na_2^+CO_3^{2-} + H_2O + CO_2(g)$$

Organic Chemistry

Organic chemistry may be defined simply as the chemistry of the compounds of carbon. Since Friedrich Wöhler synthesized urea in 1828, chemists have synthesized thousands of carbon compounds in the areas of dyes, plastics, textile fibers, medicines, and drugs. The number of organic compounds has been estimated to be in the neighborhood of a million and is constantly increasing.

The carbon atom (atomic number 6) has four electrons in its outermost energy level, which show a tendency to be shared (electronegativity of 2.5) in covalent bonds. By this means, carbon bonds to other carbons, hydrogens, halogens, oxygen, and other elements to form the many compounds of organic chemistry.

Hydrocarbons

Hydrocarbons, as the name implies, are compounds containing only carbon and hydrogen in their structure. The simplest hydrocarbon is methane, CH_4. As previously mentioned, this type of formula, which shows the kinds of atoms and their respective numbers, is called an *empirical* formula. In organic chemistry this is not sufficient to identify the compound it is used to represent. For example, the empirical formula C_2H_6O can denote either an ether or an ethyl alcohol. For this reason, a *structural* formula is used to indicate how the atoms are arranged in the molecule. The ether of C_2H_6O looks like

$$
\begin{array}{ccc}
\text{H} & & \text{H} \\
| & & | \\
\text{H}-\text{C}-\text{O}-\text{C}-\text{H} \\
| & & | \\
\text{H} & & \text{H}
\end{array}
$$

whereas the ethyl alcohol is represented by the structural formula

$$
\begin{array}{cc}
\text{H} & \text{H} \\
| & | \\
\text{H}-\text{C}-\text{C}-\text{OH} \\
| & | \\
\text{H} & \text{H}
\end{array}
$$

To avoid ambiguity, structural formulas are more often used than empirical formulas in organic chemistry. The structural formula of methane is

$$
\begin{array}{c}
\text{H} \\
| \\
\text{H}-\text{C}-\text{H} \\
| \\
\text{H}
\end{array}
$$

Alkane Series (Saturated)

Methane is the first member of a hydrocarbon series called the alkanes (or paraffin series). The general formula for this series is C_nH_{2n+2}, where n is the number of carbons in the molecule. Table 18 provides some essential information about this series. Since many other organic structures use the stem of the alkane names, you should learn these names and structures well. Notice that as the number of carbons in the chain increases, the boiling point also increases. The first four alkanes are gases at room temperature; the subsequent compounds are liquid and then become more viscous with increasing length of the chain.

Since the chain is increased by a carbon and two hydrogens in each subsequent molecule, the alkanes are referred to as a *homologous* series.

The alkanes are found in petroleum and natural gas. They are usually extracted by fractional distillation, which separates the compounds by varying the temperature so that each vaporizes at its respective boiling point. Methane, which forms 90% of natural gas, can also be prepared in the laboratory by heating soda lime (containing NaOH) with sodium acetate:

$$NaC_2H_3O_2 + NaOH \rightarrow CH_4(g) + Na_2CO_3$$

TABLE 18

THE ALKANES

Name	Formula	Number of Structural Isomers	Structure	Boiling Point (°C)
Methane	CH_4	1	H—C—H (with H above and below)	−162
Ethane	C_2H_6	1	H—C—C—H (with H's)	−89
Propane	C_3H_8	1	H—C—C—C—H (with H's)	−42
n-Butane	C_4H_{10}	2	H—C—C—C—C—H (with H's)	0
n-Pentane	C_5H_{12}	3	H—C—C—C—C—C—H (with H's)	36
n-Hexane	C_6H_{14}	5	$CH_3—CH_2—CH_2—CH_2—CH_2—CH_2$	69
n-Heptane	C_7H_{16}	7	$CH_3—CH_2—CH_2—CH_2—CH_2—CH_2—CH_3$	98
n-Octane	C_8H_{18}	18	$CH_3—CH_2—CH_2—CH_2—CH_2—CH_2—CH_2—CH_3$	126
n-Nonane	C_9H_{20}	35	$CH_3—CH_2—CH_2—CH_2—CH_2—CH_2—CH_2—CH_2—CH_3$	151
n-Decane	$C_{10}H_{22}$	75	$CH_3—CH_2—CH_2—CH_2—CH_2—CH_2—CH_2—CH_2—CH_2—CH_3$	174

(Methane through n-Butane marked "Gas at room temperatures"; n-Pentane through n-Decane marked "Liquid state")

When the alkanes are burned with sufficient air, the compounds formed are CO_2 and H_2O. An example is

$$2C_2H_6 + 7O_4 \rightarrow 4CO_2(g) + 6H_2O(g)$$

The alkanes can be reacted with halogens so that hydrogens are replaced by a halogen atom:

$$H—C—H + Br_2 \rightarrow H—C—Br + HBr$$

Some common substitution compounds of methane are

$$
\begin{array}{ccc}
\text{H} & \text{H} & \text{Cl} \\
| & | & | \\
\text{H}-\text{C}-\text{Cl} & \text{Cl}-\text{C}-\text{Cl} & \text{Cl}-\text{C}-\text{Cl} \\
| & | & | \\
\text{H} & \text{Cl} & \text{Cl}
\end{array}
$$

| methyl chloride | chloroform | carbon tetrachloride |
| monochloromethane | trichloromethane | tetrachloromethane |

Naming alkane substitutions When an alkane hydrocarbon has an end hydrogen removed, it is referred to as an alkyl substituent or group. The respective name of each is the alkane name with *-ane* replaced by *-yl*.

Alkane	Alkyl Group	Compound
methane	methyl	methyl bromide

$$
\begin{array}{ccc}
\text{H} & \text{H} & \text{H} \\
| & | & | \\
\text{H}-\text{C}-\text{H} & \text{H}-\text{C}- & \text{H}-\text{C}-\text{Br} \\
| & | & | \\
\text{H} & \text{H} & \text{H}
\end{array}
$$

| butane | butyl | butyl chloride |

$$
\begin{array}{ccc}
\text{H H H H} & \text{H H H H} & \text{H H H H} \\
| \ | \ | \ | & | \ | \ | \ | & | \ | \ | \ | \\
\text{H}-\text{C}-\text{C}-\text{C}-\text{C}-\text{H} & \text{H}-\text{C}-\text{C}-\text{C}-\text{C}- & \text{H}-\text{C}-\text{C}-\text{C}-\text{C}-\text{Cl} \\
| \ | \ | \ | & | \ | \ | \ | & | \ | \ | \ | \\
\text{H H H H} & \text{H H H H} & \text{H H H H}
\end{array}
$$

One method of naming a substitution product is to use the alkyl substituent or group name for the respective chain and the halide as shown above.

The IUPAC system uses the name of the longest carbon chain as the parent chain. The carbon atoms are numbered in the parent chain to indicate where branching or substitution takes place. The direction of numbering is chosen so that the lowest numbers possible are given to the side chains. The complete name of the compound is arrived at by first naming the attached group, each one being prefixed by the number of the carbon to which it is attached, and then the parent alkane. If a particular group appears more than once, the appropriate prefix (*di-*, *tri-*, and so on) is used to indicate how many times it appears. A carbon atom number must be used to indicate the position of each such group. If two or more of the same group are attached to the same carbon atom, the number of the carbon atom is repeated. If two or more different substituted groups are in a name, they are arranged alphabetically.

Example 1

2,2-Dimethylbutane

Solution:

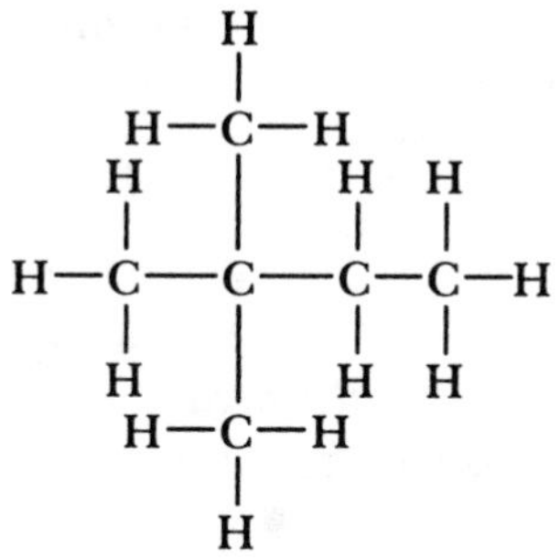

```
                H
                |
            H—C—H
        H   |   H   H
        |   |   |   |
    H—C—C—C—C—H
        |   |   |   |
        H   |   H   H
            H—C—H
                |
                H
```

Example 2

1,1-Dichloro-3-ethyl-2,4-dimethylpentane

Solution:

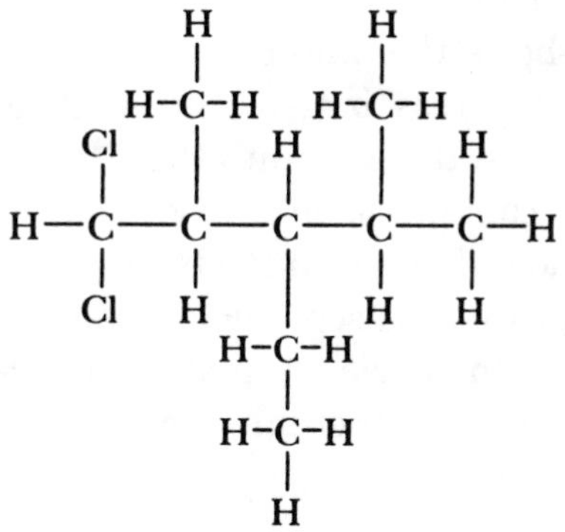

```
            H           H
            |           |
        H-C-H       H-C-H
    Cl  |       H   |       H
    |   |       |   |       |
  H—C—C—C—C—C—H
    |   |       |   |   |
    Cl  H       |   H   H
                H-C-H
                |
                H-C-H
                |
                H
```

Example 3

2-Iodo-2-methylpropane

Solution:

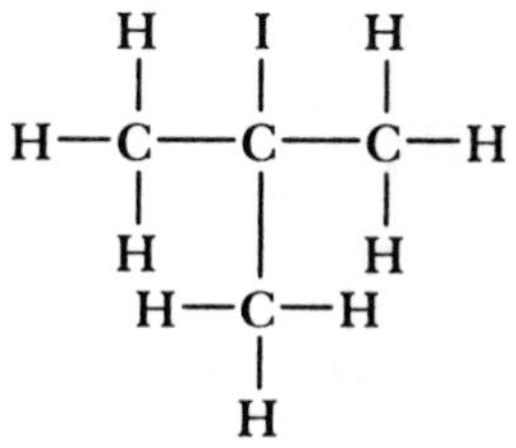

```
    H   I   H
    |   |   |
  H—C—C—C—H
    |   |   |
    H   |   H
        H—C—H
        |
        H
```

Cycloalkanes   Starting with propane in the alkane series, it is possible to obtain a ring form by attaching the two chain ends. This reduces the number of hydrogens by two.

This hydrocarbon is called cyclopropane.

Since all the members of the alkane series have single covalent bonds, this series and all such structures are said to be *saturated*.

If the hydrocarbon molecule contains double or triple covalent bonds, it is referred to as *unsaturated*.

Properties and Uses of Alkanes

Properties for some straight-chain alkanes are listed in Table 17. The trends in these properties can be explained by examining the structure of alkanes. The carbon-hydrogen bonds of alkanes are nonpolar. The only forces of attraction between nonpolar molecules are weak intermolecular forces, or London dispersion forces. These forces increase as the mass of a molecule increases.

The table also shows the physical states of alkanes. Smaller alkanes exist as gases at room temperature, whereas larger ones exist as liquids. Gasoline and kerosene consist mostly of liquid alkanes. It is not until there are 17 carbons in the chain that the solid form occurs. Paraffin wax contains solid alkanes.

Making use of the difference in the boiling point of mixtures of the liquid alkanes found in petroleum, it is possible to separate the various components through *fractional distillation*. This is the major industrial process used in refining petroleum into gasoline, kerosene, lubricating oils, and several other minor components.

Alkene Series (Unsaturated)

The alkene series has a double covalent bond between two adjacent carbon atoms. The general formula of this series is C_nH_{2n}. In naming these compounds, the suffix of the alkane is replaced by *-ene*. Two examples:

ethene (common name: ethylene)

propene (common name: propylene)

If the double bond occurs on an interior carbon, the chain is numbered so that the position of the double bond is designated by the lowest possible number assigned to the first doubly bonded carbon. For example:

$$
\begin{array}{ccccccc}
& \text{H} & \text{H} & & & \text{H} & \\
& | & | & & & | & \\
\text{H}-\text{C}-&\text{C}-&\text{C}=&\text{C}-&\text{C}-&\text{H} \\
& | & | & | & | & | \\
& \text{H} & \text{H} & \text{H} & \text{H} & \text{H} \\
\end{array}
\qquad \text{2-pentene}
$$

The bonding is more complex in the double covalent bond than in the single bonds in the molecule. Using the orbital pictures of the atom, we can show this as

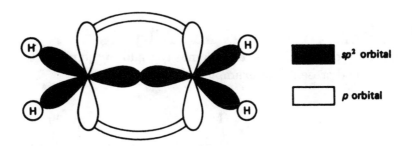

	sp^2 orbital
	p orbital

The two p lobes attached above and below constitute *one* bond called a pi (π) bond.

The sp^2 orbital bonds between the carbons and with each hydrogen are referred to as sigma (σ) bonds.

Alkyne Series (Unsaturated)

The alkyne series has a triple covalent bond between two adjacent carbons. The general formula of this series is C_nH_{2n-2}. In naming these compounds, the alkane suffix is replaced by *-yne*. A few examples:

$$
\text{H}-\text{C}\equiv\text{C}-\text{H} \qquad \text{ethyne (common name: acetylene)}
$$

$$
\begin{array}{c}
\text{H} \\
| \\
\text{H}-\text{C}\equiv\text{C}-\text{C}-\text{H} \\
| \\
\text{H} \\
\end{array}
\qquad \text{propyne}
$$

The orbital structure can be shown as

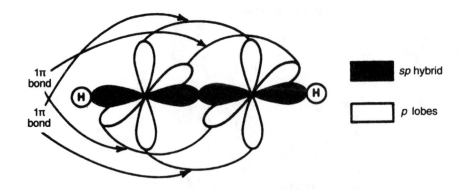

The bonds formed by the *p* orbitals and the one bond between the *sp* orbitals make up the triple bond.

Additions to Alkenes and Alkynes

Unsaturates of the alkenes and alkynes can add to their structures by breaking the double or triple bond present. For example,

$$\underset{\text{H}}{\overset{\text{H}}{\text{H—C}}}{=}\underset{\text{H}}{\overset{\text{H}}{\text{C—H}}} + \text{Br}_2 \longrightarrow \text{H}{-}\underset{\text{H}}{\overset{\text{Br}}{\text{C}}}{-}\underset{\text{H}}{\overset{\text{Br}}{\text{C}}}{-}\text{H}$$

1,2-dibromoethane
(ethylene dibromide)

The 1,2- means that on the first and second carbons in the chain a bromine atom is bonded. 1,1-dibromoethane is

$$\text{H}{-}\underset{\text{Br}}{\overset{\text{Br}}{\text{C}}}{-}\underset{\text{H}}{\overset{\text{H}}{\text{C}}}{-}\text{H}$$

An addition to ethyne could be

$$\text{H—C}{\equiv}\text{C—H} + 2\text{Br}_2 \longrightarrow \text{H}{-}\underset{\text{Br}}{\overset{\text{Br}}{\text{C}}}{-}\underset{\text{Br}}{\overset{\text{Br}}{\text{C}}}{-}\text{H}$$

Alkadienes have two double covalent bonds in each molecule. The *-ene* indicates a double bond, and the *di-* indicates two such bonds. Names of this type are also derived from the alkanes, and so butadiene has four carbons and two double bonds.

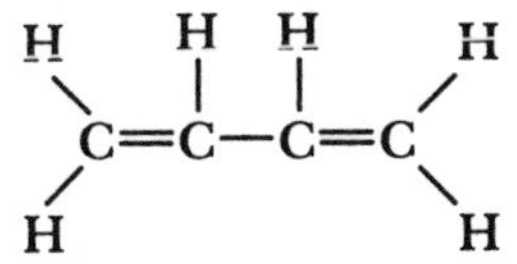

(1,3-butadiene is a more precise name and indicates that the double bonds follow the first and third carbons.)

This compound is used in making synthetic rubber.

Alicyclic Hydrocarbons

All the hydrocarbons discussed to this point have been *aliphatic*, that is, open-chain hydrocarbons with either straight or branched structures. Hydrocarbons in which the carbon atoms are arranged in a closed ring structure are cyclic hydrocarbons. The two main groups of cyclic hydrocarbons are the *alicyclic* hydrocarbons and the *aromatic* hydrocarbons. Aromatic hydrocarbons are discussed in the next section.

In the alicyclic hydrocarbons, the carbon atoms are linked by single or double bonds and are arranged in a closed structure. As previously shown with alkanes, the cycloalkanes begin with cyclopropane.

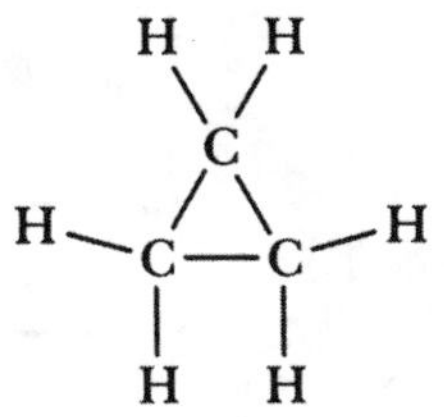

cyclopropane

Similarly, cycloalkenes contain one double bond between a pair of carbon atoms in the ring structure. Cycloalkadienes contain two double bonds between two different pairs of carbon atoms. Some examples are

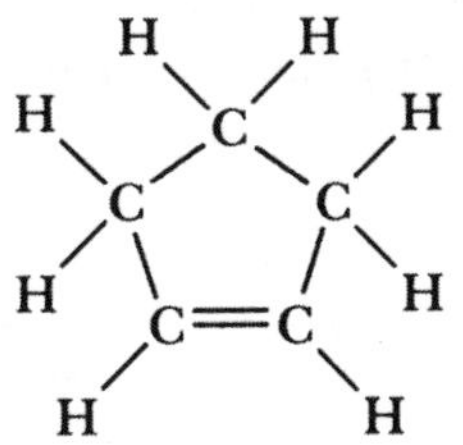

cylopentene

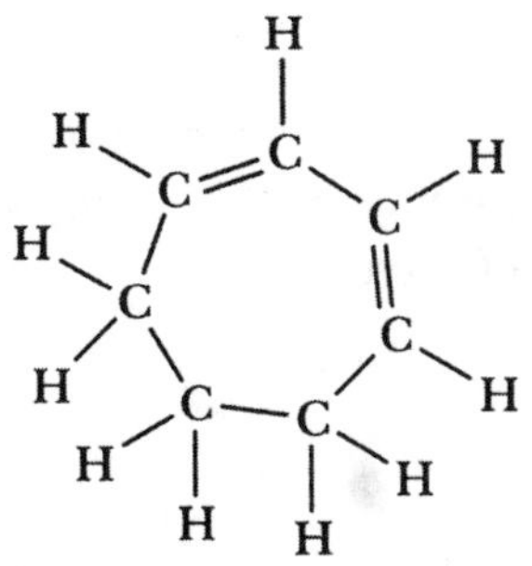

1,3-cycloheptadiene

Aromatic Hydrocarbons

The aromatic compounds are unsaturated ring structures. The basic formula of this series is C_nH_{2n-6}, and the simplest compound is benzene (C_6H_6). The benzene structure is a resonance structure represented as

<table>
<tr>
<td>

H
|
C
H—C   C—H
H—C   C—H
C
|
H

↔

H
|
C
H—C   C—H
H—C   C—H
C
|
H

</td>
<td>

Note: The carbon-to-carbon bonds are neither single nor double bonds but hybrid bonds. This is called *resonance*.

</td>
</tr>
</table>

The orbital structure can be represented as

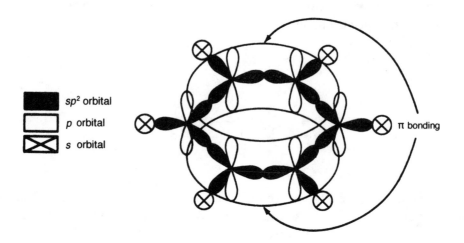

■ sp^2 orbital
□ p orbital
⊠ s orbital

π bonding

Most of the aromatics have an aroma, thus the name *aromatic*.

The C_6H_5 group is a substituent called phenyl. This is the benzene structure with one hydrogen missing. If the phenyl substituent adds a methyl group, the compound is called toluene or methylbenzene:

Abbreviation for the benzene structure

or

Two other members of the benzene series and their structures are

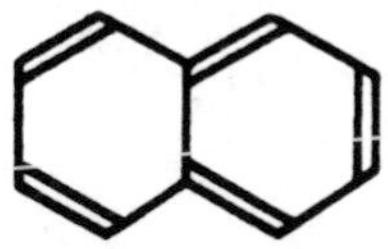

$C_{10} H_8$ naphthalene

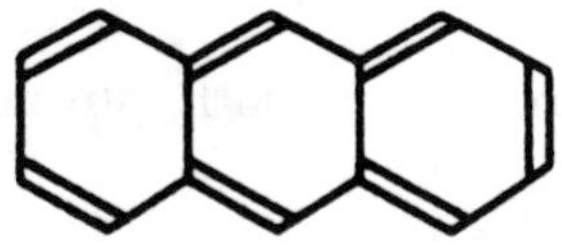

$C_{14} H_{10}$ anthracene

The IUPAC system of naming benzene derivatives, like that for chain compounds, involves numbering the carbon atoms in the ring in order to pinpoint the locations of the side chains. However, if only two groups are substituted in the benzene ring, the compound formed will be a benzene derivative having three possible isomeric forms. In such cases, the prefixes *orth*-, *meta*-, and *para*-, abbreviated as *o*-, *m*-, and *p*-, may be used to name the isomers. In the ortho structure, the two substituted groups are located on adjacent carbon atoms. In the meta structure, they are separated by one carbon atom. In the para structure, they are separated by two carbon atoms.

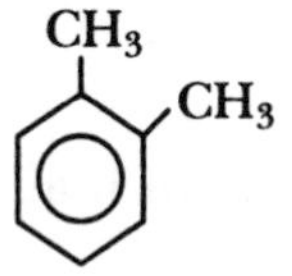

1,2-dimethylbenzene	1,3-dimethylbenzene	1,4-dimethylbenzene
o-xylene	*m*-xylene	*p*-xylene

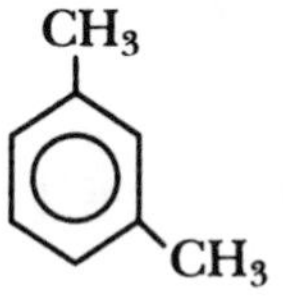

Isomers

Many of the chain hydrocarbons can have the same formula, but their structures may differ. For example, butane is the first compound that can have two different structures or *isomers* for the same formula.

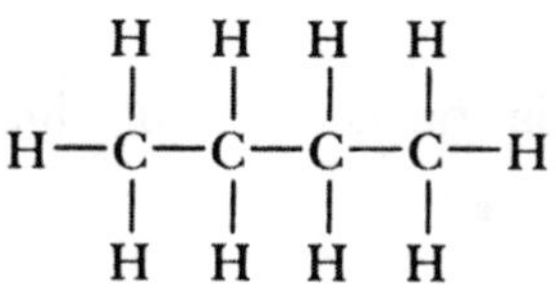

n-butane

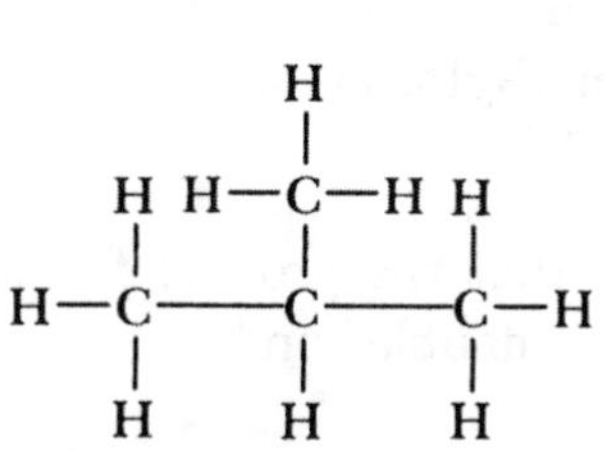

isobutane

This isomerization can be shown by the equation

$$CH_3 - CH_2 - CH_2 - CH_3 \xrightarrow[70-100°C]{AlCl_2} CH_3 - \overset{\overset{\displaystyle CH_3}{|}}{CH} - CH_3$$

butane   isobutane

The isomers have different properties, both physical and chemical, from the normal structure.

Changing Hydrocarbons

Cracking

Under proper conditions of temperature and pressure and often in the presence of a catalyst, long chains of hydrocarbons may be made into more useful smaller molecules. This process is called cracking. An example is

$$C_{16}H_{34} \rightarrow C_8H_{18} + C_8H_{16}$$

hexadecane   octane   octene

Notice that the products formed by cracking a saturated hydrocarbon of the alkanes are a saturated hydrocarbon and an unsaturated alkene.

Alkylation

Alkylation is the combining of a saturated alkane with an unsaturated alkene. An example is

$$C_4H_{10} + C_4H_8 \xrightarrow{H_2SO_4} C_8H_{18}$$

isobutane   isobutylene   isooctane
2-methylpropane   2-methylpropene   2,2,4-trimethylpentane

Notice that the product formed by alkylation is a useful chain hydrocarbon without the previous double bond.

Polymerization

The combination of two or more unsaturated molecules to form a larger chain molecule is called *polymerization*. An example is

$$
\underset{\text{2-methylpropene}}{CH_3 - \overset{\overset{\displaystyle CH_3}{|}}{C} = CH_2} + \underset{\text{2-methylpropene}}{CH_3 - \overset{\overset{\displaystyle CH_3}{|}}{C} = CH_2} \longrightarrow CH_3 - \overset{\overset{\displaystyle CH_3}{|}}{\underset{\underset{\displaystyle CH_3}{|}}{C}} - CH_2 - \overset{\overset{\displaystyle CH_3}{|}}{C} = CH_2
$$

2,4,4-trimethylpentene-1

In the product name, 2,4,4-trimethylpentene-1, the first three numbers indicate the carbons that have the trimethyl (three methyl substituents) attached—one each. The -1 at the end indicates that the first carbon has the double bond attaching it to the next carbon. In numbering compounds of the alkenes and alkynes, always number the carbon atoms from the end that has the double or triple bond.

Hydrogenation

Hydrogenation is the process of adding hydrogens to an unsaturated hydrocarbon in the presence of a suitable catalyst. It is often used to make liquid unsaturated fats into solid saturated fats. An example for hydrocarbons is

$$
\underset{\text{2-monomethylpropene-1}}{CH_3 - \overset{\overset{\displaystyle CH_3}{|}}{C} = CH_2} + H_2 \longrightarrow \underset{\text{2-monomethylpropane}}{CH_3 - \overset{\overset{\displaystyle CH_3}{|}}{\underset{\underset{\displaystyle H}{|}}{C}} - CH_3}
$$

Dehydrogenation

Dehydrogenation is the removal of hydrogens from a chain molecule in the presence of a catalyst to form an unsaturated molecule. An example is

$$
\underset{\text{butane}}{CH_3 - CH_2 - CH_2 - CH_3} \rightarrow \underset{\text{1-butene}}{CH_2 = CH - CH_2 - CH_3} + H_2
$$

Aromatization

Hydrocarbons of the alkane series having six or more carbons in the chain can be made to form an aromatic hydrocarbon with the loss of hydrogen. This must be done at high temperatures and with a suitable catalyst. An example is

$$
\underset{\text{heptane}}{CH_3 - CH_2 - CH_2 - CH_2 - CH_2 - CH_2 - CH_3} \rightarrow
$$

toluene

Fischer-Tropsch Process

This process was developed to make gasoline from natural gas, coal, or lignite by converting the original carbon compound to water gas. For example,

$$2CH_4 + O_2 \rightarrow 2CO + 4H_2$$
water gas

This water gas is then enriched with hydrogen and reacted at high temperatures with a catalyst to form a mixture of hydrocarbons, one of which is gasoline.

Hydrocarbon Derivatives

Alcohols

The simplest alcohols are alkanes that have one or more hydrogen atom replaced by the hydroxyl group, —OH. This is called its functional group.

Methanol

Methanol is the simplest alcohol. Its structure is

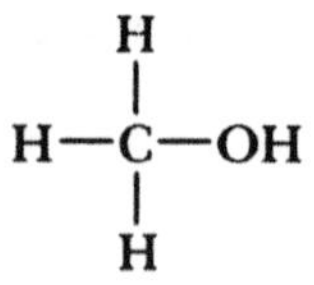

H
|
H—C—OH   methanol or wood alcohol
|
H

Laboratory preparation

$$CH_3Cl \quad + \quad NaOH \rightarrow NaCl + CH_3OH$$
chloromethane   methanol
(methyl chloride)   (methyl alcohol)

Industrial preparation   For many years methanol, wood alcohol, was obtained by the destructive distillation of wood. Today it is made by the hydrogenation of oxides of carbon in the presence of a suitable catalyst:

$$CO + 2H_2 \xrightarrow[\substack{300-400°C \\ 200\,atm.}]{catalyst} CH_3OH$$

Properties and uses   Methanol is a colorless, flammable liquid with a boiling point of 65°C. It is miscible with water, is exceedingly poisonous, and can cause blindness if taken internally. It can be used as a fuel, as a solvent, and as a denaturant to make denatured ethyl alcohol, unsuitable for drinking.

Ethanol

Ethanol is the best known and most used alcohol. Its structure is

$$
\begin{array}{ccc}
& \text{H} & \text{H} \\
& | & | \\
\text{H} - & \text{C} - \text{C} & - \text{OH} \quad \text{ethanol} \\
& | & | \\
& \text{H} & \text{H}
\end{array}
$$

(Notice that the alcohol names are derived from the alkane names by replacing the *e* with -*ol*.)

Its common names are ethyl alcohol and grain alcohol.

Laboratory preparation

$$
\text{ethane} \xrightarrow{\text{Cl}_2} \text{chloroethane} \xrightarrow{\text{NaOH}} \text{ethanol}
$$

Simple sugars can be converted to ethanol by the action of an enzyme (zymase) found in yeast.

$$C_6H_{12}O_6 \rightarrow 2C_2H_5OH + 2CO_2(g)$$
$$\text{sugar} \qquad \text{ethanol}$$

After fermentation, the alcohol is distilled off. Grains, potatoes, and other starch plants, which can be treated with an acid-water solution to form sugars, can be converted into ethyl alcohol or "whiskey."

Industrial preparation The usual process for producing industrial alcohol consists of treating ethyene (or ethylene) with concentrated H_2SO_4 and then hydrolyzing the resulting ethyl hydrogen sulfate to ethanol.

$$C_2H_4 \xrightarrow{H_2SO_4} C_2H_5HSO_4 \xrightarrow{H_2O} C_2H_5OH + H_2SO_4$$

Properties and uses Ethanol is a colorless flammable liquid with a boiling point of 78°C. It is miscible with water and is a good solvent for a wide variety of substances (these solutions are often referred to as "tinctures"). It is used as an antifreeze because of its low freezing point, –115°C, and for making acetaldehyde and ether.

Other Alcohols

Isomeric alcohols have similar formulas but different properties because of their differences in structure. If the —OH is attached to an end carbon, the alcohol is called a primary alcohol. If attached to a "middle" carbon, it is called a secondary alcohol. Some examples:

$$\underset{\substack{\text{1-propanol}\\ \text{(propyl alcohol)}}}{H-\overset{\displaystyle H}{\underset{\displaystyle H}{C}}-\overset{\displaystyle H}{\underset{\displaystyle H}{C}}-\overset{\displaystyle H}{\underset{\displaystyle H}{C}}-OH} \quad \xleftrightarrow{\text{isomers}} \quad \underset{\substack{\text{2-propanol}\\ \text{(isopropyl alcohol)}}}{H-\overset{\displaystyle H}{\underset{\displaystyle H}{C}}-\overset{\displaystyle H}{\underset{\displaystyle OH}{C}}-\overset{\displaystyle H}{\underset{\displaystyle H}{C}}-H}$$

(The number in front of the name indicates to which carbon the —OH ion is attached.)

$$CH_3-CH_2-CH_2-\overset{\displaystyle H}{\underset{\displaystyle H}{C}}-OH \quad \xleftrightarrow{\text{isomers}} \quad CH_3-CH_2-\overset{\displaystyle H}{\underset{\displaystyle OH}{C}}-CH_3$$

1-butanol	2-butanol
(sometimes called *n*-butanol or normal butanol) butyl alcohol	isobutyl alcohol

Phenols are alcohols derived from the aromatic hydrocarbons. Some important examples are

Structure	Name	Properties or Uses
⬡—OH	Phenol Carbolic acid Hydroxybenzene	Slightly acidic, extremely corrosive, poisonous. Used to make synthetic resins, plastics, drugs, dyes, and photographic developers. Good disinfectant.

Other alcohols with more than one —OH group:

Structure	Name	Properties or Uses
$H-\overset{\displaystyle H}{\underset{\displaystyle OH}{C}}-\overset{\displaystyle H}{\underset{\displaystyle OH}{C}}-H$	ethylene glycol 1,2-ethanediol	A colorless liquid, high boiling point, low freezing point. Used as permanent antifreeze in automobiles.
$H-\overset{\displaystyle H}{\underset{\displaystyle OH}{C}}-\overset{\displaystyle H}{\underset{\displaystyle OH}{C}}-\overset{\displaystyle H}{\underset{\displaystyle OH}{C}}-H$	glycerine glycerol 1,2,3-propanetriol	Colorless liquid, odorless, viscous, sweet taste. Used to make nitroglcerine, resins for paint, and cellophane.

Aldehydes

The functional group of an aldehyde is the $-C{\displaystyle {O \atop H}}$, formyl group. The general formula is RCHO, where R represents a hydrocarbon substituent.

Preparation from an alcohol Aldehydes can be prepared by the oxidation of an alcohol. This can be done by inserting a hot copper wire into the alcohol. A typical reaction is

$$\underset{\substack{\text{methanol} \\ \text{(methyl alcohol)}}}{H-\overset{\displaystyle H}{\underset{\displaystyle H}{C}}-OH} + [O] \xrightarrow{\substack{\text{mild oxidizing} \\ \text{agent}}} \left[H-\overset{\displaystyle H}{\underset{\displaystyle OH}{C}}-OH \right] \rightarrow \underset{\substack{\text{methanal} \\ \text{(formaldehyde)}}}{H-C{\displaystyle {O \atop H}}} + H_2O$$

The middle structure is an intermediate structure, but since two hydroxyl groups do not stay attached to the same carbon, it changes to the aldehyde by a water molecule "breaking away."

The aldehyde name is derived from the alcohol name by dropping the *-ol* and adding *-al*. Ethanol forms ethanal (acetaldehyde) in the same manner.

Organic Acids or Carboxylic Acids

The functional group of an organic acid is $-C{\displaystyle {O \atop OH}}$, the carboxyl group. The general formula is R—COOH.

Preparation from an aldehyde Organic acids can be prepared by the mild oxidation of an aldehyde. The simplest acid is methanoic acid, which is present in ants, bees, and other insects. A typical reaction is

$$\underset{\substack{\text{methanal} \\ \text{(formaldehyde)}}}{H-C{\displaystyle {O \atop H}}} + [O] \rightarrow \underset{\substack{\text{methanoic acid} \\ \text{(formic acid)}}}{H-C{\displaystyle {O \atop OH}}}$$

Notice that in the IUPAC system the name is derived from the alkane stem by adding *-oic*.
Ethanal can be oxidized to ethanoic acid:

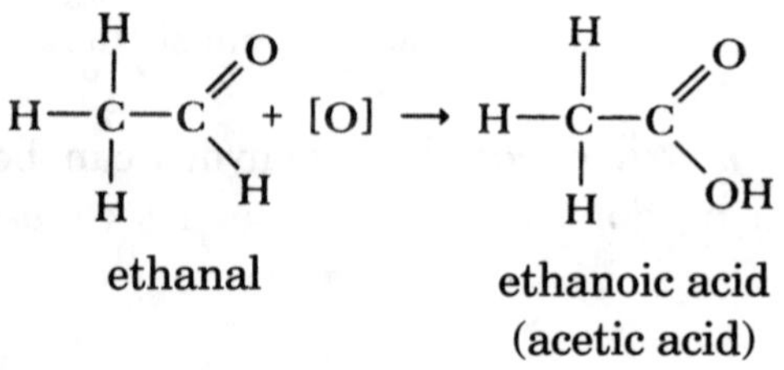

ethanal   ethanoic acid
(acetic acid)

Acetic acid, as ethanoic acid is commonly called, is a mild acid which in the concentrated form is called glacial acetic acid. Glacial acetic is used in many industrial processes, such as making cellulose acetate. Vinegar is a 4% to 8% solution of acetic acid which can be made by fermenting alcohol.

$$C_2H_5OH + O_2 \rightarrow CH_3COOH + H_2O$$

ethanol   ethanoic acid
(acetic acid)

One of the aromatic acids is benzoic acid with a carboxyl group replacing one of the hydrogens:

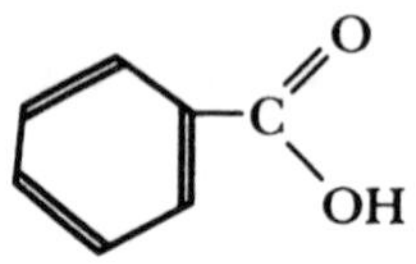

Complex Organic Acids

There are some more complex organic acids that are important to mention. These are listed in Table 19.

TABLE 19

SOME COMPLEX ORGANIC ACIDS

Acid	Source	Structure	Notes
Palmitic	Fats	$C_{15}H_{31}$—C $\overset{\text{O}}{\underset{\text{OH}}{\Bigg<}}$	Saturated
Stearic	Fats	$C_{17}H_{35}$—C $\overset{\text{O}}{\underset{\text{OH}}{\Bigg<}}$	Saturated
Oleic	Fats	$C_{17}H_{33}$—C $\overset{\text{O}}{\underset{\text{OH}}{\Bigg<}}$	Unsaturated—has a double bond in the chain
Citric	Citrus fruits	COOH—CH_2—COHCOOH—CH_2—COOH	
Lactic	Sour milk	CH_3—CHOH—COOH	
Oxalic	Rhubarb	COOH—COOH	
Salicylic	Oil of wintergreen		

Summary of Oxygen Derivatives

$$\text{R—H} \rightarrow \underset{\substack{\text{chlorine} \\ \text{substitution} \\ \text{product}}}{\text{R—Cl}} \rightarrow \underset{\text{alcohol}}{\text{R—OH}} \rightarrow \underset{\text{aldehyde}}{\text{R}^1\text{CHO}} \rightarrow \underset{\text{acid}}{\text{R}^1\text{—COOH}}$$

Functional Group

hydrocarbon   (ending -*ol*) (ending -*al*) (ending -*oic*)

Note: R^1 indicates a hydrocarbon chain different from R by one less carbon in the chain. An actual example using ethane is

$$\underset{\text{ethane}}{C_2H_6} \xrightarrow{Cl_2} \underset{\substack{\text{monochloro-} \\ \text{ethane}}}{C_2H_5Cl} \xrightarrow{NaOH} \underset{\text{ethanol}}{C_2H_5OH} \xrightarrow{[O]} \underset{\text{ethanal}}{CH_3CHO} \xrightarrow{[O]} \underset{\substack{\text{ethanoic} \\ \text{acid}}}{CH_3COOH}$$

Ketones

When a secondary alcohol is slightly oxidized, it forms a compound having the functional group $\begin{smallmatrix} R-C-R^1 \\ \| \\ O \end{smallmatrix}$, called a ketone. The R^1 indicates that this group need not be the same as R. An example is

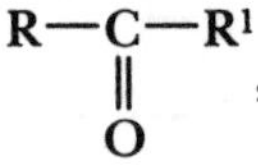

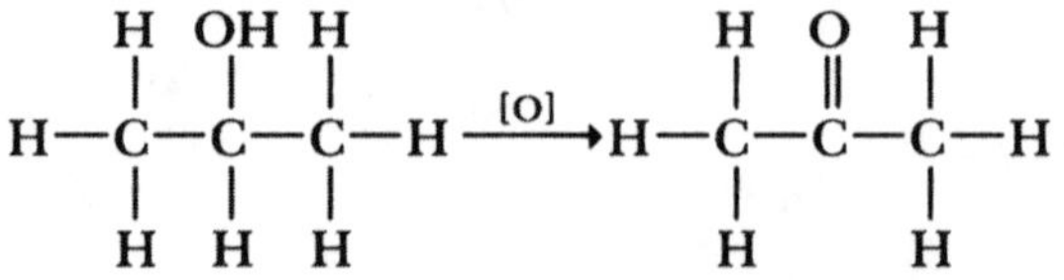

2-propanol	2-propanone
(isopropyl alcohol)	(acetone)

In the IUPAC method the name of the ketone has the ending *-one* with a digit indicating the carbon that has the double-bonded oxygen. Another method of designating a ketone is to name the radicals on either side of the ketone structure and use the word *ketone*. In the preceding reaction, the product would be dimethyl ketone.

Ethers

When a primary alcohol, such as ethanol, is dehydrated with sulfuric acid, an ether forms. The functional group is $R-O-R^1$, in which R^1 may be the same hydrocarbon group, as shown below in the first example, or a different hydrocarbon group, as shown in the second example.

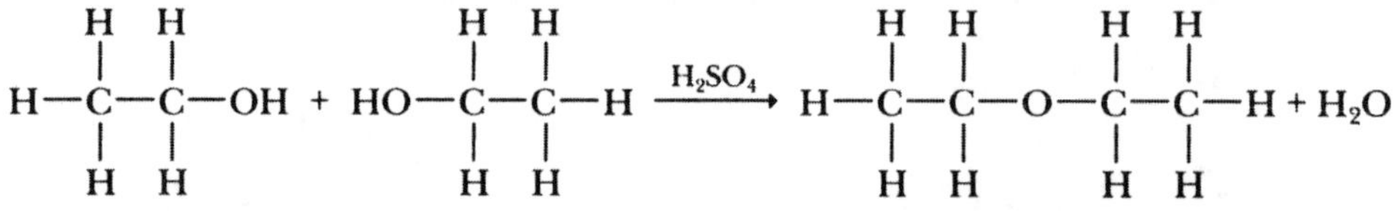

ethanol	ethanol	ethoxyethane
		(ethyl ether or diethyl ether)

Another ether with unlike groups, $R-O-R^1$:

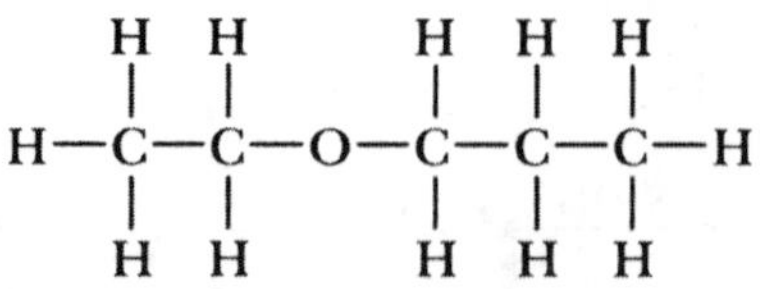

ethoxypropane
(ethyl propyl ether)

In the IUPAC method the ether name is made up of the first substituent's stem, then *-oxy*, and then the alkane name for the second substituents.

Amines and Amino Acids

The group NH_2^- is found in the amide ion or the amino group. Under the proper conditions, this ion can replace a hydrogen in a hydrocarbon compound. The resulting compound is called an *amine*. Some examples are

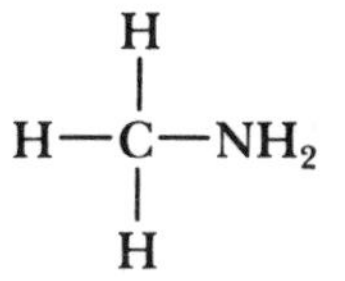

methylamine

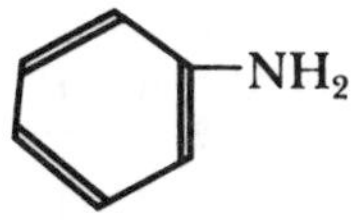

aniline

In *amides*, the NH_2^- group replaces a hydrogen in the carboxyl group. When naming amides, the *-ic* of the common name or the *-oic* of the IUPAC name of the parent acid is replaced by *-amide*. See below.

Amino acids are organic acids that contain one or more amino groups. The simplest uncombined amino acid is glycine, or amino acetic acid, NH_2CH_2—COOH. More than 20 amino acids are known, about half of which are necessary in the human diet because they are needed to make up the body proteins.

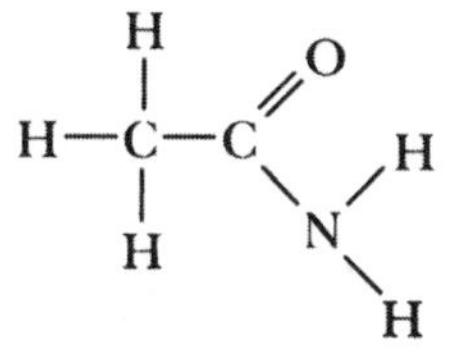

acetamide
(ethanamide)

Esters

Esters are often compared to inorganic salts because their preparations are similar. To make a salt, you react the appropriate acid and base. To make an ester, you react the appropriate organic acid and alcohol. For example,

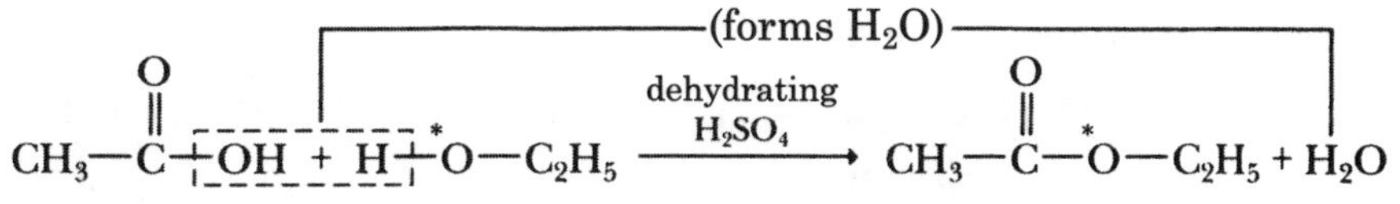

ethanoic
acid

ethanol

ethyl ethanoate
(ethyl acetate)

The name is made up of the alkyl substituent of the alcohol and the acid name, in which *-ic* is replaced with *-ate*.

The general equation is

$$\overset{*}{RO} - H + R^1 CO - OH \rightarrow R^1 CO\overset{*}{O} - R + HOH$$

$$\text{alcohol} \qquad \text{acid} \qquad\qquad \text{ester}$$

Esters usually have a sweet smell and are used in perfumes and flavor extracts.

SOME COMMON ESTERS

Name	Formula	Characteristic Odor
Ethyl butyrate	$C_3H_7COO \cdot C_2H_5$	Pineapple
Amyl acetate	$CH_3COO \cdot C_5H_{11}$	Banana, pear
Methyl salicylate	$C_6H_4(OH)COO \cdot CH_3$	Wintergreen
Amyl valerate	$C_4H_9COO \cdot C_5H_{11}$	Apple
Octyl acetate	$CH_3COO \cdot C_8H_{17}$	Orange
Methyl anthranilate	$C_3H_4(NH_2)COO \cdot CH_3$	Grape

Some esters found in the seeds of plants and the bodies of animals are fats. Stearin is such an ester.

$$
\begin{array}{lll}
C_{17}H_{35} - COOH & HO - CH_2 & C_{17}H_{35}COO - CH_2 \\
C_{17}H_{35} - COOH + HO - CH \longrightarrow 3\,H_2O + & C_{17}H_{35}COO - CH \\
C_{17}H_{35} - COOH & HO - CH_2 & C_{17}H_{35}COO - CH_2
\end{array}
$$

$$\text{stearic acid} \qquad \text{glycerol} \qquad\qquad \text{glycerol stearate}$$
$$\text{(stearin)}$$

Some other examples.

$$\text{Olein or glyceryl oleate} \quad = (C_{17}H_{35}COO)_3C_3H_5$$
$$\text{Butyrin or glyceryl butyrate} = (C_3H_7COO)_3C_3H_5$$

Soap was once made by housewives using a fat (like stearin) and lye (sodium hydroxide). This saponification reaction is

$$(C_{17}H_{35}COO)_3C_3H_5 + 3NaOH \rightarrow 3C_{17}H_{35}COONa + C_3H_5(OH)_3$$
$$\text{stearin} \qquad\quad \text{lye} \qquad\quad \text{soap} \qquad \text{glycerine}$$
$$\text{(sodium stearate)}$$

Carbohydrates

Carbohydrates are made up of carbon, hydrogen, and oxygen. Usually the hydrogen-to-oxygen ratio is $2 : 1$. The simple carbohydrates are more-or-less sweet and are referred to as saccharides (meaning "sweet").

Monosaccharides and Disaccharides

The monosaccharides are simple compounds having six carbons in the formula, but their properties differ because of structural differences. This is shown in dextrose and levulose, which have the same formula (are isomers) but different properties.

dextrose (glucose)   levulose (fructose)

$$
\begin{array}{cc}
\begin{array}{l}
\mathrm{H} \\
| \\
\mathrm{C}\!=\!\mathrm{O} \\
| \\
\mathrm{H}\!-\!\mathrm{C}\!-\!\mathrm{OH} \\
| \\
\mathrm{HO}\!-\!\mathrm{C}\!-\!\mathrm{H} \\
| \\
\mathrm{H}\!-\!\mathrm{C}\!-\!\mathrm{OH} \\
| \\
\mathrm{H}\!-\!\mathrm{C}\!-\!\mathrm{OH} \\
| \\
\mathrm{H}\!-\!\mathrm{C}\!-\!\mathrm{OH} \\
| \\
\mathrm{H}
\end{array}
&
\begin{array}{l}
\mathrm{H} \\
| \\
\mathrm{H}\!-\!\mathrm{C}\!-\!\mathrm{OH} \\
| \\
\mathrm{C}\!=\!\mathrm{O} \\
| \\
\mathrm{HO}\!-\!\mathrm{C}\!-\!\mathrm{H} \\
| \\
\mathrm{H}\!-\!\mathrm{C}\!-\!\mathrm{OH} \\
| \\
\mathrm{H}\!-\!\mathrm{C}\!-\!\mathrm{OH} \\
| \\
\mathrm{H}\!-\!\mathrm{C}\!-\!\mathrm{OH} \\
| \\
\mathrm{H}
\end{array}
\end{array}
$$

Not as sweet as ordinary sugar (sucrose); does not need to be digested.	Twice as sweet as dextrose; must be changed to dextrose in the body before available for body use.

The sugars that contain an aldehyde structure ($-\mathrm{C}\!\!\stackrel{\scriptstyle H}{\diagup}\!\!=\!\mathrm{O}$), such as dextrose, or a group that readily changes to this structure act as mild reducing agents and are called *reducing sugars*. They can be identified by their ability to reduce copper(II) hydroxide, in Fehling's solution or Benedict's solution, to copper(I) oxide which is recognized by its brick-red color. This test is often used to detect the disease known as diabetes.

The "double" sugars, those having twelve carbons in their structures, are called disaccharides. Some of these are sucrose, lactose (found in milk), and maltose.

Sucrose can be converted to simple sugars by treating a solution of it with a little acid and boiling. This process is called *inversion*, and the products, dextrose and levulose, are called *invert* sugars. The general name for this type of reaction with disaccharides is *hydrolysis*.

$$C_{12}H_{22}O_{11} + H_2O \rightarrow C_6H_{12}O_6 + C_6H_{12}O_6$$

sucrose   dextrose   levulose

Polysaccharides

These complex carbohydrates are made up of some multiple of $C_6H_{10}O_5$. This is shown in the general formula $(C_6H_{10}O_5)_n$, in which n represents the variable multiple.

This group includes starch, dextrin, and cellulose.

Polymers

Polymers are very large organic compounds made of repeating units. The term *polymer* comes from two Greek roots: *poly* meaning "many," and *mer* meaning "part." The repeating units in a polymer are called *monomers*. (*Mono* means "one" in Greek.) You can compare a polymer to a long string of beads, and a monomer to an individual bead.

Almost all living organisms make and use different polymers. Plants use glucose as a monomer to form the polymers starch, an important food source, and cellulose, an important structural compound in plants and the principal component of paper. Different amino acids link together to form proteins, which are also polymers. Depending on the sequence of amino acids, the protein might be the hair on your head, a muscle in your arm, or an enzyme that helps you to digest food.

One of the first completely synthetic polymers was nylon. Other synthetic polymers have a wide variety of different properties and uses. Polyethylene is a lightweight, inexpensive polymer used to make such items as trash bags and plastic containers. Polyvinyl chloride is used as plastic wrap because it can be made into a thin film that adheres well to itself. Polymethylmethacrylate is a polymer valued for its transparency and resistance to shattering. It is used as a substitute for glass. Another well-known polymer is Teflon, which is used as a nonstick finish on metal cookware.

Polymers can be classified by the way they behave when heated. A *thermoplastic polymer* melts when heated and can be reshaped many times. A *thermosetting polymer* does not melt when heated but keeps its original shape. The molecules of a linear polymer are free to move. They slide back and forth against each other easily when heated, and so they are called thermoplastics. The molecules of a branched polymer contain side chains that prevent the molecules from sliding across each other easily. More heat is required to melt a branched polymer than a linear polymer but they are still likely to be thermoplastic. In cross-linked polymers adjacent molecules have formed bonds with each other. Individual molecules are not able to slide past each other when heated, and so they retain their shape when heated and are thermosetting polymers.

The two principal methods of synthesizing polymers are addition polymerization and condensation polymerization.

An addition polymer is a polymer formed by chain addition reactions between monomers containing a double bond. Molecules of ethene can polymerize with each other to form polyethene, commonly called polyethylene.

$$n\,CH_2{=}CH_2 \longrightarrow (-CH_2{-}CH_2{-})_n$$
$$\text{(catalyst)}$$

$$\text{ethene} \qquad\qquad \text{polyethylene}$$

The letter n shows that the addition reaction can be repeated multiple times to form a polymer n monomers long. This can be repeated hundreds of times.

A condensation polymer is a polymer formed by condensation reactions. Monomers of condensation polymers must contain two functional groups. This allows each monomer to link with two other monomers by condensation reactions. Condensation polymers are usually copolymers with two monomers in alternating order.

One example of a condensation polymer is shown on the following page. The hexanediamines with two amine groups react with adipic acids with two carboxyl groups to form nylon 66 and water.

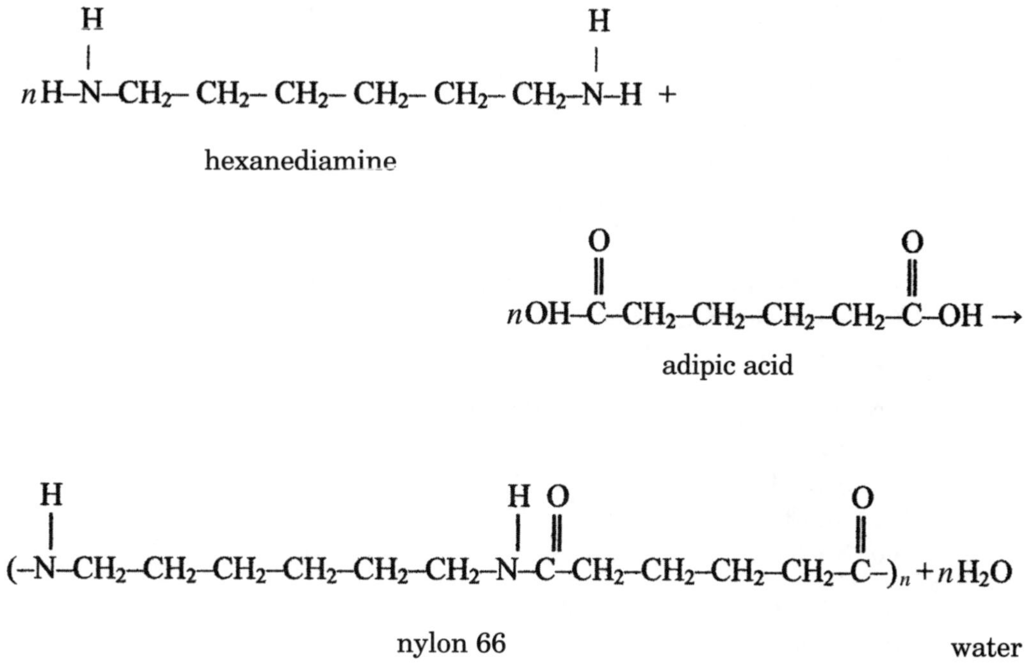

$$n \text{ H–N–CH}_2\text{– CH}_2\text{– CH}_2\text{– CH}_2\text{– CH}_2\text{– CH}_2\text{–N–H } +$$
(with H on the N atoms)

hexanediamine

$$n \text{ OH–C–CH}_2\text{–CH}_2\text{–CH}_2\text{–CH}_2\text{–C–OH } \rightarrow$$
(with O double-bonded to each C)

adipic acid

$$(\text{–N–CH}_2\text{–CH}_2\text{–CH}_2\text{–CH}_2\text{–CH}_2\text{–CH}_2\text{–N–C–CH}_2\text{–CH}_2\text{–CH}_2\text{–CH}_2\text{–C–})_n + n\,\text{H}_2\text{O}$$
(with H on the N atoms and O double-bonded to each C)

nylon 66   water

The product contains two kinds of monomers, an adipic acid monomer and a hexanediamine monomer. This copolymer is known as nylon 66 because each of the monomers contains six carbon atoms. It is the most widely used of all synthetic polymers.

Chapter 14 Review

1. Carbon atoms usually
   (A) lose 4 electrons
   (B) gain 4 electrons
   (C) form 4 covalent bonds
   (D) share the 2 electrons in the first principal energy level

2. Coke is produced from bituminous coal by
   (A) cracking
   (B) synthesis
   (C) substitution
   (D) destructive distillation

3. The usual method for preparing carbon dioxide in the laboratory is
   (A) heating a carbonate
   (B) fermentation
   (C) reacting an acid and a carbonate
   (D) burning carbonaceous materials

4. The precipitate formed when carbon dioxide is bubbled into limewater is
   (A) $CaCl_2$
   (B) H_2CO_3
   (C) CaO
   (D) $CaCO_3$

5. The "lead" in a lead pencil is
   (A) bone black
   (B) graphite and clay
   (C) lead oxide
   (D) lead peroxide

6. A decolorizer used in sugar refining is
   (A) bone black
   (B) chlorine water
   (C) hydrogen peroxide
   (D) sulfur dioxide

7. A common ingredient of baking powder is
   (A) NaCl
   (B) $NaHCO_3$
   (C) Na_2CO_3
   (D) NaOH

8. The first and simplest alkane is
   (A) ethane
   (B) methane
   (C) C_2H_2
   (D) methene
   (E) CCl_4

9. Slight oxidation of a primary alcohol gives
   (A) a ketone
   (B) an organic acid
   (C) an ether
   (D) an aldehyde
   (E) an ester

10. The characteristic group of the organic ester is
   (A) —CO—
   (B) —COOH
   (C) —CHO
   (D) —O—
   (E) —COO—

11. Fermentation of glucose gives
   (A) CO_2 and H_2O
   (B) CO and alcohol
   (C) CO_2 and CH_3OH
   (D) CO and C_2H_5OH
   (E) CO_2 and C_2H_5OH

12. The organic acid that can be made from ethanol is
   (A) acetic acid
   (B) formic acid
   (C) C_3H_7OH
   (D) found in bees and ants
   (E) butanoic acid

13. An ester can be prepared by the reaction of
    (A) two alcohols
    (B) an alcohol and an aldehyde
    (C) an alcohol and an organic acid
    (D) an organic acid and an aldehyde
    (E) an acid and a ketone

14. Phenol is a derivative of an
    (A) alkane
    (B) alkene
    (C) aliphatic (chain) hydrocarbon
    (D) aromatic hydrocarbon
    (E) alkyne

15. Sucrose is a
    (A) reducing sugar
    (B) monosaccharide
    (C) sugar substitute with a ketone group
    (D) disaccharide
    (E) sugar with an aldehyde group

16. Compounds that have the same composition but differ in structural formulas
    (A) are used for substitution products
    (B) are called isomers
    (C) are called polymers
    (D) have the same properties
    (E) are usually alkanes

17. Ethene is the first member of the
    (A) alkane series
    (B) saturated hydrocarbons
    (C) alkyne series
    (D) unsaturated hydrocarbons
    (E) aromatic hydrocarbons

18. Fehling's solution gives a positive test with
    (A) glucose
    (B) sucrose
    (C) copper(I) oxide
    (D) starch

Answers

1. **C**	5. **B**	9. **D**	13. **C**	17. **D**
2. **D**	6. **A**	10. **E**	14. **D**	18. **A**
3. **C**	7. **B**	11. **E**	15. **D**	
4. **D**	8. **B**	12. **A**	16. **B**	

Terms You Should Know

alcohol
aldehyde
alkane
alkene
alkylation
alkyne
amine
amino acid
aromatics
aromatization
carbohydrate
cracking
dehydrogenation
destructive distillation

ester
fullerenes
Fischer-Tropsch process
hydrocarbon
hydrogenation
isomer
ketone
octane
photosynthesis
polymerization
polymers
polysaccharide
Solvay process

Chapter 15

NUCLEONICS

Radioactivity

The discovery of radioactivity came in 1896 (less than 2 months after Wilhelm Röntgen announced the discovery of X rays). Röntgen had pointed out that the X rays came from a spot on a glass tube where a beam of electrons, in his experiments, was hitting, and that this spot simultaneously showed strong fluorescence. It occurred to Henri Becquerel and others that X rays might in some way be related to fluorescence (emission of light when exposed to some exciting agency) and to phosphorescence (emission of light *after* the exciting agency is removed).

Becquerel accordingly tested a number of phosphorescent substances to determine whether they emitted X rays while phosphorescing. He had no success until he tried a compound of uranium. Then he found that the uranium compound, whether or not it was allowed to phosphoresce by exposure to light, continuously emitted something that could penetrate lightproof paper and even thicker materials.

Becquerel determined that the compounds of uranium and the element itself produced *ionization* in the surrounding air. Thus either the ionizing effect, as indicated by the rate of discharge of a charged electroscope, or the degree of darkening of a photographic plate, could be used to measure the intensity of the invisible emission. Moreover, the emission from the uranium was continuous, perhaps even permanent, and required no energy from any external source. Yet, probably because of the current interest and excitement over X rays, Becquerel's work received little attention until early in 1898 when Marie and Pierre Curie entered the picture. (Pierre Curie was one of Becquerel's colleagues in Paris.)

Searching for the source of the intense radiation in uranium ore, they used tons of it to isolate very small quantities of two new elements, radium and polonium, both radioactive. Along with Becquerel, the Curies shared the Nobel Prize in Physics in 1903. Unfortunately, Pierre died soon afterward in a tragic accident involving a horse-drawn street cart. Marie Curie carried on their work and was chosen to succeed him at the Sorbonne as the first woman professor at that university. In 1911 she received a second Nobel Prize for her work in the chemistry of the new elements, radium and polonium. She founded the Radium Institute in Paris and devoted much of her time to the application of radioactivity to different mediums.

The Nature of Radioactive Emissions

While the early separation experiments were in progress, an understanding was slowly being gained about the nature of the spontaneous emission from the various radioactive elements. Becquerel thought at first that there were simply X rays, but *three* different kinds of radioactive emission, now called *alpha particles*, *beta particles*, and *gamma rays*, were soon found. We now know that alpha particles are positively charged particles of helium nuclei, beta particles are streams of high-speed electrons, and gamma rays are high-energy radiations similar to X rays. The emission of these three types of radiation is depicted below:

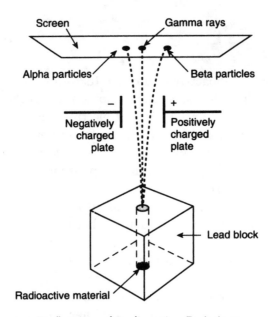

Deflection of Radioactive Emissions

The important characteristics of each type of radiation can be summarized as follows:

Alpha particle   (helium nucleus $_2^4$He) Positively charged, +2

1. Ejection reduces the atomic number by 2 and the atomic weight by 4 amu.
2. High energy, relatively low velocity.
3. Range: about 5 cm in air.
4. Shielding needed: stopped by the thickness of a sheet of paper, skin.
5. Interactions: produces about 100,000 ionizations per centimeter; repelled by the positively charged nucleus; attracts electrons but does not capture them until its speed is much reduced.
6. An example: Thorium-230 has an unstable nucleus and undergoes radioactive decay through alpha emission. The nuclear equation that describes this reaction is

$$_{90}^{230}\text{Th} \rightarrow {}_2^4\text{He} + {}_{88}^{226}\text{Ra}$$

In a decay reaction like this, the initial element (thorium-230) is called the parent nuclide and the resulting element (radium-226) is called the daughter nuclide.

Beta particle   (fast electron) Negatively charged, –1

1. Ejected when a *neutron* decays into a proton and an electron.
2. High velocity, low energy.
3. Range: about 12 m.
4. Shielding needed: stopped by 1 cm of aluminum or the thickness of the average book.
5. Interactions: weak due to high velocity but produces about 100 ionizations per centimeter.
6. An example: Protactinium-234 is a radioactive nuclide that undergoes beta emission. The nuclear equation is

$$^{234}_{91}\text{Pa} \rightarrow \,^{234}_{92}\text{U} + \,^{0}_{-1}\text{e}$$

Gamma radiation   (electromagnetic radiation identical with light; high energy) No charge

1. Beta particles and gamma rays are usually emitted together; after a beta is emitted, a gamma ray follows.
2. Arrangement in nucleus is unknown. Same velocity as visible light.
3. Range: no specific range.
4. Shielding needed: about 13 cm of lead.
5. Interactions: weak of itself; gives energy to electrons, which then perform the ionization.

Methods of Detection of Alpha, Beta, and Gamma Rays

All methods of detection of these radiations rely on their ability to ionize. There are a number of methods in common use.

Photographic plate   The fogging of a photographic emulsion led to the discovery of radioactivity. If this emulsion is viewed under a high-power microscope, it is seen that beta and gamma rays cause the silver bromide grains to develop in a scattered fashion. However, alpha particles, owing to the dense ionization they produce, leave a definite track of exposed grains in the emulsion. Hence not only is the alpha particle detected, but also its range (in the emulsion) can be measured. Special emulsions are capable of showing the beta-particle tracks.

Film pages are used to measure radiation exposure of people working in a radiation environment.

Scintillation counter   A fluorescent screen (e.g., ZnS) will show the presence of electrons and X rays, as we have already seen. If the screen is viewed with a magnifying eyepiece, small flashes of light, called scintillations, will be observed. By observing the scintillations, one not only can detect the presence of alpha particles, but also can actually count them.

The cloud chamber   One of the most useful instruments for detecting and measuring radiation is the Wilson cloud chamber. Its operation depends on the well-known fact that moisture tends to condense around ions (the probable explanation for some formation of clouds in the sky). If an enclosed region of air is saturated with water vapor (this is always the case if water is present) and the air is cooled suddenly, it becomes supersaturated, that is, it contains for the instant more water vapor than it can hold permanently, and a fog of water droplets develops around the ions in the chamber.

For example, and alpha particle traveling through such a supersaturated atmosphere supplies a trail of ions on which water droplets will condense. This trail is thus made visible. These trails are usually photographed by a camera that operates whenever a piston moves downward causing the air to expand and cool.

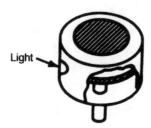

Cloud Chamber

The bubble chamber This device utilizes a liquid such as ether, ethyl alcohol, pentane, or propane, which is superheated to well above the boiling point. When the pressure is released quickly, the liquid is in a highly unstable condition, ready to boil violently. An ionizing particle passing through the liquid at this instant leaves a trail of tiny bubbles which may be photographed.

The electroscope One of the simplest and one of the first used in work on radioactivity consists of two thin gold leaves which are charged by a battery with the same type of charge. Radioactive materials produce ions in proportion to their radioactivity. A negatively charged electroscope becomes discharged when ions in the air take electrons from the electroscope, causing the leaves to gradually collapse.

Electroscope

The Geiger counter   This instrument is perhaps the most widely used at the present time for measuring individual radiation. It consists of a fine wire of tungsten mounted along the axis of a tube that contains a gas at reduced pressure. A difference of potential of about 1000 V is applied in such a way as to make the metal tube negative with respect to the wire. The voltage is high enough so that the electrons produced are accelerated by the electric field. Near the wire, where the field is strongest, the accelerated particles produce more ions—positive ones going to the walls and negative ones being collected by the wire. Any particle that will produce an ion gives rise to the same avalanche of ions, and so the type of particle cannot be identified. However, each individual particle can be detected.

Decay Series and Transmutations

The nuclei of uranium, radium, and other radioactive elements are continually disintegrating. It should be emphasized that spontaneous disintegration produces the gas known as radon. The time it takes for half of the material to disintegrate is called its *half-life*.

For example, for radium, we know that on the average half of all the radium nuclei present will have disintegrated to radon in 1590 years. In another 1590 years, half of this remainder will decay, and so on. When a radium atom disintegrates, it loses an alpha particle, which upon gaining two electrons eventually becomes a neutral helium atom. The remainder of the atom becomes radon.

Such a conversion of an element to a new element (because of a change in the number of protons) is called a *transmutation*. This transmutation can be produced artificially by bombarding the nuclei of a substance with various particles from a particle accelerator such as the cyclotron.

The following uranium-radium disintegration series shows how a radioactive atom can change when it loses each kind of particle. Note that the atomic number is shown by a subscript ($_{92}$U), and the isotopic mass by a superscript (^{238}U). The alpha particle is shown by the Greek symbol α, and the beta particle by β.

$$^{238}_{92}\text{U} \xrightarrow{-\alpha} {}^{234}_{90}\text{Th} \xrightarrow{-\beta} {}^{234}_{91}\text{Pa} \xrightarrow{-\beta} {}^{234}_{92}\text{U} \xrightarrow{-\alpha} {}^{230}_{90}\text{Th} \xrightarrow{-\alpha}$$

$$^{226}_{88}\text{Ra} \xrightarrow{-\alpha} {}^{222}_{86}\text{Rn} \xrightarrow{-\alpha} {}^{218}_{84}\text{Po} \xrightarrow{-\alpha} {}^{214}_{82}\text{Pb} \xrightarrow{-\beta} {}^{214}_{83}\text{Bi} \xrightarrow{-\beta}$$

$$^{214}_{84}\text{Po} \xrightarrow{-\alpha} {}^{214}_{82}\text{Pb} \xrightarrow{-\beta} {}^{210}_{83}\text{Bi} \xrightarrow{-\beta} {}^{210}_{84}\text{Po} \xrightarrow{-\alpha} {}^{206}_{82}\text{Pb(stable)}$$

This can be shown graphically in the radioactive decay series on the next page.

Particle Lost	Weight Change	Atomic Number Change
Alpha (α)	Loses 4 amu	Decreases by 2
Beta (β)	None	Increases by 1

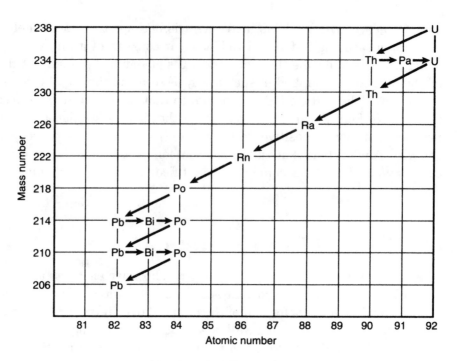

Uranium-238 Decay Series

The stability of an atom seems to be related to binding energy. The binding energy is the amount of energy released when a nucleus is formed from its component particles. If you add the mass of the components and compare this sum to the actual mass of the nucleus formed, there will be a small difference in these figures. This difference in mass can be converted to its energy equivalent using Einstein's equation $E = mc^2$, where E is the energy, m is the mass, and c is the velocity of light. It is this energy that is called the binding energy.

It has been found that the lightest and heaviest elements have the smallest binding energy per nuclear particle and thus are less stable than the elements of intermediate atomic masses, which have the greatest binding energy.

The relationship of an even or odd number of protons to the number of neutrons affects the stability of a nucleus. Many stable nuclei have even numbers of protons and neutrons, while stability is less frequent in nuclei that have an even number of protons and an odd number of neutrons, or vice versa. Only a few stable nuclei are known that have odd numbers of protons and neutrons.

Radioactive Dating

A helpful adaptation of radioactive decay is its use in determining the age of substances such as rocks and relics. Because ^{14}C has a half-life of about 5700 years and occurs in the remains of organic materials, it has been useful in dating these materials. A small percentage of CO_2 in the atmosphere contains ^{14}C. The stable isotope of carbon is ^{12}C. ^{14}C is a beta emitter and decays to form ^{14}N:

$$^{14}_{6}C \rightarrow {}^{14}_{7}N + {}^{0}_{-1}e$$

In any living organism, the ratio of ^{14}C to ^{12}C is the same as it is in the atmosphere because of the constant interchange of materials between organism and surroundings. When an organism dies, this interaction stops and the ^{14}C gradually decays to nitrogen. By comparing the relative amounts of ^{14}C and ^{12}C in the remains, the age of the organism can be established. Since ^{14}C has a half-life of 5700 years, if there was 5 g of ^{14}C at first and now there is only half as much, 2.5 g, its age would be 5700 years. In other words, old wood emits half as much beta radiation per gram of carbon as that emitted by living plant tissues. This method was used to determine the age of the Dead Sea Scrolls (about 1900 years) and has been found to be in agreement with several other dating techniques that have been employed.

Nuclear Energy

Since Einstein predicted that matter could be converted to energy over a half-century ago, scientists have tried to unlock this energy source. Einstein's prediction was verified in 1932 by Sir John Cockcroft and Ernest Walton, who produced small quantities of helium and a tremendous amount of energy from the reaction

$$^{7}_{3}Li + ^{1}_{1}H \rightarrow 2^{4}_{2}He + energy$$

In fact, the energy released was almost exactly the amount calculated from Einstein's equation.

In 1942 Enrico Fermi and his co-workers discovered that a sustained chain reaction of fission could be controlled to produce large quantities of energy. This led to the development of the nuclear fission bomb, which brought World War II to an end.

Conditions for Fission

When fissionable material like ^{235}U is bombarded with a "slow" neutron, fission occurs, giving off different fission products. An example of such a reaction is shown in Figure 42. As long as one of the released neutrons produces another reaction, the chain reaction will continue. If each fission starts more than one reaction, the process becomes tremendously powerful in a very short time. This occurs in the atomic bomb.

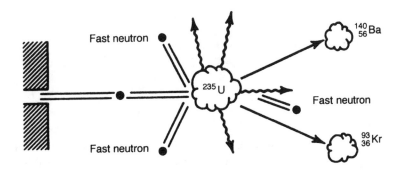

Figure 42. Fission Chain Reaction.

To obtain the neutron "trigger" for this reaction, neutrons must be slowed down so that they do not pass through the nucleus without effect. This slowing down or moderating is best done by letting fast neutrons collide with many relatively light atoms such as hydrogen, deuterium, and carbon. Graphite, paraffin, ordinary water, and "heavy water" (containing deuterium instead of ordinary hydrogen) are all suitable moderators.

Nuclear fission can be made to occur in an uncontrolled explosion or in a controlled nuclear reactor. In both cases enough fissionable material must be present so that, once the reaction starts, it can at least sustain itself. This amount of fissionable material is called the *critical mass*. In the bomb, the number of reactions increases tremendously, whereas in a reactor the rate of fissions is controlled.

The typical nuclear reactor or "pile" is made up of the following kinds of material:

1. Fissionable material—sustains the chain reaction.
2. Moderator—slows down fission neutrons.
3. Control rods (cadmium or boron steel rods)—absorb excess neutrons and control rate of reaction.
4. Concrete encasement—provides shielding from radiation.

See the model diagram of a nuclear reactor below.

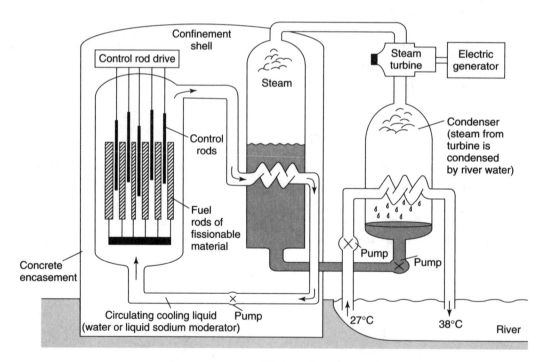

Nuclear Reactor

There are many variations of this basic reactor as reactors become more efficient. Many U.S. cities now receive some of their electrical power from a nuclear power station.

Methods of Obtaining Fissionable Material

The natural abundance of ^{235}U is extremely small: about 0.0005% of the earth's crust, with 0.7% natural uranium. Therefore the problem is to separate this isotope from the rest of the more abundant isotope. Two methods have been used to isolate ^{235}U: the electromagnetic process and gaseous diffusion.

The electromagnetic process is based on the principle of the mass spectrograph. Gaseous ions with similar charges but different masses are passed between the poles of a magnet. The lighter ones are deflected more than the heavier ones and are thus collected in separate compartments. These large mass separators are called calutrons and were used extensively to collect the first uranium-235 isotopes.

The diffusion process was found to be faster and less expensive than the electromagnetic process. This process makes use of the gas diffusion principle that light molecules diffuse faster than heavier ones. When uranium is converted into gaseous uranium hexafluoride, UF_6, and allowed to diffuse through miles of porous partitions under reduced pressure, the lighter $^{235}UF_6$ will diffuse faster than the $^{238}UF_6$. Thus the gas coming through first will be richer in ^{235}U than when it started. The ^{235}U is then retrieved from the fluoride compound for use in atomic devices or reactors.

Fusion

Fusion is the opposite of fission in that instead of a nucleus being split, two nuclei join to form a new nucleus with a great amount of released energy. The sun's energy results from such a fusion reaction, in which four hydrogen protons are eventually made into one helium nucleus with the liberation of a large amount of energy. The reactions are

$$^1_1H + ^1_1H \rightarrow ^2_1H + ^0_1e$$
$$^1_1H + ^2_1H \rightarrow ^3_2He$$
$$^3_2He + ^3_2He \rightarrow ^4_2He + 2^1_1H$$
$$^3_2He + ^1_1H \rightarrow ^4_2He + ^0_1e$$

So far, nuclear fusion has been produced only in an uncontrolled explosion, the hydrogen bomb. Scientists in every major country of the world are attempting to find a means of controlling a sustained fusion reaction.

In the meantime, nuclear fission is being adapted to produce energy for electrical power stations, merchant ships, aircraft carriers, and submarines. Experimentation is proceeding on the nuclear aircraft engine and a nuclear rocket engine.

Radiation Exposure

Nuclear radiation can transfer its energy to the electrons of atoms or molecules and cause ionization. The *roentgen* is a unit used to measure nuclear radiation. It is equal to the amount of radiation that produces 2×10^9 ion pairs when it passes through 1 cm^3 of dry air. Ionization can damage living tissue. Radiation damage to human tissue is measured in *rems* (roentgen equivalent, man). One rem is the quantity of ionizing radiation that does as much damage to human tissue as is done by 1 roentgen of high-voltage X rays. Cancer and genetic effects caused by DNA mutations are long-term radiation damage to living tissue. DNA can be mutated directly by interaction with radiation or indirectly by interaction with previously ionized molecules.

Everyone is exposed to environmental background radiation. Average exposure for people living in the United States is estimated to be about 0.1 rem per year. At high altitudes, people have increased exposure because of increased cosmic-ray levels.

Radon-222 trapped inside home basements may also cause increased exposure. Because it is a gas, it can move up from the soil into homes through cracks and holes in the foundation.

New Subatomic Particles

As scientists continue the search for more information about the atom and its makeup, they have been especially interested in the subatomic particles. In the 1960s, with more powerful accelerators available to investigate the atom, they found more than 100 different subatomic particles. More recently, research has suggested a new order that brings a simplicity to the particles that compose matter. By studying the tracks made on photographic plates by the particles that result from nuclear collision, Murray Gell-Mann at the California Institute of Technology and Zweig at CERN in Switzerland arrived at a new way of accounting for these subatomic particles. Their theory is that all small particles are combinations of still smaller particles called quarks. A quark is a particle with a fractional charge of either one third or two thirds the charge of an electron. This theory simplifies the multitude of particles.

At first three quarks were postulated, but the number has grown to six. The usual subatomic particles presented in a first-year chemistry book account for only two quarks. Each quark is described or named by its particular characteristic, which is referred to as its "flavor." The following table lists the six quarks and their fractional charges.

SIX KINDS OF QUARKS

Name	Electric Charge
Up	$\frac{2}{3}$
Down	$-\frac{1}{3}$
Charmed	$\frac{2}{3}$
Strangeness	$-\frac{1}{3}$
Top (truth)	$\frac{2}{3}$
Bottom (beauty)	$-\frac{1}{3}$

The two quarks described as "up" and "down" are thought to constitute most of the ordinary matter in the universe. Scientists have never isolated individual quarks; they attribute their failure to the magnitude of the binding force that holds the quarks together.

The electron is now believed to be composed of smaller elementary particles called *leptons*. This is also true of the neutrino, which is similar in mass to the electron but has a neutral charge.

Chapter 15 Review

1. Radioactive changes differ from ordinary chemical changes because radioactive changes
   (A) involve changes in the nucleus
   (B) are explosive
   (C) absorb energy
   (D) release energy

2. Isotopes of uranium have different
   (A) atomic numbers
   (B) atomic masses
   (C) numbers of planetary electrons
   (D) numbers of protons

3. Pierre and Marie Curie discovered
   (A) oxygen
   (B) hydrogen
   (C) chlorine
   (D) radium

4. The number of protons in the nucleus of an atom of atomic number 32 is
   (A) 4
   (B) 32
   (C) 42
   (D) 73

5. Atoms of ^{235}U and ^{238}U differ in structure by three
   (A) electrons
   (B) isotopes
   (C) neutrons
   (D) protons

6. In a hydrogen bomb, hydrogen is converted into
   (A) uranium
   (B) helium
   (C) barium
   (D) plutonium

7. The use of radioactive isotopes has already produced promising results in the treatment of certain types of
   (A) cancer
   (B) heart disease
   (C) pneumonia
   (D) diabetes

8. The emission of a beta particle results in a new element with the atomic number
   (A) increased by 1
   (B) increased by 2
   (C) decreased by 1
   (D) decreased by 2

9. A substance used as a moderator in a nuclear reactor is
   (A) marble
   (B) hydrogen
   (C) tritium
   (D) graphite

10. A self-sustaining nuclear fission chain reaction depends upon the release of
    (A) protons
    (B) neutrons
    (C) electrons
    (D) alpha particles

11. Einstein's formula for the Law of Conservation of Mass and Energy states that
    $E =$ _____.

12. The energy of the sun is thought to be produced by the fusion of hydrogen atoms into _____.

13. Could a mass spectrograph (or calutron) be used to separate ^{54}Mn from ^{54}Cr? _____.

14. The release of energy from the combination of two nuclei into one is called _____.

15. Write the equation for the first observed artificial transmutation by Rutherford: _____.

Matching

Choose an answer from this list and match it with the appropriate numbered statement below.

A. Crooke	H. gamma ray
B. alpha particle	I. gaseous diffusion
C. beta particle	J. neptunium
D. Becquerel	K. plutonium
E. Einstein	L. ^{235}U
F. chain reaction	M. ^{238}U
G. deuteron	

16. Helium nucleus: _____

17. Stable isotope of uranium: _____

18. One method of separating isotopes: _____

19. Most penetrating ray from radioactive decay: _____

20. Mass-energy conversion: _____

Answers

1. **A**	3. **D**	5. **C**	7. **A**	9. **D**
2. **B**	4. **B**	6. **B**	8. **A**	10. **B**

11. mc^2

12. He

13. No

14. fusion

15. $^{14}_{7}\text{N} + ^{4}_{2}\text{He} \rightarrow ^{17}_{8}\text{O} + ^{1}_{1}\text{H}$

16. B

17. M

18. I

19. H

20. E

Terms You Should Know

alpha particle	Geiger counter
beta particle	half-life
bubble chamber	lepton
cloud chamber	phosphorescence
critical mass	quark
electroscope	radon
fission	rem
fluorescence	roentgen
fusion	transmutation
gamma ray	

Chapter 16

REPRESENTATIVE LABORATORY SETUPS

New Technology in the Laboratory

Laboratory setups vary from school to school depending on whether the lab is equipped with macro- or microscale equipment. Microlabs use specialized equipment that allows lab work to be done on a much smaller scale. The basic principles are the same as when using full-sized equipment, but microscale equipment lowers the cost of materials, results in less waste, and poses less danger. The examples in this book are of macroscale experiments.

Along with learning to use microscale equipment, most labs require a student to learn how to use technological tools to assist in experiments. The most common are:

Gravimetric balance with direct readings to thousandths of a gram instead of a triple-beam balance

pH meters that give pH readings directly instead of using indicators

Spectrophotometer that measures the percentage of light transmitted at specific frequencies so that the molarity of a sample can be determined without doing a titration

Computer-assisted labs that use probes to take readings, e.g., temperature and pressure, so that programs available for computers can print out a graph of the relationship of readings taken over time.

Some Basic Setups

Figures 43 through 54 show some of the basic laboratory setups used in beginning chemistry. The purpose of these diagrams is to review the basic techniques of assembling equipment with regard to some knowledge about the reactants and products involved.

In the equations, the letters in parentheses have the following meanings:

(s) = solid
(ℓ) = liquid
(g) = gas

1 **Preparation of a gaseous product, nonsoluble in water, by water displacement from solid reactants.**

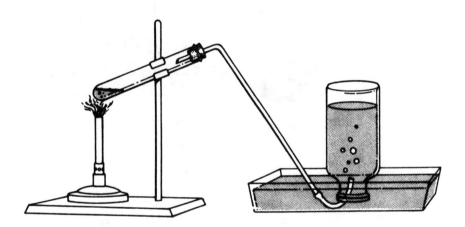

Figure 43. Preparation of Oxygen.

Preparation of oxygen (O_2).

$$2KClO_3(s) + MnO_2(s) \rightarrow 2KCl + 3O_2(g) + MnO_2$$

2 **Preparation of a gaseous product, nonsoluble in water, by water displacement from at least one reactant in solution.**

Note: Purpose of the thistle tube shown in Figure 48:
   a. Introduction of more liquid without "opening" the reacting vessel.
   b. Safety valve to indicate blocked delivery tube by the rise of liquid in the thistle tube.

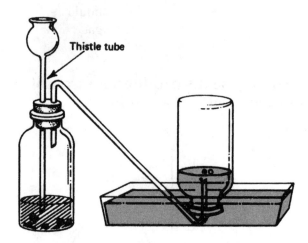

Figure 44. Preparation of Carbon Dioxide, Nitric Oxide, and Hydrogen.

Preparation of carbon dioxide (CO_2).

$$CaCO_3(s) + 2HCl(\ell) \rightarrow CaCl_2 + H_2O + CO_2(g)$$

Preparation of nitric oxide (NO).

$$3Cu(s) + 8 \text{ dilute } HNO_3(\ell) \rightarrow 3Cu(NO_3)_2 + 4H_2O + 2NO(g)$$

Preparation of hydrogen (H_2).

$$Zn(s) + 2HCl(\ell) \rightarrow ZnCl_2 + H_2(g)$$

3 | **Preparation of a gaseous product heavier than air that can best be collected by the upward displacement of air.**

Using a thistle tube with a stopcock to control the flow of a liquid reactant:

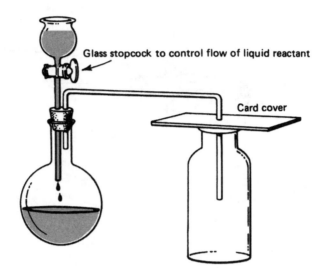

Figure 45. Preparation of Sulfur Dioxide and Hydrogen Sulfide.

Preparation of sulfur dioxide (SO_2).

$$Na_2SO_3(s) + H_2SO_4(\ell) \rightarrow Na_2SO_4 + H_2O + SO_2(g)$$

strong, irritating odor

Preparation of hydrogen sulfide (H_2S).

$$2HCl(\ell) + FeS(s) \rightarrow FeCl_2 + H_2S(g)$$

rotten egg odor

4 **Preparation of a gaseous product that is soluble in water and lighter than air by the downward displacement of air.**

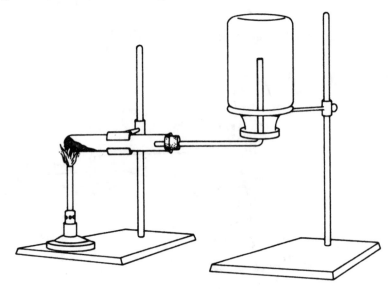

Figure 46. Preparation of Ammonia.

Preparation of ammonia (NH_3).

$$2NH_4Cl(s) + Ca(OH)_2(s) \rightarrow CaCl_2 + 2H_2O + 2NH_3(g)$$

5 **Distillation of a liquid.**

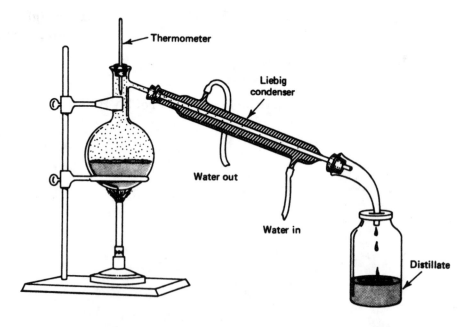

Figure 47. Distillation of a Liquid.

Removes dissolved impurities, which remain in the flask. Does not remove volatile materials, which vaporize and pass over into the distillate.

6 **Preparation of a gaseous product, not dissolved by water, by means of electrolysis.**

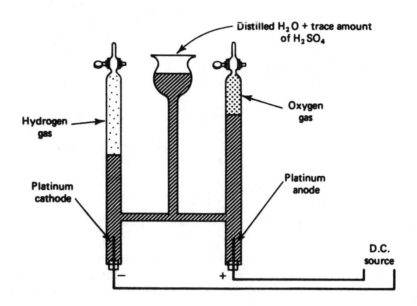

Figure 48. Electrolysis Setup.

Anode reaction:	$H_2O(\ell) \rightarrow \frac{1}{2}O_2(g) + 2H^+ + 2e^-$
Cathode reaction:	$2H_2O(\ell) + 2e^- \rightarrow H_2(g) + 2OH^-$
Cell reaction:	$3H_2O(\ell) \rightarrow \frac{1}{2}O_2(g) + H_2(g) + 2H_2O(\ell)$
or	$H_2O(\ell) \rightarrow \frac{1}{2}O_2(g) + H_2(g)$

---

7   ## Separation of a mixture by chromatography.

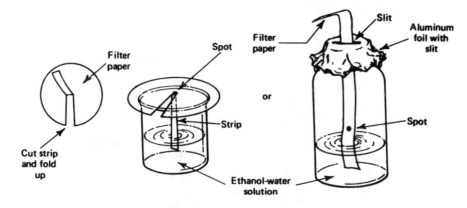

Figure 49. Chromatography Setup.

Chromatography is a process used to separate parts of a mixture. The component parts separate as the solvent carrier moves past the spot by capillary action. Because of variations in solubility, attraction to the filter paper, and density, each fraction moves at a different rate. Once separation occurs, the fractions are either identified by their color or removed for other tests. A usual example is the use of Shaeffer Skrip Ink No. 32, which separates into yellow, red, and blue streaks of dyes.

8   ## Measuring potentials in electrochemical cells.

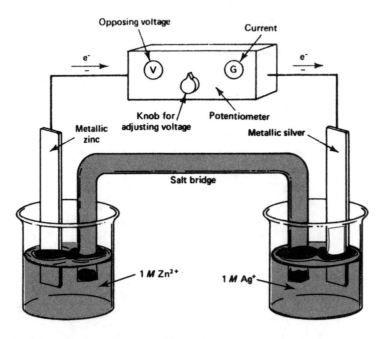

Figure 50. Potentiometer Setup for Measuring Potential.

The voltmeter in this zinc-silver electrochemical cell reads approximately 1.56 V. This means that the Ag to Ag^+ half-cell has 1.56 V more electron-attracting ability than the Zn to Zn^{2+} half-cell. If the potential of the zinc half-cell is known, the potential of the silver half-cell can be determined by adding 1.56 V to the potential of the zinc half-cell. In a setup like this, only the difference in potential between two half-cells can be measured.

9 | Titration setup.

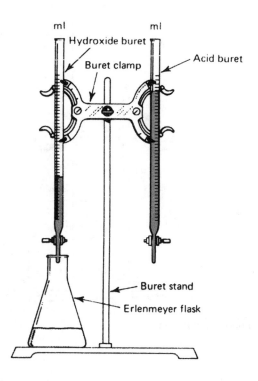

Figure 51. Titration Using Burets.

The titration of a NaOH solution of unknown concentration in one buret with 0.1 M HCl in the other.

Introduce approximately 15 mL of NaOH into the flask and add an indicator such as litmus or phenolphthalein. Add the HCl slowly with constant swirling. When a color change occurs and is retained, record the amount used. To find the molarity of the NaOH use the formula

$$M_{acid} \times V_{acid} = M_{base} \times V_{base}$$

10	**Replacement of hydrogen by a metal.**

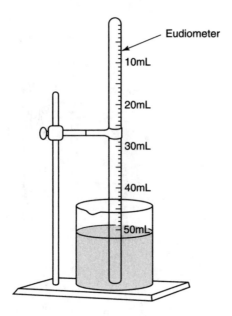

Figure 52. Eudiometer Apparatus.

Measure the mass of a strip of magnesium with an analytical balance to the nearest 0.001 g. Using a strip with a mass of about 0.040 g produces about 40 mL of H_2. Pour 5 mL of concentrated HCl into the eudiometer and slowly fill the remainder with water. Try to minimize mixing. Lower the coil of Mg strip into the tube, invert it, and lower it to the bottom of the battery jar. After the reaction is complete, measure the volume of the gas released and calculate the mass of hydrogen replaced by the magnesium. (Refer to Chapter 5 for a discussion of gas laws.)

11 ## Drying gases to remove water vapor.

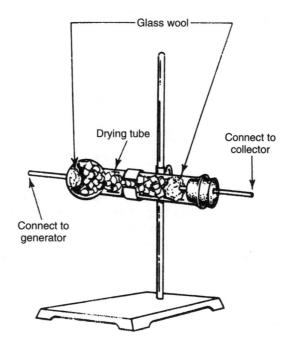

Figure 53. CaCl₂ Drying Tube.

Calcium chloride is hydroscopic and absorbs water vapor as the gas passes through this tube.

12 ## Measuring temperature for phase change and heating curves.

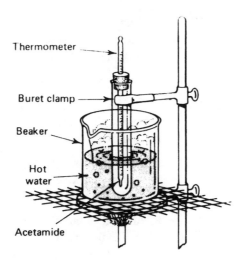

Figure 54. Heating Curve and Phase Change Setup.

This setup can be used with either acetamide or paradichlorobenzene in the test tube. By keeping a careful time and temperature observation chart, you can obtain data to plot a time-temperature graph and can note the effect of a phase change on this plot.

Summary of Qualitative Tests

I. Identification of Gases

Gas	Test	Result
Ammonia NH_3	1. Smell cautiously. 2. Test with litmus. 3. Expose to HCl fumes.	1. Sharp odor. 2. Red litmus turns blue. 3. White fumes form (NH_4Cl).
Carbon dioxide CO_2	1. Pass through limewater, $Ca(OH)_2$.	1. White precipitate (ppt.) forms, $CaCO_3$.
Carbon monoxide CO	1. Burn it and pass product through limewater, $Ca(OH)_2$.	1. White ppt. forms, $CaCO_3$.
Hydrogen H_2	1. Allow it to mix with some air and then ignite. 2. Burn it and then trap product.	1. Gas explodes. 2. Burns with blue flame—product H_2O turns cobalt chloride paper from blue to pink.
Hydrogen chloride HCl	1. Smell cautiously. 2. Exhale over the gas. 3. Dissolve in water and test with litmus. 4. Add $AgNO_3$ to the solution.	1. Choking odor. 2. Vapor fumes form. 3. Blue litmus turns red. 4. Forms white ppt.
Hydrogen sulfide H_2S	1. Smell cautiously. 2. Test with moist lead acetate paper.	1. Rotten egg odor. 2. Turns brown-black (PbS).
Nitric oxide NO	1. Expose to the air.	1. Colorless gas turns reddish brown.
Nitrous oxide N_2O	1. Insert glowing splint. 2. Add nitric oxide gas.	1. Bursts into flame. 2. Remains colorless.
Oxygen O_2	1. Insert glowing splint. 2. Add nitric oxide gas.	1. Bursts into flame. 2. Turns reddish brown.
Sulfur dioxide SO_2	1. Smell cautiously. 2. Allow it to bubble into purple potassium permanganate solution.	1. Choking odor. 2. Solution becomes colorless.

II. Identification of Negative Ions

Ion	Test	Result
Acetate $C_2H_3O_2^-$	Add conc. H_2SO_4 and warm gently.	Odor of vinegar released.
Bromide Br^-	Add chlorine water and some CCl_4; shake.	Reddish-brown color concentrated in CCl_4 layer.
Carbonate CO_3^-	Add HCl acid; pass released gas through limewater.	White cloudy ppt. forms.
Chloride Cl^-	1. Add silver nitrate solution. 2. Then add nitric acid, later followed by ammonium hydroxide.	1. White ppt. forms. 2. Ppt. insoluble in HNO_3 but dissolves in NH_4OH.
Hydroxide OH^-	Test with red litmus paper.	Turns blue.
Iodide I^-	Add chlorine water and some CCl_4; shake.	Purple color concentrated in CCl_4 layer.
Nitrate NO_3^-	Add freshly made ferrous sulfate sol., and then conc. H_2SO_4 carefully down the side of the tilted tube.	Brown ring forms at junction of layers.
Nitrite NO_2^-	Add dilute H_2SO_4.	Brown fumes (NO_2) released.
Sulfate SO_4^-	Add sol. of $BaCl_2$, then HCl.	White ppt. forms; insoluble in HCl.
Sulfide S^{2-}	Add HCl and test gas released with lead acetate paper.	Gas, with rotten egg odor, turns paper brown-black.
Sulfite SO_3^{2-}	Add HCl and pass gas into purple $KMnO_4$ sol.	Solution turns colorless.

III. Identification of Positive Ions

Ion	Test	Result
Ammonium NH_4^+	Add strong base (NaOH); heat gently.	Odor of ammonia.
Ferrous Fe^{2+}	Add sol. of potassium ferricyanide, $K_3Fe(CN)_6$.	Dark blue ppt. forms (Turnball's blue).
Ferric Fe^{3+}	Add sol. of potassium ferrocyanide, $K_4Fe(CN)_6$.	Dark blue ppt. forms (Prussian blue).
Hydrogen H^+	Test with blue litmus paper.	Turns red.

IV. Qualitative Tests for Metals

Flame Tests

Carefully clean a platinum wire by dipping it into dilute HNO_3 and heating in a Bunsen flame. Repeat until the flame is colorless. Dip heated wire into the substance being tested (either solid or solution) and then hold it in the hot outer part of the Bunsen flame.

Compound of	Color of Flame
Na	Yellow
K	Violet (use cobalt-blue glass to screen out Na impurities)
Li	Crimson
Ca	Orange-red
Ba	Green
Sr	Bright red

Hydrogen Sulfide Tests

Bubble hydrogen sulfide gas through the solution of a salt of the metal being tested. Check color of the precipitate formed.

Compound of	Color of Sulfide Precipitate
Lead (Pb)	Brown-black (PbS)
Copper (Cu)	Black (CuS)
Silver (Ag)	Black (Ag_2S)
Mercury (Hg)	Black (HgS)
Nickel (Ni)	Black (NiS)
Iron (Fe)	Black (FeS)
Cadmium (Cd)	Yellow (CdS)
Arsenic (As)	Light yellow (As_2S_3)
Antimony (Sb)	Orange (Sb_2S_3)
Zinc (Zn)	White (ZnS)
Bismuth (Bi)	Brown (Bi_2S_3)

PRACTICE TESTS IN CHEMISTRY

General Information

Because one of the most prominent achievement tests in chemistry is the College Board SAT II: Chemistry test, the review tests in this section are modeled after this test. The types of questions and the areas covered reflect those used in past and present achievement tests in chemistry.

The College Board SAT II: Chemistry test is planned to test the principles and concepts drawn from the factual material found largely in inorganic chemistry and, to a much lesser extent, that found in organic chemistry. Only a few questions are asked concerning industrial or analytical chemistry.

According to a description of the SAT II: Chemistry test by the College Entrance Examination Board, it includes kinetic molecular theory and the three states of matter; atomic theory and structure and the periodic table; nuclear reactions; quantitative relations as applied to chemical formulas and equations; chemical bonding and molecular structure, and their relations to properties; the nature of chemical reactions, including acid-base reactions, oxidation-reduction reactions, ionic reactions, and other chemical changes occurring in solution; energy changes accompanying chemical reactions; interpretation of chemical equilibria and reaction rates; solution phenomena; electrochemistry, nuclear chemistry, and radiochemistry; physical and chemical properties of the more familiar metals, transition elements, and nonmetals and of the more familiar compounds; understanding and interpretation of laboratory procedures and observations.

You will be provided with a periodic table to use during the test. All necessary information regarding atomic numbers and atomic masses is given on the chart. Plan to use a chart similar to the one provided at the beginning of the first practice test. It will provide the atomic information needed for the practice tests.

The directions on Practice Tests 1 through 5 are identical to those used on the present College Board SAT II: Chemistry test. Record your answers by blackening the corresponding spaces on the answer sheets provided.

Basic Topics and Abilities Tested

The following charts show the content of the test and the levels of thinking skills tested:

	Topics	Percent of Test (approx.)	Number of Questions (approx.)
Structure	1. Atomic theory and structure; periodic relationships 2. Nuclear reactions 3. Chemical bonding and molecular structure	25	21
States of Matter	1. Kinetic molecular theory of gases and the gas laws 2. Liquids, solids, and phase changes 3. Solutions; concentration units, solubility, conductivity, and colligative properties	15	13
Reaction Types	1. Acids and bases 2. Oxidation-reduction; electrochemical cells 3. Precipitation	14	12
Stoichiometry	Mole concept, Avogadro's number, empirical and molecular formulas, stoichiometric calculations, percentage composition, limiting reagents	12	10
Equilibrium and Reaction Rates	Equilibrium; mass action expressions, ionic equilibria, Le Châtelier's Principle; factors affecting rates of reaction	7	6
Thermodynamics	Energy changes in chemical reactions; physical processes, Hess's Law, randomness	6	5
Descriptive Chemistry	Physical and chemical properties of elements and their more familiar compounds, chemical reactivity and products of chemical reactions, simple examples from organic and environmental chemistry	13	11
Laboratory	Equipment, measurement, procedures, observations, safety, calculations, interpretation of results	7	6

Note: Each test contains approximately five questions on equation balancing and/or predicting products of chemical reactions. These are distributed among the various content categories.

Thinking Skills Tested	Percent of Test (approx.)
Recalling fundamental concepts, specific pieces of information, and basic terminology (low-level skill)	20
Showing a *comprehension of the basics* and the *ability to apply this information* in a rather straightforward manner to questions, situations, and the solution of qualitative or quantitative problem-oriented questions (medium-level skill)	45
Using the ability to *analyze* information and/or situations and to *synthesize* the knowledge learned to *evaluate* how and what ideas or relationships should be used to draw conclusions or to solve problems (high-level skill)	35

The first chart gives you a general overview of the content of the test. Your knowledge of the topics and your skills in recalling, applying, and synthesizing this knowledge are evaluated through 85 multiple-choice questions. This material is that generally covered in an introductory course in chemistry at a level suitable for college preparation. While every test covers the topics listed, different aspects of each topic are stressed from year to year. Add to this the differences that exist in high school courses with respect to the percentage of time devoted to each major topic and to the specific subtopics covered, and you may find that there are questions on topics with which you have little or no familiarity.

Each of the sample tests in this book is constructed to match closely the distribution of topics shown in the preceding chart so that you will gain a feel for the makeup of the actual test. After each test, a chart will show you which questions relate to each topic. This will be very helpful to you in planning your review because you can identify the areas on which you need to concentrate in your studies. Another chart enables you to see which chapters correspond to the various topic areas.

What Types of Questions Appear on the Test?

There are three general types of questions on the SAT II: Chemistry Test. They are matching questions, true/false and relationship questions, and multiple-choice questions. This section will discuss each type of question and give specific examples of how to answer each type. You should learn the directions for each type so that you will not waste time becoming familiar with them on the test day. The directions in this section are identical to those that are on the test.

Type 1: Matching Questions in Part A

In each of these questions, you are given five lettered choices to be used to answer all the questions in that set. The choices may be in the form of statements, pictures, graphs, experimental findings, equations, or specific situations. Answering a question may be as simple as recalling information or as difficult as analyzing the information given to establish what you need to do qualitatively or quantitatively to synthesize your answer. The directions for this type of question specifically state that a choice can be used once, more than once, or not at all in each set.

PART A

<u>Directions:</u> Each set of lettered choices below refers to the numbered statements or formulas immediately following it. Select the one lettered choice that best fits each statement or formula and then fill in the corresponding oval on the answer sheet. A choice may be used once, more than once, or not at all in each set.

<u>Example</u>

<u>Questions 1–3</u> refer to the following graphs:

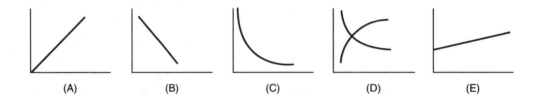

| (A) | (B) | (C) | (D) | (E) |

1. The graph that best shows the relationship of volume to temperature for an ideal gas while the pressure is held constant
2. The graph that best shows the relationship of volume to pressure for an ideal gas while the temperature is held constant
3. The graph that best shows the relationship of the number of grams of solute that is soluble in 100 grams of water at varying temperatures if the solubility begins as a small quantity and increases slowly as the temperature is increased.

These three questions require you to recall the basic gas laws and the graphic depiction of the relationship expressed in each law as well as how solubility can be shown graphically.

To answer the first question, you must recognize that the relationship between gas volume and changes in temperature is a direct relationship that is depicted by graphing Charles' Law, $V_1/T_1 = V_2/T_2$. The only graph that shows this type of direct relationship with the appropriate slope is (A).

To answer the second question you need to understand that Boyle's Law states that the pressure of a gas specimen is inversely proportional to its volume at constant temperature. Mathematically this means that pressure (P) times volume (V) is constant, or $P_1V_1 = P_2V_2$. This inversely proportional relationship is accurately depicted in (C). Although (B) shows the x axis increasing as the y axis decreases, it is not the graph for an inverse proportion.

The third question requires that you have knowledge about solubility curves and can translate the solubility relationship given in words to the graph shown as choice (E).

Type 2: True/False and Relationship Questions in Part B

On the actual Chemistry Test, this type of question must be answered in a special section of your answer sheet labeled "chemistry" at the lower left-hand corner of page 2. These questions will be numbered beginning with 101. They consist of a statement or assertion in column I and, on the other side of the word BECAUSE, another statement or assertion in column II. Your first task is to determine if each of the statements is true or false and record your answer for each respectively in the answer blocks for column I and column II by darkening either the Ⓣ oval or the Ⓕ oval. Then you must use your reasoning skills and your understanding of the topic to determine if there is a cause-and-effect relationship between the two statements.

Here are the directions and two examples of a relationship analysis question.

PART B

Directions: Each question below consists of two statements, I in the left-hand column and II in the right-hand column. For each question, determine whether statement I is true or false *and* if statement II is true or false and fill in the corresponding T or F ovals on your answer sheet. *Fill in oval CE only if statement II is a correct explanation of statement I.*

Sample Answer Grid:

CHEMISTRY* Fill in oval CE only if
II is correct explanation of I.

	I	II	CE*
101.	Ⓣ Ⓕ	Ⓣ Ⓕ	◯

Example 1

101. When 2 liters of oxygen gas reacts completely with 2 liters of hydrogen gas, the limiting factor is the volume of the oxygen   BECAUSE   the coefficients in the balanced equation of a gaseous reaction give the volume relationship of the reacting gases.

The reaction that takes place is

$$2H_2 + O_2 \rightarrow 2H_2O$$

The coefficients of this gaseous reaction show that 2 liters of hydrogen react with 1 liter of oxygen. This leaves 1 liter of unreacted oxygen. The limiting factor is the quantity of hydrogen.

Knowing how to solve this quantitative relationship shows that statement I is not true. However, statement II is a true statement of the relationship of coefficients in a balanced equation for a gaseous chemical reaction. So the answer blocks should be completed like this:

	I	II	CE*
101.	T (F)	T F	○

Example 2

101. Water is a good solvent of ionic and polar compounds

BECAUSE

the water molecule has polar properties caused by the factors involved in the bonding of the hydrogen and oxygen atoms.

Statement I is true because water is known to be such a good solvent that it is sometimes referred to as the universal solvent. This property is attributed mostly to its polar structure. Your knowledge of the polar covalent bond between the oxygen and hydrogen atoms and the angular orientation of the hydrogens with 105° between them contribute to the establishment of a permanent dipole moment. This feature also gives rise to a high degree of hydrogen bonding. These properties combine to make water a powerful solvent for both polar and ionic compounds. Knowing these concepts and how substances go into solution, you know that statement I is true and that the assertion in statement II explains statement I. There is a cause-and-effect relationship between the two statements. Therefore the answers should be marked like this:

	I	II	CE*
101.	(T) F	(T) F	●

Type 3: Multiple-Choice Questions in Part C

These five-choice questions are usually written as questions but are sometimes written as incomplete statements. You are given five suggested answers or completions. You must select the one that is best in each case and record your choice in the appropriate oval. In some questions you are asked to select the one inappropriate answer. Such questions contain a word in capital letters such as NOT, LEAST, or EXCEPT.

In some of these questions, you may be asked to make an association between a graphic, pictorial, or mathematical representation and a word explanation or problem. The solution may involve solving a scientific problem by interpreting these representations. In some cases the same representation may be used for a series of two or more questions. In no case, however, is the answer to one question necessary for answering a subsequent question correctly. Each question in the set is independent of the others.

PART C

Directions: Each of the questions or incomplete statements below is followed by five suggested answers or completions. Select the one that is best in each case and then fill in the corresponding oval on the answer sheet.

Example 1

40. In this graphic representation of a chemical reaction, which arrow depicts the **activation energy?**

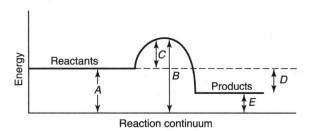

To answer this question you need to know how to interpret the energy levels in this graphic representation of energy level changes along the time continuum of the reaction. The activation energy is the minimum energy required for a chemical reaction to take place. The reactant molecules come together, and chemical bonds are stretched, broken, and formed in producing the products. During this process the energy of the system increases to a maximum and then decreases to the energy of the products. The activation energy is the difference between the maximum energy and the energy of the reactants. Choice (C) in the graphic depiction shows this energy barrier that has to be overcome for the reaction to proceed. This oval should be darkened.

Example 2

41. If the molecular mass of NH_3 is 17, what is the density of this compound at STP?
    (A) 0.25 g/L
    (B) 0.76 g/L
    (C) 1.25 g/L
    (D) 3.04 g/L
    (E) 9.11 g/L

The solution of this quantitative problem depends on the application of several principles. One principle is that the molar mass of a gas expressed in grams will occupy 22.4 liters at standard temperature and pressure (STP). The other is that the density of a gas at STP is the mass of one liter of the gas. Therefore 17 grams of ammonia occupy 22.4 liters, and 1 liter of this is 17 g/22.4 L, or 0.76 g/L. The correct mathematical answer is (B).

Example 3

Some questions in this part will give you a question followed by three or four bits of information labeled by Roman numerals I through III or IV. One or more of these statements may correctly answer the question. You must select from the five lettered choices the one that best answers the question.

45. Which bond(s) is(are) ionic?

    I. H—Cl(g)
   II. S—Cl(g)
  III. Cs—F(g)

  (A) I, II, and III
  (B) I and II only
  (C) II and III only
  (D) I only
  (E) III only

To determine the type of bonding that exists in these three substances, you must use your knowledge of ionic bonds and of how they are formed instead of other types of bonds. You must also use your knowledge of the relationship between the electronegativity of individual elements and the position of an element in the periodic table. The first two choices (I and II) are compounds formed between elements that do not differ enough in electronegativity to cause the formation of an ionic bond. This can be confirmed by checking the position of the elements in the periodic table and how electronegativity varies with respect to the position of the elements in the periodic table. Because cesium fluoride, choice III, consists of elements that appear in the lower right corner and the upper left corner of the periodic table, they have a sufficient difference in electronegativity values that an ionic bond can be predicted between them. Of the choices given, only (E) would be a correct answer.

CHEMISTRY TEST

Material in the following table may be useful in answering the questions in this examination.

Periodic Table of the Elements

Use this periodic table when taking a practice test

1.0080 H 1																		4.003 He 2
6.939 Li 3	9.012 Be 4											10.811 B 5	12.01115 C 6	14.007 N 7	15.999 O 8	18.998 F 9	20.183 Ne 10
22.990 Na 11	24.312 Mg 12											26.981 Al 13	28.086 Si 14	30.974 P 15	32.064 S 16	35.453 Cl 17	39.948 Ar 18
39.102 K 19	40.08 Ca 20	44.956 Sc 21	47.90 Ti 22	50.942 V 23	51.996 Cr 24	54.938 Mn 25	55.847 Fe 26	58.933 Co 27	58.71 Ni 28	63.54 Cu 29	65.37 Zn 30	69.72 Ga 31	72.59 Ge 32	74.922 As 33	78.96 Se 34	79.909 Br 35	83.80 Kr 36
85.47 Rb 37	87.62 Sr 38	88.905 Y 39	91.22 Zr 40	92.906 Nb 41	95.94 Mo 42	(99) Tc 43	101.07 Ru 44	102.91 Rh 45	106.4 Pd 46	107.87 Ag 47	112.40 Cd 48	114.82 In 49	118.69 Sn 50	121.75 Sb 51	127.60 Te 52	126.90 I 53	131.30 Xe 54
132.90 Cs 55	137.34 Ba 56	57-71	178.49 Hf 72	180.95 Ta 73	183.85 W 74	186.21 Re 75	190.2 Os 76	192.2 Ir 77	195.09 Pt 78	196.97 Au 79	200.59 Hg 80	204.37 Tl 81	207.19 Pb 82	208.98 Bi 83	(210) Po 84	(210) At 85	(222) Rn 86
223 Fr 87	(226) Ra 88	89-103	Unq 104	Unp 105	Unh 106	Uns 107	Uno 108	Une 109	110	111	112	113	114	115	116	117	118

138.91 La 57	140.12 Ce 58	140.91 Pr 59	144.24 Nd 60	(147) Pm 61	150.35 Sm 62	151.96 Eu 63	157.25 Gd 64	158.92 Tb 65	162.50 Dy 66	164.93 Ho 67	167.26 Er 68	168.93 Tm 69	173.04 Yb 70	174.97 Lu 71
227 Ac 89	232.04 Th 90	(231) Pa 91	238.03 U 92	(237) Np 93	(242) Pu 94	(243) Am 95	(247) Cm 96	(249) Bk 97	(251) Cf 98	(254) Es 99	(253) Fm 100	(256) Md 101	(254) No 102	(257) Lr 103

Answer Sheet for Practice Test 1

Before you attempt any of the Practice Tests, read the information in the Introduction section. When you understand the material there and are aware of the types of questions on the test and their respective instructions, you are ready to take a Practice Test.

Remember that you have one hour and that you may not use a calculator. Use the periodic table given in front of this test and record your answers in the appropriate spaces on the answer sheet.

Determine the correct answer for each question. Then, using a No. 2 pencil, blacken completely the oval containing the letter of your choice. (This answer sheet is modeled on the SAT II: Chemistry test answer sheet.)

1. Ⓐ Ⓑ Ⓒ Ⓓ Ⓔ
2. Ⓐ Ⓑ Ⓒ Ⓓ Ⓔ
3. Ⓐ Ⓑ Ⓒ Ⓓ Ⓔ
4. Ⓐ Ⓑ Ⓒ Ⓓ Ⓔ
5. Ⓐ Ⓑ Ⓒ Ⓓ Ⓔ
6. Ⓐ Ⓑ Ⓒ Ⓓ Ⓔ
7. Ⓐ Ⓑ Ⓒ Ⓓ Ⓔ
8. Ⓐ Ⓑ Ⓒ Ⓓ Ⓔ
9. Ⓐ Ⓑ Ⓒ Ⓓ Ⓔ
10. Ⓐ Ⓑ Ⓒ Ⓓ Ⓔ
11. Ⓐ Ⓑ Ⓒ Ⓓ Ⓔ
12. Ⓐ Ⓑ Ⓒ Ⓓ Ⓔ
13. Ⓐ Ⓑ Ⓒ Ⓓ Ⓔ
14. Ⓐ Ⓑ Ⓒ Ⓓ Ⓔ
15. Ⓐ Ⓑ Ⓒ Ⓓ Ⓔ
16. Ⓐ Ⓑ Ⓒ Ⓓ Ⓔ
17. Ⓐ Ⓑ Ⓒ Ⓓ Ⓔ
18. Ⓐ Ⓑ Ⓒ Ⓓ Ⓔ
19. Ⓐ Ⓑ Ⓒ Ⓓ Ⓔ
20. Ⓐ Ⓑ Ⓒ Ⓓ Ⓔ
21. Ⓐ Ⓑ Ⓒ Ⓓ Ⓔ
22. Ⓐ Ⓑ Ⓒ Ⓓ Ⓔ
23. Ⓐ Ⓑ Ⓒ Ⓓ Ⓔ

ON THE ACTUAL CHEMISTRY TEST, THE FOLLOWING TYPE OF QUESTION MUST BE ANSWERED ON A SPECIAL SECTION (LABELED "CHEMISTRY") AT THE LOWER LEFT-HAND CORNER OF PAGE 2 OF YOUR ANSWER SHEET. THESE QUESTIONS WILL BE NUMBERED BEGINNING WITH 101 AND MUST BE ANSWERED ACCORDING TO THE FOLLOWING DIRECTIONS.

CHEMISTRY* Fill in oval CE only if II is a correct explanation of I.

	I	II	CE*
101.	Ⓣ Ⓕ	Ⓣ Ⓕ	◯
102.	Ⓣ Ⓕ	Ⓣ Ⓕ	◯
103.	Ⓣ Ⓕ	Ⓣ Ⓕ	◯
104.	Ⓣ Ⓕ	Ⓣ Ⓕ	◯
105.	Ⓣ Ⓕ	Ⓣ Ⓕ	◯
106.	Ⓣ Ⓕ	Ⓣ Ⓕ	◯
107.	Ⓣ Ⓕ	Ⓣ Ⓕ	◯
108.	Ⓣ Ⓕ	Ⓣ Ⓕ	◯
109.	Ⓣ Ⓕ	Ⓣ Ⓕ	◯
110.	Ⓣ Ⓕ	Ⓣ Ⓕ	◯
111.	Ⓣ Ⓕ	Ⓣ Ⓕ	◯
112.	Ⓣ Ⓕ	Ⓣ Ⓕ	◯
113.	Ⓣ Ⓕ	Ⓣ Ⓕ	◯
114.	Ⓣ Ⓕ	Ⓣ Ⓕ	◯
115.	Ⓣ Ⓕ	Ⓣ Ⓕ	◯
116.	Ⓣ Ⓕ	Ⓣ Ⓕ	◯

ON THE ACTUAL CHEMISTRY TEST, THE REMAINING QUESTIONS MUST BE ANSWERED BY RETURNING TO THE SECTION OF YOUR ANSWER SHEET YOU STARTED FOR CHEMISTRY.

40. Ⓐ Ⓑ Ⓒ Ⓓ Ⓔ	56. Ⓐ Ⓑ Ⓒ Ⓓ Ⓔ	71. Ⓐ Ⓑ Ⓒ Ⓓ Ⓔ
41. Ⓐ Ⓑ Ⓒ Ⓓ Ⓔ	57. Ⓐ Ⓑ Ⓒ Ⓓ Ⓔ	72. Ⓐ Ⓑ Ⓒ Ⓓ Ⓔ
42. Ⓐ Ⓑ Ⓒ Ⓓ Ⓔ	58. Ⓐ Ⓑ Ⓒ Ⓓ Ⓔ	73. Ⓐ Ⓑ Ⓒ Ⓓ Ⓔ
43. Ⓐ Ⓑ Ⓒ Ⓓ Ⓔ	59. Ⓐ Ⓑ Ⓒ Ⓓ Ⓔ	74. Ⓐ Ⓑ Ⓒ Ⓓ Ⓔ
44. Ⓐ Ⓑ Ⓒ Ⓓ Ⓔ	60. Ⓐ Ⓑ Ⓒ Ⓓ Ⓔ	75. Ⓐ Ⓑ Ⓒ Ⓓ Ⓔ
45. Ⓐ Ⓑ Ⓒ Ⓓ Ⓔ	61. Ⓐ Ⓑ Ⓒ Ⓓ Ⓔ	76. Ⓐ Ⓑ Ⓒ Ⓓ Ⓔ
46. Ⓐ Ⓑ Ⓒ Ⓓ Ⓔ	62. Ⓐ Ⓑ Ⓒ Ⓓ Ⓔ	77. Ⓐ Ⓑ Ⓒ Ⓓ Ⓔ
47. Ⓐ Ⓑ Ⓒ Ⓓ Ⓔ	63. Ⓐ Ⓑ Ⓒ Ⓓ Ⓔ	78. Ⓐ Ⓑ Ⓒ Ⓓ Ⓔ
48. Ⓐ Ⓑ Ⓒ Ⓓ Ⓔ	64. Ⓐ Ⓑ Ⓒ Ⓓ Ⓔ	79. Ⓐ Ⓑ Ⓒ Ⓓ Ⓔ
49. Ⓐ Ⓑ Ⓒ Ⓓ Ⓔ	65. Ⓐ Ⓑ Ⓒ Ⓓ Ⓔ	80. Ⓐ Ⓑ Ⓒ Ⓓ Ⓔ
50. Ⓐ Ⓑ Ⓒ Ⓓ Ⓔ	66. Ⓐ Ⓑ Ⓒ Ⓓ Ⓔ	81. Ⓐ Ⓑ Ⓒ Ⓓ Ⓔ
51. Ⓐ Ⓑ Ⓒ Ⓓ Ⓔ	67. Ⓐ Ⓑ Ⓒ Ⓓ Ⓔ	82. Ⓐ Ⓑ Ⓒ Ⓓ Ⓔ
52. Ⓐ Ⓑ Ⓒ Ⓓ Ⓔ	68. Ⓐ Ⓑ Ⓒ Ⓓ Ⓔ	83. Ⓐ Ⓑ Ⓒ Ⓓ Ⓔ
53. Ⓐ Ⓑ Ⓒ Ⓓ Ⓔ	69. Ⓐ Ⓑ Ⓒ Ⓓ Ⓔ	84. Ⓐ Ⓑ Ⓒ Ⓓ Ⓔ
54. Ⓐ Ⓑ Ⓒ Ⓓ Ⓔ	70. Ⓐ Ⓑ Ⓒ Ⓓ Ⓔ	85. Ⓐ Ⓑ Ⓒ Ⓓ Ⓔ
55. Ⓐ Ⓑ Ⓒ Ⓓ Ⓔ		

PRACTICE TEST 1

Note: For all questions involving solutions and/or chemical equations, assume that the system is in water unless otherwise stated.

PART A

Directions: Each set of lettered choices refers to the numbered statements or formulas immediately following it. Select the one lettered choice that best fits each statement or formula and then fill in the corresponding oval on the answer sheet. A choice may be used once, more than once, or not at all in each set.

Questions 1–9

PERIODIC TABLE (ABBREVIATED)

^3Li				(D)	^{10}Ne
	(A)		(C)		
(B)	^{20}Ca				(E)

1. The most electronegative element

2. The element with a possible oxidation number of –2

3. The element that would react in a one-to-one ratio with (D)

4. The element with the smallest ionic radius

5. The element with the smallest first ionization potential

6. The element with a complete p orbital as its outermost energy level

7. A member of the alkali-metals family

8. A noble gas

9. The element that would react most actively when placed in water to form a strong base

Questions 10–12 refer to the following heating curve for water:

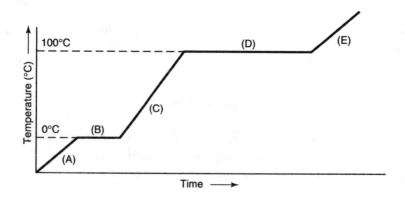

10. In which part of the curve is the state of H_2O only a solid?

11. When is the heat to change the state of H_2O greater?

12. Where is the temperature of H_2O changing at 1°C/cal/g?

Questions 13–15 refer to the following diagram:

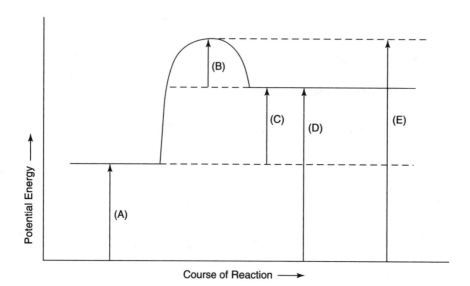

13. Indicates the forward activation energy of this reaction

14. Indicates the portion of the curve that would be directly affected by the addition of a catalyst

15. Indicates the difference between the activation energies for the reverse and forward reactions and equals the energy change in the reaction.

Questions 16–18

    (A) Iron
    (B) Gold
    (C) Sodium
    (D) Argon
    (E) Uranium

16. A metal element that resists reaction with acids

17. A monoatomic element with filled p orbitals

18. A transition element that occurs when the inner $3d$ orbital is partially filled

Questions 19–20

    (A) Rhombic sulfur
    (B) Monoclinic sulfur
    (C) Sulfur trioxide
    (D) Sulfate
    (E) Sulfite

19. A substance that exhibits a resonance structure

20. A product formed from a base reacting with sulfurous acid

Questions 21–23

    (A) Dilute
    (B) Concentrated
    (C) Unsaturated
    (D) Saturated
    (E) Supersaturated

21. The condition, unrelated to quantities, that indicates that the rate going into solution is equal to the rate coming out of solution

22. The condition that exists when a water solution that has been at equilibrium and is saturated is heated to a higher temperature with a higher solubility but no additional solute is added

23. The descriptive term that indicates there is a large quantity of solute, compared to the amount of solvent, in a solution

PART B

ON THE ACTUAL CHEMISTRY TEST, THE FOLLOWING TYPE OF QUESTION MUST BE ANSWERED ON A SPECIAL SECTION (LABELED "CHEMISTRY") AT THE LOWER LEFT-HAND CORNER OF PAGE 2 OF YOUR ANSWER SHEET. THESE QUESTIONS WILL BE NUMBERED BEGINNING WITH 101 AND MUST BE ANSWERED ACCORDING TO THE FOLLOWING DIRECTIONS.

<u>Directions</u>: Each question below consists of two statements, I in the left-hand column and II in the right-hand column. For each question, determine whether statement I is true or false <u>and</u> if statement II is true or false and fill in the corresponding T or F ovals on your answer sheet. <u>Fill in oval CE only if statement II is a correct explanation of statement I.</u>

Sample Answer Grid:

CHEMISTRY * Fill in oval CE only if
II is a correct explanation of I.

	I	II	CE*
101.	T F	T F	◯

101. Nonmetallic oxides are usually acid anhydrides   BECAUSE   nonmetallic oxides form acids when placed in water.

102. When HCl gas and NH$_3$ gas come into contact, a white smoke forms   BECAUSE   the NH$_3$ and HCl react to form a white solid, ammonium chlorate.

103. The reaction of barium chloride and sodium sulfate does not go to completion   BECAUSE   the compound barium sulfate is formed as an insoluble precipitate

104. When two elements react exothermically to form a compound, the compound should be relatively stable   BECAUSE   the release of energy from a combustion reaction indicates that the compound formed is at a lower energy level than the reactants and thus relatively stable.

105. The ion of a nonmetallic atom is larger in radius than the atom   BECAUSE   when a nonmetallic ion is formed, it gains electrons in the outer orbital and thus increases the size of the electron cloud around the nucleus.

106. Oxidation and reduction occur together   BECAUSE   in redox reactions electrons must be gained and lost.

107. Decreasing the atmospheric pressure on a pot of boiling water causes it to stop boiling

BECAUSE changes in pressure are directly related to the boiling point of water.

108. The reaction of hydrogen with oxygen to form water is an exothermic reaction

BECAUSE water molecules have polar covalent bonds.

109. Atoms of different elements can have the same mass number

BECAUSE the atoms of each element have a characteristic number of protons in the nucleus.

110. The transmutation decay of $^{238}_{92}$U can be shown as $^{238}_{92}$U $\rightarrow$ $^{234}_{90}$Th + $^{4}_{2}$He

BECAUSE the transmutation of $^{238}_{92}$U is accompanied by the release of a beta particle.

111. $^{13}_{12}$C and $^{12}_{12}$C are isotopes of the element carbon

BECAUSE isotopes of an element have the same number of protons in the nucleus.

112. The Cu^{2+} ion needs to be oxidized to form Cu metal

BECAUSE oxidation is a gain of electrons.

113. The volume of a gas at 1000°C and 600 mm Hg pressure will be decreased at STP

BECAUSE decreasing temperature and increasing pressure will cause the volume to decrease.

114. The pH of a 0.01-*M* solution of HCl will be 2

BECAUSE dilute HCl dissociates into two essentially ionic particles.

115. Nuclear fusion on the sun converts hydrogen to helium with a release of energy

BECAUSE some mass is converted to energy in a solar fusion.

116. The "bullet" usually used to initiate the fusion of ^{235}U is a neutron

BECAUSE capture of the neutron by the ^{235}U nucleus causes an unstable condition that leads to its disintegration.

PART C

Directions: Each of the questions or incomplete statements below is followed by five suggested answers or completions. Select the one that is best in each case and then fill in the corresponding oval on the answer sheet.

40. What is the approximate formula mass of $Ca(NO_3)_2$?
    (A) 70
    (B) 82
    (C) 102
    (D) 150
    (E) 164

41. In this reaction: $XClO_3 + A \rightarrow XCl + O_2(g) + A$, which substance is the catalyst?
    (A) X
    (B) $XClO_3$
    (C) A
    (D) XCl
    (E) O_2

42. The normal electron configuration for ethyne (acetylene) is
    (A) H:C::C:H
    (B) H:C̈:C̈:H
    (C) H·C:::C·H
    (D) H:C: :C:H
    (E) H:Ċ:Ċ:H

43. According to the Kinetic Molecular Theory, molecules increase in kinetic energy when they
    (A) are mixed with other molecules at lower temperature
    (B) are frozen into a solid
    (C) are condensed into a liquid
    (D) are heated to a higher temperature
    (E) collide with each other in a container at lower temperature

44. How many atoms are represented in the formula $Ca_3(PO_4)_2$?
    (A) 5
    (B) 8
    (C) 9
    (D) 12
    (E) 13

45. All of the following have covalent bonds EXCEPT
    (A) HCl
    (B) CCl_4
    (C) H_2O
    (D) CsF
    (E) CO_2

46. Which of the following is (are) the weakest attractive force?
    (A) van der Waals forces
    (B) coordinate covalent bonding
    (C) covalent bonding
    (D) polar covalent bonding
    (E) ionic bonding

47. Which of these resembles the molecular structure of the water molecule?

(A)

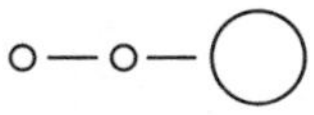

(B)

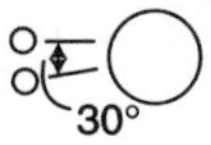

(C)

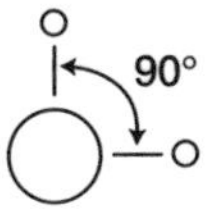

(D)

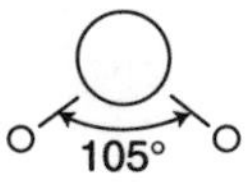

(E)

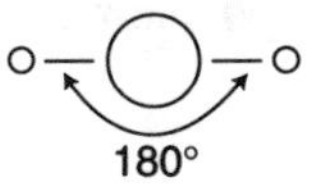

48. The two most important considerations in deciding whether a reaction will occur spontaneously are
(A) the stability and state of the reactants
(B) the energy gained and the heat evolved
(C) a negative value for enthalpy and a positive value for entropy
(D) a positive value for enthalpy and a negative value for entropy
(E) the endothermic energy and the structure of the products

49. The reaction of an acid like HCl and a base like NaOH always
(A) forms a precipitate
(B) forms a volatile product
(C) forms an insoluble salt and water
(D) forms a sulfate salt and water
(E) forms a salt and water

50. The oxidation number of sulfur in H_2SO_4 is
(A) +2
(B) +3
(C) +4
(D) +6
(E) +8

51. Balance this reaction by using the oxidation-reduction method of electron exchange:

$$KMnO_4 + H_2SO_3 \rightarrow K_2SO_4 + MnSO_4 + H_2SO_4 + H_2O$$

Which of the following partial equations is the correct reduction half-reaction for the *balanced* equation?
(A) $5SO_3^{2-} + 5H_2O \rightarrow 5SO_4^{2-} + 2H^+ + 10e^-$
(B) $2MnO_4^- + 16H^+ + 10e^- \rightarrow 2Mn^{2+} + 8H_2O$
(C) $SO_3^{2-} \rightarrow SO_4^{2-} + 2e^-$
(D) $SO_3^{2-} + 2H^+ \rightarrow SO_4^{2-} + H_2O + 2e^-$
(E) $Mn^{7+} \rightarrow Mn^{2+} + 5e^-$

52. Which of the following, when placed into water, will test as an acid solution?

    I. $HCl(g) + H_2O$
   II. Excess $H_3O^+ + H_2O$
  III. $CuSO_4(s) + H_2O$

(A) I only
(B) III only
(C) I and II only
(D) II and III only
(E) I, II, and III

53. The property of matter that is independent of its surrounding conditions and position is
(A) volume
(B) density
(C) mass
(D) weight
(E) state

54. Where are the highest ionization energies found in the periodic table?
(A) upper left corner
(B) lower left corner
(C) upper right corner
(D) lower right corner
(E) middle of transition elements

55. Which of the following pairs of compounds can be used to illustrate the Law of Multiple Proportions?
(A) NO and NO_2
(B) CH_4 and CO_2
(C) ZnO_2 and $ZnCl_2$
(D) NH_4 and NH_4Cl
(E) H_2O and HCl

56. In the equilibrium reaction $A + B \rightleftharpoons AB + heat$ (in a closed container), how could the forward reaction rate be increased?

    I. By increasing the concentration of AB
   II. By increasing the concentration of A
  III. By removing some of product AB

(A) I only
(B) III only
(C) I and III only
(D) II and III only
(E) I, II, and III

57. In the reaction of sodium with water, the balanced equation has which of the following coefficients?
   I. 1
   II. 2
   III. 3

   (A) I only
   (B) III only
   (C) I and II only
   (D) II and III only
   (E) I, II, and III

58. If 10 L of CO gas react with sufficient oxygen to completely react, how many liters of CO_2 gas are formed?
   (A) 5
   (B) 10
   (C) 15
   (D) 20
   (E) 40

59. If 49 g of H_2SO_4 react with 80 g of NaOH, how much reactant will be left over after the reaction is complete?
   (A) 24.5 g H_2SO_4
   (B) none of either compound
   (C) 20 g NaOH
   (D) 40 g NaOH
   (E) 60 g NaOH

60. If the molar concentration of Ag^+ ions in 1 liter of a saturated water solution of silver chloride is 1.38×10^{-5} mol/L, what is the K_{sp} of this solution?
   (A) 0.34×10^{-10}
   (B) 0.69×10^{-5}
   (C) 1.90×10^{-10}
   (D) 2.76×10^{-5}
   (E) 2.76×10^{-10}

61. If the density of a diatomic gas is 1.43 g/L, what is its molar mass?
   (A) 16 g
   (B) 32 g
   (C) 48 g
   (D) 64 g
   (E) 14.3 g

62. From 2 mol of $KClO_3$ how many liters of O_2 can be produced by decomposition of all the $KClO_3$?
   (A) 11.2
   (B) 22.4
   (C) 33.6
   (D) 44.8
   (E) 67.2

63. Which value best determines whether a reaction is spontaneous?
    (A) change in Gibbs free energy, ΔG
    (B) change in entropy, ΔS
    (C) change in kinetic energy, ΔKE
    (D) change in enthalpy, ΔH
    (E) change in heat of formation, ΔH^0

Questions 64–68 refer to the following experimental setup and data:

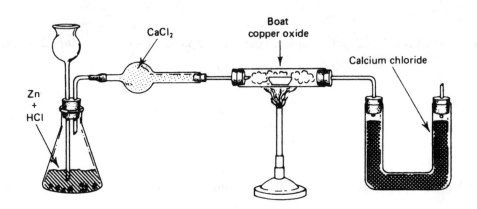

Recorded data:

Before: CuO + porcelain boat = 62.869 g
    $CaCl_2$ + U-tube = 80.483 g

After: Porcelain boat + contents = 54.869 g
   $CaCl_2$ + U-tube   = 89.483 g

Reactions: $Zn + 2HCl \rightarrow ZnCl_2 + H_2$
   then $H_2 + CuO \rightarrow H_2O + Cu$

64. What type of reaction occurred in the porcelain boat?
    (A) electrolysis
    (B) double displacement
    (C) reduction and oxidation
    (D) decomposition or analysis
    (E) combination or synthesis

65. Why was the $CaCl_2$ tube placed between the generator and the tube containing the porcelain boat?
    (A) to absorb evaporated HCl
    (B) to absorb evaporated H_2O
    (C) to slow down the gases released
    (D) to absorb the evaporated Zn particles
    (E) to remove the initial air that passes through the tube

66. How many grams of hydrogen were used in the formation of the water that was a product?
    (A) 1
    (B) 2
    (C) 4
    (D) 8
    (E) 9

67. What conclusion can you draw from this experiment?
    (A) Hydrogen diffuses faster than oxygen.
    (B) Hydrogen is lighter than oxygen.
    (C) The molar mass of oxygen is 32 g/mol.
    (D) Water is a triatomic molecule with polar characteristics.
    (E) Water is formed from hydrogen and oxygen in a ratio of 1 : 8 by weight.

68. Which of the following is an observation rather than a conclusion?
    (A) A substance is an acid if it changes litmus paper from blue to red.
    (B) A gas is lighter than air if it escapes from a bottle left mouth upward.
    (C) The gas H_2 forms an explosive mixture with air.
    (D) Air is mixed with hydrogen gas and ignited; it explodes.
    (E) An oil liquid is immiscible with water because it separates into a layer above the water.

69. When HCl fumes and NH_3 fumes are introduced into opposite ends of a long, dry glass tube, a white ring forms in the tube. Which answer explains this phenomenon?
    (A) NH_4Cl forms.
    (B) HCl diffuses faster.
    (C) NH_3 diffuses faster.
    (D) The ring occurs closer to the end into which HCl was introduced.
    (E) The ring occurs in the middle of the tube.

70. The correct formula for calcium hydrogen sulfate is
    (A) CaH_2SO_4
    (B) $CaHSO_4$
    (C) $Ca(HSO_4)_2$
    (D) Ca_2HSO_4
    (E) $Ca_2H_2SO_4$

71. Ten grams of sodium hydroxide is dissolved in enough water to make 1 L of solution. What is the molarity of the solution?
    (A) 0.25 M
    (B) 0.5 M
    (C) 1 M
    (D) 1.5 M
    (E) 4 M

72. For a saturated solution of salt in water, which statement is true?
    (A) All dissolving has stopped.
    (B) Crystals begin to grow.
    (C) An equilibrium has been established.
    (D) Crystals of the solute will continue to dissolve.
    (E) The solute is exceeding its solubility.

73. In which of the following series is the pi bond present in the bonding structure?

    I. Alkane
    II. Alkene
    III. Alkyne

    (A) I only
    (B) III only
    (C) I and III only
    (D) II and III only
    (E) I, II, and III

Questions 74–76 refer to the following setup:

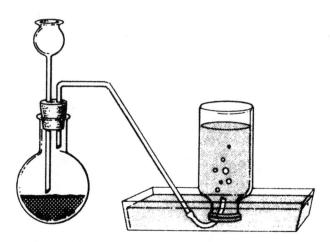

74. Why could you NOT use this setup for preparing H_2 if the generator contained Zn and vinegar?
    (A) Hydrogen would not be produced.
    (B) The setup of the generator is improper.
    (C) The generator must be heated with a burner.
    (D) The delivery tube setup is wrong.
    (E) The gas cannot be collected over water.

75. In a proper laboratory setup for collecting a gas by water displacement, which of these gases could NOT be collected over H_2O because of its solubility?
   (A) CO_2
   (B) NO
   (C) O_2
   (D) NH_3
   (E) CH_4

76. What is the approximate percentage of oxygen in the formula mass of $Ca(NO_3)_2$?
   (A) 28
   (B) 42
   (C) 58
   (D) 96
   (E) 164

77. For the reaction $N_2O_4(g) \rightleftharpoons 2NO_2(g)$, the K_{eq} expression is

   (A) $K_{eq} = \dfrac{[N_2O_4]}{[NO_2]}$

   (B) $K_{eq} = \dfrac{[N_2O_4]}{[NO_2]^2}$

   (C) $K_{eq} = \dfrac{[NO_2]}{[N_2O_4]}$

   (D) $K_{eq} = \dfrac{[NO_2]^2}{[N_2O_4]}$

   (E) $K_{eq} = \dfrac{[N_2O_4]^2}{[NO_2]}$

78. What is the K_{eq} for the above reaction if at equilibrium the concentration of N_2O_4 is 4×10^{-2} mol/L and that of NO_2 is 2×10^{-2} mol/L?
   (A) 1×10^{-2}
   (B) 2×10^{-2}
   (C) 2×10^{-2}
   (D) 4×10^{-4}
   (E) 8×10^{-2}

79. How much water, in liters, must be added to 0.5 L of 6 M HCl to make it 2 M?
   (A) 0.33
   (B) 0.5
   (C) 1
   (D) 1.5
   (E) 2

80. Four grams of hydrogen are ignited with 4 g of oxygen. How many grams of water can be formed?
    (A) 0.5
    (B) 2.5
    (C) 4.5
    (D) 8
    (E) 36

81. Which structure is an ester?

(A)
```
      H       H
      |       |
  H—C—O—C—H
      |       |
      H       H
```

(B)
```
      H  H  H
      |  |  |
  H—C—C—C—OH
      |  |  |
      H  H  H
```

(C)
```
      H  H      O
      |  |     //
  H—C—C—C
      |  |     \
      H  H      O—H
```

(D)
```
      H  H      O
      |  |     //
  H—C—C—C
      |  |     \
      H  H      H
```

(E)
```
      H  H  O       H
      |  |  ||       |
  H—C—C—C—O—C—H
      |  |           |
      H  H           H
```

82. What piece of apparatus can be used to introduce more liquid into a reaction and also serve as a pressure valve?
    (A) stopcock
    (B) pinchcock
    (C) thistle tube
    (D) flask
    (E) condenser

83. Which formulas could represent the empirical formula and the molecular formula of a given compound?
    (A) CH_2O and $C_4H_6O_4$
    (B) CHO and $C_6H_{12}O_6$
    (C) CH_4 and C_5H_{12}
    (D) CH_2 and C_3H_6
    (E) CO and CO_2

84. The reaction Fe $\rightarrow$ Fe^{2+} + 2e$^-$ (+0.44 volt) would occur spontaneously with which of the following?

    I. Pb $\rightarrow$ Pb^{2+} + 2e$^-$ (+0.13 volt)
    II. Cu $\rightarrow$ Cu^{2+} + 2e$^-$ (−0.34 volt)
    III. 2Ag + 2e$^-$ $\rightarrow$ 2Ag^0 (+0.80 volt)

    (A) I only
    (B) III only
    (C) I and III only
    (D) II and III only
    (E) I, II, and III

85. 2Na(s) + Cl$_2$(g) $\rightarrow$ 2NaCl(s) + 822 kJ
    How much heat is released by the above reaction if 0.5 mole of sodium reacts completely with chlorine?
    (A) 205 kJ
    (B) 411 kJ
    (C) 822 kJ
    (D) 1644 kJ
    (E) 3288 kJ

STOP
IF YOU FINISH BEFORE ONE HOUR IS UP, YOU MAY GO BACK
TO CHECK YOUR WORK OR COMPLETE UNANSWERED QUESTIONS.

Answers and Explanations for Test 1

1. **(D)** The most electronegative element (F) would be found in the upper right corner; the noble gases are exceptions at the far right.

2. **(C)** Elements in the group with (C) have a possible oxidation number of –2.

3. **(B)** Elements in the group with (B) react in a 1 : 1 ratio with elements in the group with (D) since one has an electron to lose and the other one needs an electron to complete its outer energy level.

4. **(A)** (A) loses two electrons to form an ion whose remaining electrons, being close to the nucleus, are pulled in closer because of the unbalanced 2+ charge.

5. **(B)** Since (B) has only one electron in the outer 4s orbital, it can more easily be removed than can an electron from the 3s orbital of (A), which is closer to the positive nucleus.

6. **(E)** All Group VIII elements have a complete p orbital as the outer energy level. This explains why these elements are "inert."

7. **(B)** The Group IA elements comprise the alkali-metals family.

8. **(E)** The Group VIII elements are the inert or noble gases.

9. **(B)** The alkali metals react with water to form a strong base.

10. **(A)** H_2O is ice in part A.

11. **(D)** H_2O changes state at parts B and D. The heat of vaporization at D (540 cal/g) is greater than the heat of fusion (80 cal/g) at B.

12. **(C)** Water is heating at 1°C/cal/g in part C.

13. **(B)** This is the energy needed to start the reaction.

14. **(B)** The addition of a catalyst affects only the part indicated by B, by either raising or lowering the energy needed to commence the reaction.

15. **(C)** The net energy released is the endothermic quantity indicated by C.

16. **(B)** Gold is known as a noble metal because of its resistance to acids. Aqua regia, a mixture of HNO_3 and HCl, will react with gold.

17. **(D)** Helium is the only element that is monoatomic in the molecular form.

18. **(A)** Iron has five electrons in the d orbitals, which are partially filled.

19. **(C)** Only sulfur trioxide has a resonance structure (as shown here):

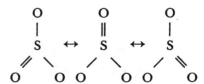

20. **(E)** Sulfurous acid reacts with a base to form a sulfite salt.

21. **(D)** The condition described is the equilibrium that exists at saturation.

22. **(C)** With the increased temperature more solute may go into solution; therefore the solution is now unsaturated.

23. **(B)** The term *concentrated* means that there is a large amount of solute in the solvent.

101. **(T, T, CE)** Nonmetallic oxides are usually acid anhydrides, and they form acids in water.

102. **(T, F)** The white smoke formed is ammonium chloride, not ammonium chlorate.

103. **(F, T)** The reaction does go to completion since barium sulfate is a precipitate.

104. (**T, T, CE**) The product of an exothermic reaction is relatively stable because it is at a lower energy level than the reactants.

105. (**T, T, CE**) Both statements are true, and the reason explains the assertion.

106. (**T, T, CE**) Both the statements are true, and the reason explains the assertion.

107. (**F, T**) The statement is false while the reason is true. Decreasing the pressure on a boiling pot will only cause it to boil more vigorously.

108. (**T, T**) The statements are true, but the reason doesn't explain the assertion.

109. (**T, T**) The statements are true, but the reason doesn't explain the assertion.

110. (**T, F**) The transmutation equation is true and shows the release of an alpha particle, not a beta particle.

111. (**T, T, CE**) The two configurations of carbon are isotopes because they have the same atomic number, which means that they have the same number of protons in the nucleus.

112. (**F, F**) Both the statement and the reason are false.

113. (**T, T, CE**) To go from 100°C to 0°C decreases the volume as the gas gets colder; therefore, the temperature fraction expressed in kelvins must be $\frac{273}{373}$ to decrease the volume. To go from 600 mm to 760 mm of pressure increases the pressure, thus causing the gas to contract. The fraction must then be $\frac{600}{760}$ to cause the volume to decrease. You could use the formula

$$V_2 = V_1 \times \frac{T_2}{T_1} \times \frac{P_1}{P_2}$$

114. (**T, T**) Since HCl is a strong acid and ionizes completely in a dilute solution of water, the $[H^+]$ (H_3O^+ is the same thing) or molar concentration of a 0.01-M concentration is 1×10^{-2} mol/L.

$$pH = -\log[H^+]$$
$$pH = -\log[1 \times 10^{-2}]$$
$$pH = -(-2) = 2$$

The pH is 2, but the reason, although true, does not explain the statement.

115. (**T, T, CE**) Both statements are true, and the reason explains the assertion.

116. (**F, T**) The statement should refer to fission, not fusion. The reason is true.

40. (**E**) The total formula mass is

$$\begin{aligned} Ca &= 40 \\ 2N &= 28 \\ 6O &= \underline{96} \\ \text{Total} & 164 \end{aligned}$$

41. (**C**) The catalyst, by definition, is not consumed in the reaction and ends up in its original form as one of the products.

42. (**D**) The ethyne molecule is the first member of the acetylene series with a general formula of C_nH_{2n-2}. It contains a triple bond between the two C atoms: H:C::C:H.

43. (**D**) Heating molecules increases their kinetic energy.

44. (**E**) 3Ca + 2P + 8O = 13 atoms.

45. (**D**) Cesium and fluorine are from the most electropositive and electronegative portions, respectively, of the periodic chart and thus form an ionic bond by cesium giving an electron to fluorine to form the respective ions.

46. (**A**) Van der Waals forces are the weak attraction of the nuclear positive charge of one atom to the negative electron field of an adjacent atom. They are much weaker than the others named.

47. (**D**) The molecular structure of water is a polar covalent compound with the hydrogens 105° apart.

48. **(C)** The most important considerations for a spontaneous reaction are (1) that the reaction is exothermic with a negative enthalpy so that once started it tends to continue on its own because of the energy released and (2) that reactions tend to go to the highest state of randomness shown by a positive entropy value.

49. **(E)** Normal H^+ acids and OH^- bases form water and a salt—not necessarily a soluble salt.

50. **(D)** All compounds have a charge of 0. H usually has +1, and O usually has –2, so

$$\begin{array}{ll} H_2 & = +2 \\ O_4 & = -8 \\ \underline{S} & = \ x \\ \text{Total} = & 0 \end{array}$$

$$+2 - 8 + x = 0$$
$$x = -2 + 8 = +6$$

51. **(B)** In the balanced equation the two half reactions are

$$5SO_3^{2-} + 5H_2O \rightarrow 5SO_4^{2-} + 10H^+ + 10e^-$$
$$\underline{2MnO_4^- + 16H^+ + 10e^- \rightarrow 2Mn^{2+} + 8H_2O}$$
$$2KMnO_4 + 5H_2SO_3 \rightarrow$$
$$K_2SO_4 + 2MnSO_4 + 2H_2SO_4 + 3H_2O$$

52. **(E)** HCl and H_3O^+ give acid solutions, as does $CuSO_4$, when it hydrolyzes in water. I, II, and III are correct.

53. **(C)** Mass is a constant and is not dependent on position or surrounding conditions.

54. **(C)** The complete outer energy level of electrons of the smallest inert gases has the highest ionization potential.

55. **(A)** Only NO and NO_2 fit the definition of the Law of Multiple Proportions, in which one substance stays the same and the other varies in units of whole integers.

56. **(D)** Increasing the concentration of one or both of the reactants and removing some of the product formed would cause the forward reaction to increase in rate to try to regain the equilibrium condition. II and III are correct.

57. **(C)** I and II are correct. The equation is $2Na + 2H_2O \rightarrow 2NaOH + H_2(g)$. The coefficients include a 1 and a 2.

58. **(B)** $2CO + O_2 \rightarrow 2CO_2$ indicates 2 vol of CO react with 1 vol of O_2 to form 2 vol of CO_2. Therefore 10 L of CO form 10 L of CO_2.

59. **(D)** $H_2SO_4 + 2NaOH \rightarrow 2H_2O + Na_2SO_4$ is the equation for this reaction. 1 mol H_2SO_4 = 98 g. 1 mol NaOH = 40 g. Then 49 g of H_2SO_4 = $\frac{1}{2}$ mol H_2SO_4. The equation shows that 1 mol of sulfuric reacts with 2 mol of sodium hydroxide or a ratio of 1 : 2. Therefore, $\frac{1}{2}$ mol of sulfuric reacts with 1 mol of sodium hydroxide in this reaction. We found 1 mol of NaOH to equal 40 g. Since 80 g of NaOH is given, 40 g of it will remain after the reaction has gone to completion.

60. **(C)** The $K_{sp} = [Ag^+][Cl^-] = 1.9 \times 10^{-10}$. The molar concentration of both Ag^+ and Cl^- is 1.38×10^{-5} mol/L.

$$\begin{aligned} K_{sp} &= [Ag^+][Cl^-] \\ &= [1.38 \times 10^{-5}][1.38 \times 10^{-5}] \\ &= 1.9 \times 10^{-10} \end{aligned}$$

61. **(B)** If 1.43 g is the weight of 1 L, then the weight of 22.4 L, which is the gram-molecular volume of a gas at STP, will give the gram-molecular weight. So 22.4 L × 1.43 g/L = 32 g, the gram-molecular mass.

62. **(E)** The equation is

$$2KClO_3 \rightarrow 2KCl + 3O_2$$

This shows that 2 mol of $KClO_3$ yields 3 mol of O_2. Three mol of $O_2 = 3 \times 22.4$ L = 67.2 L of O_2.

63. **(A)** Gibbs free energy combines the overall energy changes and the entropy change. The formula is $\Delta G = \Delta H - T\,\Delta S$. Only if ΔG is negative will the reaction be spontaneous in the forward direction.

64. **(C)** The CuO was reduced while the H_2 was oxidized, forming H_2O + Cu. The reaction is

$$CuO + H_2 \rightarrow H_2O + Cu$$

65. **(B)** The purpose of this $CaCl_2$ tube is to absorb any water evaporated from the generator. If it were not present, some water vapor would pass through to the final drying tube and cause the weight of water gained there to be larger than it should be from the reaction only.

66. **(A)** Since 8 g of O_2 was lost by the CuO, the weight of water gained by the U-tube came from 8 g of O_2 and 1 g of H_2, to make 9 g of water that it absorbed.

67. **(E)** The weight ratio of water is 1 g of H_2 to 8 g of O_2 or 1 : 8. The other statements are true but are not conclusions from this experiment.

68. **(D)** This is the only observation of the group. All the others are conclusions. Remember that an observation is only what you see, smell, taste, or measure with a piece of equipment.

69. **(A)** The phenomenon is the formation of the white ring, which is NH_4Cl. Although measuring the distance traveled by each gas could be used to verify Graham's Law of Gaseous Diffusion, this was not asked. The relationship is that the diffusion rate is inversely proportional to the square root of the gas's molecular weight.

70. **(C)** Since Ca^{2+} and HSO_4^- combine, the formula is $Ca(HSO_4)_2$.

71. **(A)** NaOH is 40 g/mol. 10 g is 10/40 or 0.25 mol in 1 L or 0.25 M.

72. **(C)** A saturated solution represents a condition where the solute is going into solution as rapidly as some solute is coming out of the solution.

73. **(D)** II and III are correct since the double-bonded carbons in the alkene and the triple-bonded carbons in the alkyne series have pi bonds.

74. **(B)** The thistle tube is not below the level of the liquid in the generator, and the gas would escape into the air. (Vinegar is an acid and would produce hydrogen.)

75. **(D)** NH_3 is very soluble and could not be collected in this manner. All others are not sufficiently soluble to hamper this method of collection.

76. **(C)** The formula mass is the total of (Ca = 40) + (2N = 28) + (6O = 96), or 164. Since oxygen is 96 amu of the 164 total, the percentage is 96/164 × 100% = 58.5%.

77. **(D)** The K_{eq} expression is made up of the concentration(s) of the products over those of the reactants, with the coefficients becoming exponents. So

$$K_{eq} = \frac{[NO_2]^2}{[N_2O_4]}$$

78. **(A)** $K_{eq} = \dfrac{[2 \times 10^{-2}]^2}{[4 \times 10^{-2}]} = \dfrac{4 \times 10^{-4}}{4 \times 10^{-2}} = 1 \times 10^{-2}$.

79. **(C)** In dilution expressions $M_1 \times V_1 = M_2 \times V_2$ can be used. Substituting $(6\,M)(0.5\,L) = (2\,M)\,(x\ \text{liters})$ gives $x = 1.5$ L total volume. Since there was 0.5 L to begin with, an additional 1 L must be added.

80. **(C)** The reaction equation and information given can be set up like this:

$$\begin{array}{ccc} \text{Given} & \text{Given} & \\ 4 \text{ g} & 4 \text{ g} & x \text{ grams} \\ 2H_2 + & O_2 \rightarrow & 2H_2O \\ 4 \text{ g} & 32 \text{ g} & 36 \text{ g} \end{array}$$

Studying this shows that the limiting element will be the 4 g of oxygen since 4 g of H_2 would require 32 g of O_2. The solution setup is

$$\frac{4 \text{ g } O_2}{32 \text{ g } O_2} = \frac{x \text{ grams } H_2O}{36 \text{ g } H_2O}$$

$$\therefore = 4.5 \text{ g } H_2O$$

81. **(E)** The functional group is $R-\overset{\overset{\displaystyle O}{\|}}{C}-R_1$. This appears only in **(E)**.

82. **(C)** The thistle tube serves both these purposes.

83. **(D)** The empirical formula is a representation of the elements in their simplest ratio. Therefore, CH_2 is the simplest ratio of the molecular formula C_3H_6.

84. **(E)** I, II, and III would occur since their reduction reactions would be −0.13, +0.34, and +0.80 volt, respectively. These numbers added to +0.44 volt separately to give a positive E^0 for the reaction with Fe.

85. **(A)** In the equation, 2 moles of Na will release 822 kJ of heat. If only 0.5 mole of Na is consumed, only one fourth as much heat will be released or $\frac{1}{4} \times 822 \text{ kJ} = 205 \text{ kJ}$.

Diagnosing Your Needs

After taking Practice Test 1, check your answers against the correct ones. Then fill in the chart below.

In the space under each equation number, place a check if you answered that question correctly.

Example

If your answer to question 5 was correct, place a check in the appropriate box.

Next, total the check marks for each section and insert the number in the designated block. Now do the arithmetic indicated and insert your percent for each area.

SUBJECT AREA	(✔) QUESTIONS ANSWERED CORRECTLY

I. Atomic Theory and Structure, including periodic relationships	6	8	105	109	110	111	47	54
☐ No. of checks ÷ 8 × 100 = _____ %								

II. Nuclear Reactions							115	116
☐ No. of checks ÷ 2 × 100 = _____ %								

III. Chemical Bonding and Molecular Structure	3	17	18	19	45	46	55	70	73
☐ No. of checks ÷ 9 × 100 = _____ %									

IV. States of Matter and Kinetic Molecular Theory of Gases	10	11	12	107	108	113	43
☐ No. of checks ÷ 7 × 100 = _____ %							

V. Solutions, including concentration units, solubility, and colligative properties	21	103	60	71	79
☐ No. of checks ÷ 5 × 100 = _____ %					

VI. Acids and Bases	20	101	102	114	52	49	69
☐ No. of checks ÷ 7 × 100 = _____ %							

VII. Oxidation-Reduction and Electrochemistry	106	112	50	51	64	84
☐ No. of checks ÷ 6 × 100 = _____ %						

SUBJECT AREA	(✔) QUESTIONS ANSWERED CORRECTLY									
VIII. Stoichiometry	40	44	57	58	59	61	62	66	76	80
▢ No. of checks ÷10 × 100 = _____%										

IX. Reaction Rates	41	48
▢ No. of checks ÷ 2 × 100 = _____%		

X. Equilibrium	22	56	72	77	78
▢ No. of checks ÷ 5 × 100 = _____%					

XI. Thermodynamics: energy changes in chemical reactions, randomness, and criteria for spontaneity	13	14	15	63	104	85
▢ No. of checks ÷ 6 × 100 = _____%						

XII. Descriptive Chemistry: physical and chemical properties of elements and their familiar compounds; organic chemistry; periodic properties	1	2	4	5	7	9
	16	23	42	53	81	83
▢ No. of checks ÷ 12 × 100 = _____%						

XIII. Laboratory: equipment, procedures, observations, safety, calculations, and interpretation of results	65	67	68	74	75	82
▢ No. of checks ÷ 6 × 100 = _____%						

Planning Your Study

The percentages give you an idea of how you have done on the various major areas of the test. Because of the limited number of questions on some parts, these percentages may not be as reliable as the percentages for parts with larger numbers of questions. However, you should now have at least a rough idea of the areas in which you have done well and those in which you need more study.

Start your study with the areas in which you are weakest. The corresponding chapters are indicated on the following page.

Subject Area	Chapters to Review
I. Atomic Theory and Structure, including periodic relationships	2
II. Nuclear Reactions	15
III. Chemical Bonding and Molecular Structure	3, 4
IV. States of Matter and Kinetic Molecular Theory of Gases	1
V. Solutions, including concentration units, solubility, and colligative properties	7
VI. Acids and Bases	11
VII. Oxidation-Reduction and Electrochemistry	12
VIII. Stoichiometry	5, 6
IX. Reaction Rates	9
X. Equilibrium	10
XI. Thermodynamics, including energy changes in chemical reactions, randomness, and criteria for spontaneity	8
XII. Descriptive Chemistry: physical and chemical properties of elements and their familiar compounds; organic chemistry; periodic properties	1, 2, 13, 14
XIII. Laboratory: equipment, procedures, observations, safety, calculations, and interpretation of results.	All lab diagrams, 16

Answer Sheet for Practice Test 2

Determine the correct answer for each question. Then, using a No. 2 pencil, blacken completely the oval containing the letter of your choice.

1. Ⓐ Ⓑ Ⓒ Ⓓ Ⓔ
2. Ⓐ Ⓑ Ⓒ Ⓓ Ⓔ
3. Ⓐ Ⓑ Ⓒ Ⓓ Ⓔ
4. Ⓐ Ⓑ Ⓒ Ⓓ Ⓔ
5. Ⓐ Ⓑ Ⓒ Ⓓ Ⓔ
6. Ⓐ Ⓑ Ⓒ Ⓓ Ⓔ
7. Ⓐ Ⓑ Ⓒ Ⓓ Ⓔ
8. Ⓐ Ⓑ Ⓒ Ⓓ Ⓔ
9. Ⓐ Ⓑ Ⓒ Ⓓ Ⓔ
10. Ⓐ Ⓑ Ⓒ Ⓓ Ⓔ
11. Ⓐ Ⓑ Ⓒ Ⓓ Ⓔ
12. Ⓐ Ⓑ Ⓒ Ⓓ Ⓔ
13. Ⓐ Ⓑ Ⓒ Ⓓ Ⓔ
14. Ⓐ Ⓑ Ⓒ Ⓓ Ⓔ
15. Ⓐ Ⓑ Ⓒ Ⓓ Ⓔ
16. Ⓐ Ⓑ Ⓒ Ⓓ Ⓔ
17. Ⓐ Ⓑ Ⓒ Ⓓ Ⓔ
18. Ⓐ Ⓑ Ⓒ Ⓓ Ⓔ
19. Ⓐ Ⓑ Ⓒ Ⓓ Ⓔ
20. Ⓐ Ⓑ Ⓒ Ⓓ Ⓔ
21. Ⓐ Ⓑ Ⓒ Ⓓ Ⓔ
22. Ⓐ Ⓑ Ⓒ Ⓓ Ⓔ
23. Ⓐ Ⓑ Ⓒ Ⓓ Ⓔ

ON THE ACTUAL CHEMISTRY TEST, THE RE-MAINING QUESTIONS MUST BE ANSWERED BY RETURNING TO THE SECTION OF YOUR AN-SWER SHEET YOU STARTED FOR CHEMISTRY.

ON THE ACTUAL CHEMISTRY TEST, THE FOLLOWING TYPE OF QUESTION MUST BE ANSWERED ON A SPECIAL SECTION (LABELED "CHEMISTRY") AT THE LOWER LEFT-HAND CORNER OF PAGE 2 OF YOUR ANSWER SHEET. THESE QUESTIONS WILL BE NUMBERED BEGINNING WITH 101 AND MUST BE ANSWERED ACCORDING TO THE FOLLOWING DIRECTIONS.

CHEMISTRY* Fill in oval CE only if II is a correct explanation of I.

	I	II	CE*
101.	Ⓣ Ⓕ	Ⓣ Ⓕ	◯
102.	Ⓣ Ⓕ	Ⓣ Ⓕ	◯
103.	Ⓣ Ⓕ	Ⓣ Ⓕ	◯
104.	Ⓣ Ⓕ	Ⓣ Ⓕ	◯
105.	Ⓣ Ⓕ	Ⓣ Ⓕ	◯
106.	Ⓣ Ⓕ	Ⓣ Ⓕ	◯
107.	Ⓣ Ⓕ	Ⓣ Ⓕ	◯
108.	Ⓣ Ⓕ	Ⓣ Ⓕ	◯
109.	Ⓣ Ⓕ	Ⓣ Ⓕ	◯
110.	Ⓣ Ⓕ	Ⓣ Ⓕ	◯
111.	Ⓣ Ⓕ	Ⓣ Ⓕ	◯
112.	Ⓣ Ⓕ	Ⓣ Ⓕ	◯
113.	Ⓣ Ⓕ	Ⓣ Ⓕ	◯
114.	Ⓣ Ⓕ	Ⓣ Ⓕ	◯
115.	Ⓣ Ⓕ	Ⓣ Ⓕ	◯
116.	Ⓣ Ⓕ	Ⓣ Ⓕ	◯

40. Ⓐ Ⓑ Ⓒ Ⓓ Ⓔ
41. Ⓐ Ⓑ Ⓒ Ⓓ Ⓔ
42. Ⓐ Ⓑ Ⓒ Ⓓ Ⓔ
43. Ⓐ Ⓑ Ⓒ Ⓓ Ⓔ
44. Ⓐ Ⓑ Ⓒ Ⓓ Ⓔ
45. Ⓐ Ⓑ Ⓒ Ⓓ Ⓔ
46. Ⓐ Ⓑ Ⓒ Ⓓ Ⓔ
47. Ⓐ Ⓑ Ⓒ Ⓓ Ⓔ
48. Ⓐ Ⓑ Ⓒ Ⓓ Ⓔ
49. Ⓐ Ⓑ Ⓒ Ⓓ Ⓔ
50. Ⓐ Ⓑ Ⓒ Ⓓ Ⓔ
51. Ⓐ Ⓑ Ⓒ Ⓓ Ⓔ
52. Ⓐ Ⓑ Ⓒ Ⓓ Ⓔ
53. Ⓐ Ⓑ Ⓒ Ⓓ Ⓔ
54. Ⓐ Ⓑ Ⓒ Ⓓ Ⓔ
55. Ⓐ Ⓑ Ⓒ Ⓓ Ⓔ

56. Ⓐ Ⓑ Ⓒ Ⓓ Ⓔ
57. Ⓐ Ⓑ Ⓒ Ⓓ Ⓔ
58. Ⓐ Ⓑ Ⓒ Ⓓ Ⓔ
59. Ⓐ Ⓑ Ⓒ Ⓓ Ⓔ
60. Ⓐ Ⓑ Ⓒ Ⓓ Ⓔ
61. Ⓐ Ⓑ Ⓒ Ⓓ Ⓔ
62. Ⓐ Ⓑ Ⓒ Ⓓ Ⓔ
63. Ⓐ Ⓑ Ⓒ Ⓓ Ⓔ
64. Ⓐ Ⓑ Ⓒ Ⓓ Ⓔ
65. Ⓐ Ⓑ Ⓒ Ⓓ Ⓔ
66. Ⓐ Ⓑ Ⓒ Ⓓ Ⓔ
67. Ⓐ Ⓑ Ⓒ Ⓓ Ⓔ
68. Ⓐ Ⓑ Ⓒ Ⓓ Ⓔ
69. Ⓐ Ⓑ Ⓒ Ⓓ Ⓔ
70. Ⓐ Ⓑ Ⓒ Ⓓ Ⓔ

71. Ⓐ Ⓑ Ⓒ Ⓓ Ⓔ
72. Ⓐ Ⓑ Ⓒ Ⓓ Ⓔ
73. Ⓐ Ⓑ Ⓒ Ⓓ Ⓔ
74. Ⓐ Ⓑ Ⓒ Ⓓ Ⓔ
75. Ⓐ Ⓑ Ⓒ Ⓓ Ⓔ
76. Ⓐ Ⓑ Ⓒ Ⓓ Ⓔ
77. Ⓐ Ⓑ Ⓒ Ⓓ Ⓔ
78. Ⓐ Ⓑ Ⓒ Ⓓ Ⓔ
79. Ⓐ Ⓑ Ⓒ Ⓓ Ⓔ
80. Ⓐ Ⓑ Ⓒ Ⓓ Ⓔ
81. Ⓐ Ⓑ Ⓒ Ⓓ Ⓔ
82. Ⓐ Ⓑ Ⓒ Ⓓ Ⓔ
83. Ⓐ Ⓑ Ⓒ Ⓓ Ⓔ
84. Ⓐ Ⓑ Ⓒ Ⓓ Ⓔ
85. Ⓐ Ⓑ Ⓒ Ⓓ Ⓔ

PRACTICE TEST 2

Note: For all questions involving solutions and/or chemical equations, assume that the system is in water unless otherwise stated.

PART A

Directions: Each set of lettered choices refers to the numbered statements or formulas immediately following it. Select the one lettered choice that best fits each statement or formula and then fill in the corresponding oval on the answer sheet. A choice may be used once, more than once, or not at all in each set.

Questions 1–4

     (A) Law of Definite Composition
     (B) High dielectric constant
     (C) van der Waals forces
     (D) Graham's Law of Diffusion (Effusion)
     (E) Triple point

1. At a particular temperature and pressure, three states of a compound may coexist.

2. Water is a good solvent.

3. The rate of movement of hydrogen gas compared to oxygen gas is 4 : 1.

4. The molecules of nitrous oxide and nitrogen dioxide differ by a multiple of the mass of one oxygen.

Questions 5–7 refer to the following diagram:

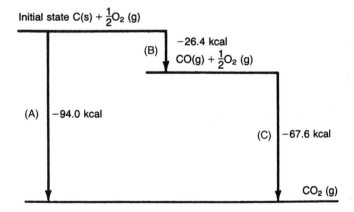

5. The ΔH of the reaction to form CO from $C + O_2$

6. The ΔH of the reaction to form CO_2 from $CO + O_2$

7. The ΔH of the reaction to form CO_2 from $C + O_2$

Questions 8–11

    (A) Hydrogen bond
    (B) Ionic bond
    (C) Polar covalent bond
    (D) Pure covalent bond
    (E) Metallic bond

8. The type of bond between atoms of potassium and chloride in a crystal of potassium chloride

9. The type of bond between the atoms in a nitrogen molecule

10. The type of bond between the atoms in a molecule of CO_2 (electronegativity difference = 1)

11. The type of bond between the atoms of calcium in a crystal of calcium

Questions 12–14 refer to the following diagram which is a phase diagram for CO_2.

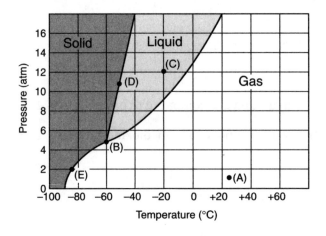

12. The point at which all three states of CO_2 can exist

13. The point at which CO_2 can exist only as a liquid

14. The point at which CO_2 can exist as a solid and a gas under 2 atmospheres of pressure

Questions 15–23 refer to the following graph, which shows the variation of the first ionization potential with respect to increasing atomic numbers.

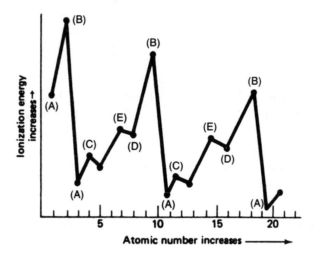

15. The atoms likely to react with water to release hydrogen

16. Nonmetals that are all found in the gaseous state at STP

17. The noble gases

18. The alkali metals

19. The half-filled condition of the p orbitals

20. The filled s orbitals with the exception of He

21. The beginning of pairing in the p orbitals

22. The most active metals

23. The filled p orbitals

PART B

ON THE ACTUAL CHEMISTRY TEST, THE FOLLOWING TYPE OF QUESTION MUST BE ANSWERED ON A SPECIAL SECTION (LABELED "CHEMISTRY") AT THE LOWER LEFT-HAND CORNER OF PAGE 2 OF YOUR ANSWER SHEET. THESE QUESTIONS WILL BE NUMBERED BEGINNING WITH 101 AND MUST BE ANSWERED ACCORDING TO THE FOLLOWING DIRECTIONS.

Directions: Each question below consists of two statements, I in the left-hand column and II in the right-hand column. For each question, determine whether statement I is true or false and if statement II is true or false and fill in the corresponding T or F ovals on your answer sheet. Fill in oval CE only if statement II is a correct explanation of statement I.

Sample Answer Grid:

CHEMISTRY * Fill in oval CE only if
II is a correct explanation of I.

	I	II	CE*
101.	ⓉⒻ	ⓉⒻ	◯

101. The structure of SO_3 is shown by using   BECAUSE SO_3 is very unstable and resonates
     more than one structural formula   between these possible structures.

102. When the ΔG value of a reaction at a   BECAUSE   when ΔG is negative, the ΔH is also
     given temperature is negative, the   negative.
     reaction occurs spontaneously

103. One mole of CO_2 has a greater mass than   BECAUSE   the molecular mass of CO_2 is greater
     one mole of H_2O than the molecular mass of H_2O.

104. Hydrosulfuric acid is often used in   BECAUSE $H_2S(aq)$ reacts with many metallic
     qualitative tests   ions to produce colorful precipitates.

105. Crystals of sodium chloride go into   BECAUSE   the sodium ion has a 1+ charge and
     solution in water as ions   the chloride ion has a 1– charge and
                                              they are hydrated by the water
                                              molecules.

106. In an equilibrium reaction, if the   BECAUSE   when a stress is applied to a reaction
     concentration of the reactants is   in equilibrium, the equilibrium shifts
     increased, the reaction will increase its   in the direction that opposes the
     forward rate   stress.

107. The $\Delta H_{reaction}$ of a particular reaction can   BECAUSE   Hess's Law conforms to the First
     be arrived at by the summation of the   Law of thermodynamics, which
     $\Delta H_{reaction}$ values of two or more reactions   states that the total energy of the
     that, added together, give the $\Delta H_{reaction}$ of   universe is a constant.
     the particular reaction

108. In a reaction that has both a forward and   BECAUSE   the reverse reaction does not begin
     a reverse reaction, $A + B \rightleftharpoons AB$, when only   until equilibrium is reached.
     A and B are introduced into a reacting
     vessel, the forward reaction rate is the
     highest at the beginning and begins to
     decrease from that point until equilibrium
     is reached

109. At equilibrium, the forward reaction and   BECAUSE   the reactants and products have
     reverse reaction stop   reached the equilibrium
                                                 concentrations.

110. The hydrid orbital form of carbon in acetylene is believed to be the *sp* form   BECAUSE C_2H_2 is a linear compound with a triple bond between the carbons.

111. The weakest of the bonds between molecules are coordinate covalent bonds   BECAUSE   coordinate covalent bonds represent the weak attractive force of the electrons of one molecule for the positively charged nucleus of another.

112. A saturated solution is not necessarily concentrated   BECAUSE   *dilute* and *concentrated* are terms that relate only to the relative amount of solute dissolved in the solvent.

113. Lithium is the most active metal in the first group of the periodic chart   BECAUSE   lithium has only one electron in the outer energy level.

114. The anions migrate to the cathode in an electrochemical reaction   BECAUSE   positively charged ions are attracted to the negatively charged cathode.

115. The atomic number of a neutral atom that has a mass of 39 and has 19 electrons is 19   BECAUSE   the number of protons in a neutral atom is equal to the number of electrons.

116. For an element with an atomic number of 17, the most probable oxidation number is +1   BECAUSE   the outer energy level of the halogen family has a tendency to add one electron to itself.

PART C

Directions: Each of the questions or incomplete statements below is followed by five suggested answers or completions. Select the one that is best in each case and then fill in the corresponding oval on the answer sheet.

40. All of the following involve a chemical change EXCEPT
    (A) the formation of HCl from H_2 and Cl_2
    (B) the color change when NO is exposed to air
    (C) the formation of steam from burning H_2 and O_2
    (D) the solidification of Crisco at low temperatures
    (E) the odor of NH_3 when NH_4Cl is rubbed together with $Ca(OH)_2$ powder

41. When most fuels burn, the products include carbon dioxide and
    (A) hydrocarbons
    (B) hydrogen
    (C) water
    (D) hydroxide
    (E) hydrogen peroxide

42. In the metric system, the prefix *kilo-* means
   (A) 10^0
   (B) 10^{-1}
   (C) 10^{-2}
   (D) 10^2
   (E) 10^3

43. How many atoms are in 1 mol of water?
   (A) 3
   (B) 54
   (C) 6.02×10^{23}
   (D) $2(6.02 \times 10^{23})$
   (E) $3(6.02 \times 10^{23})$

44. Which of the following atoms normally forms monoatomic molecules?
   (A) Cl
   (B) H
   (C) O
   (D) N
   (E) He

45. Which method is often employed in the separation of the hydrocarbons found in petroleum?
   (A) alkylation
   (B) hydrogenation
   (C) catalytic cracking
   (D) fractional distillation
   (E) polymerization

46. The complete loss of an electron of one atom to another atom with the consequent formation of electrostatic charges is said to be
   (A) a covalent bond
   (B) a polar covalent bond
   (C) an ionic bond
   (D) a coordinate covalent bond
   (E) a pi bond between p orbitals

47. In the electrolysis of water, the cathode reaction is
   (A) $2H_2O(\ell) + 2e^- \rightarrow H_2(g) + 2OH^- + O_2(g)$
   (B) $2H_2O(\ell) \rightarrow \frac{1}{2}O_2(g) + 2H^+ + 2e^-$
   (C) $2OH^- + 2e^- \rightarrow O^2(g) + H_2(g)$
   (D) $2H^+ + 2e^- \rightarrow H_2(g)$
   (E) $2H_2O(\ell) + 4e^- \rightarrow O_2(g) + 2H_2(g)$

Question 48 refers to a solution of 0.100 M acetic acid.

48. What is the H_3O^+ concentration? ($K_a = 1.8 \times 10^{-5}$)
    (A) 1.8×10^{-5}
    (B) 1.8×10^{-4}
    (C) 1.8×10^{-3}
    (D) 1.3×10^{-3}
    (E) 0.9×10^{-3}

49. If a radioactive element with a half-life of 100 years is found to have transmutated so that only 25% of the original sample remains, what is the age of the sample?
    (A) 25 years
    (B) 50 years
    (C) 100 years
    (D) 200 years
    (E) 400 years

50. What is the pH of an acetic acid solution if the $[H_3O^+] = 1 \times 10^{-4}$ mol/L?
    (A) 1
    (B) 2
    (C) 3
    (D) 4
    (E) 5

51. The polarity of water is useful in explaining which of the following?
    I.   The solution process
    II. The ionization process
    III. The high conductivity of distilled water

    (A) I only
    (B) II only
    (C) I and II only
    (D) II and III only
    (E) I, II, and III

52. When sulfur dioxide is bubbled through water, the solution will contain
    (A) sulfurous acid
    (B) sulfuric acid
    (C) hyposulfuric acid
    (D) persulfuric acid
    (E) anhydrous sulfuric acid

53. Four grams of hydrogen gas at STP contain
    (A) 6.02×10^{23} atoms
    (B) 12.04×10^{23} atoms
    (C) 12.04×10^{46} atoms
    (D) 1.2×10^{22} molecules
    (E) 12.04×10^{23} molecules

54. Analysis of a gas gave C = 85.7% and H = 14.3%. If the formula mass of this gas is 42 amu, what are the empirical formula and the true formula?
   (A) CH; C_4H_4
   (B) CH_2; C_3H_6
   (C) CH_3; C_3H_9
   (D) C_2H_2; C_3H_6
   (E) C_2H_4; C_3H_6

55. Which fraction would be used to correct a given volume of gas at 30°C to its new volume when it is heated to 60°C and the pressure is kept constant?
   (A) $\dfrac{30}{60}$

   (B) $\dfrac{60}{30}$

   (C) $\dfrac{273}{333}$

   (D) $\dfrac{303}{333}$

   (E) $\dfrac{333}{303}$

56. What would be the predicted freezing point of a solution that has 684 g of sugar (1 mol = 342 g) dissolved in 1500 g of water?
   (A) 1.86°C
   (B) –0.93°C
   (C) –1.39°C
   (D) –2.48°C
   (E) –2.79°C

57. What is the approximate pH of a 0.005-M solution of H_2SO_4?
   (A) 1
   (B) 2
   (C) 5
   (D) 9
   (E) 13

58. How many grams of NaOH are needed to make 100 g of a 5% solution?
   (A) 2
   (B) 5
   (C) 20
   (D) 40
   (E) 95

59. For the Haber process, $N_2 + 3H_2 \rightleftharpoons 2NH_3 + heat$ (at equilibrium), which of the following statements concerning the reaction rate are true?

    I.   The reaction to the right increases when the pressure is increased.
    II. The reaction to the right decreases when the temperature is increased.
    III. The reaction to the right decreases when NH_3 is removed from the chamber.

    (A) I only
    (B) II only
    (C) I and II only
    (D) II and III only
    (E) I, II, and III

60. If you titrate a $1\,M$ H_2SO_4 solution against 50 mL of $1\,M$ NaOH solution, what volume of H_2SO_4, in milliliters, will be needed for neutralization?
    (A) 10
    (B) 25
    (C) 40
    (D) 50
    (E) 100

61. How many grams of CO_2 can be prepared from 150 g of calcium carbonate reacting with an excess of hydrochloric acid solution?
    (A) 11
    (B) 22
    (C) 33
    (D) 44
    (E) 66

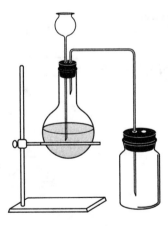

62. The above diagram represents a setup that may be used to prepare and collect
    (A) NH_3
    (B) NO
    (C) H_2
    (D) SO_3
    (E) CO_2

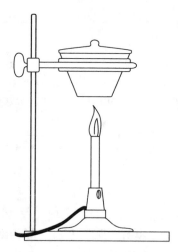

63. The lab setup shown above was used for the gravimetric analysis of the empirical formula of MgO. In synthesizing MgO from a Mg strip in the crucible, which of the following is NOT true?
A. The initial strip of Mg should be cleaned.
B. The lid of the crucible should fit tightly to exclude oxygen.
C. The heating of the covered crucible should continue until the Mg is fully reacted.
D. Cool the crucible, lid, and contents to room temperature before measuring their mass.
E. When the Mg appears to be fully reacted, partially remove the crucible lid and continue heating.

Questions 64–66 refer to the following experimental setup and data.

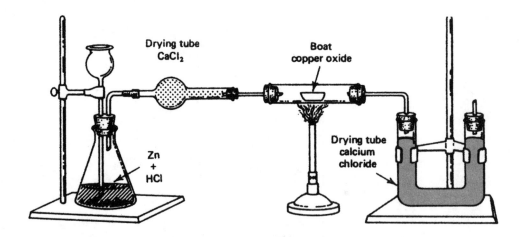

Recorded data:
    Weight of U-tube . . . . .20.36 g
    Weight of U-tube and calcium chloride before . . . . .39.32 g
    Weight of U-tube and calcium chloride after . . . . .57.32 g
    Weight of boat and copper oxide before . . . . .30.23 g
    Weight of boat and copper oxide after . . . . .14.23 g
    Weight of boat . . . . 5.00 g

64. What is the reason for the first $CaCl_2$ drying tube?
    (A) generate water
    (B) absorb hydrogen
    (C) absorb water that evaporates from the flask
    (D) decompose the water from the flask
    (E) act as a catalyst for the combination of hydrogen and oxygen

65. What conclusion can be derived from the data collected?
    (A) Oxygen was lost from the $CaCl_2$.
    (B) Oxygen was generated in the U-tube.
    (C) Water was formed from the reaction.
    (D) Hydrogen was absorbed by the $CaCl_2$.
    (E) CuO was formed in the decomposition.

66. What is the ratio of the mass of water formed to the mass of hydrogen used in the formation of water?
    (A) $1:8$
    (B) $1:9$
    (C) $8:1$
    (D) $9:1$
    (E) $8:9$

67. What is the mass, in grams, of 1 mol of $KAl(SO_4)_2 \cdot 12H_2O$?
    (A) 132
    (B) 180
    (C) 394
    (D) 474
    (E) 516

68. What mass of aluminum will be completely oxidized by 44.8 L of oxygen at STP?
    (A) 18 g
    (B) 37.8 g
    (C) 50.4 g
    (D) 72.0 g
    (E) 100.8 g

69. In general, when metal oxides react with water, they form solutions that are
    (A) acidic
    (B) basic
    (C) neutral
    (D) unstable
    (E) colored

Questions 70–72 refer to the following diagram.

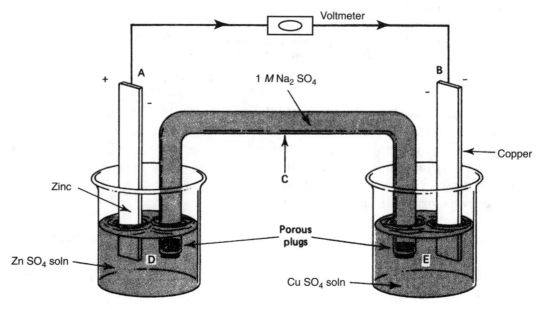

70. The oxidation reaction will occur at
    (A) A
    (B) B
    (C) C
    (D) D
    (E) E

71. The apparatus at C is called the
    (A) anode
    (B) cathode
    (C) salt bridge
    (D) ion bridge
    (E) osmotic bridge

72. The standard potentials of the metals are:
    $Zn^{2+} + 2\ e^- = Zn$ $E_0 = -0.76$ volt
    $Cu^0 = Cu^{2+} + 2e^-$ $E_0 = -0.34$ volt

    What will be the voltmeter reading for this reaction?
    (A) +1.10
    (B) −1.10
    (C) +0.42
    (D) −0.42
    (E) −1.52

73. How many liters of oxygen (STP) can you prepare from the decomposition of 213 g of sodium chlorate?
    (A) 11.2
    (B) 22.4
    (C) 44.8
    (D) 67.2
    (E) 78.4

74. In the equation $Al(OH)_3 + H_2SO_4 \rightarrow Al_2(SO_4)_3 + H_2O$, the coefficients of the balanced equation are
    (A) 1, 3, 1, 2
    (B) 2, 3, 2, 6
    (C) 2, 3, 1, 6
    (D) 2, 6, 1, 3
    (E) 1, 3, 1, 6

75. What is $\Delta H_{reaction}$ for the decomposition of sodium chlorate? [(H_f^0 values are $NaClO_3(s) = -85.7$ kcal/mol, $NaCl(s) = -98.2$ kcal/mol, $O_2(g) = 0$ kcal/mol.)]
    (A) −183.9 kcal
    (B) −91.9 kcal
    (C) +45.3 kcal
    (D) +22.5 kcal
    (E) −12.5 kcal

76. Isotopes of an element are related because which of the following is(are) the same in these isotopes?
    I.   Atomic mass
    II. Atomic number
    III. Arrangement of orbital electrons

    (A) I only
    (B) II only
    (C) I and II only
    (D) II and III only
    (E) I, II, and III

77. In the reaction of Zn with dilute HCl to form H_2, which of the following will increase the reaction rate?

    I.   Increasing the temperature
    II. Increasing the exposed surface of Zn
    III. Using a more active metal instead of Zn

    (A) I only
    (B) II only
    (C) I and III only
    (D) II and III only
    (E) I, II, and III

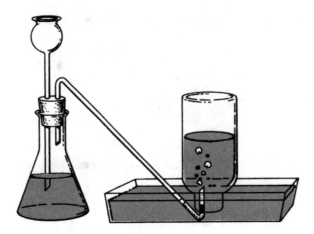

78. The laboratory setup shown above can be used to prepare a
    (A) gas lighter than air and soluble in water
    (B) gas heavier than air and soluble in water
    (C) gas soluble in water
    (D) gas insoluble in water
    (E) gas that reacts with water

79. In the reaction $CaCO_3 + 2HCl \rightarrow CaCl_2 + H_2O + CO_2$, if 4 mol of HCl is available to the reaction with an unlimited supply of $CaCO_3$, how many moles of CO_2 can be produced at STP?
    (A) 1
    (B) 1.5
    (C) 2
    (D) 2.5
    (E) 3

80. A saturated solution of $BaSO_4$ at 25°C contains 3.9×10^{-5} mol/L of Ba^{2+} ions. What is the K_{sp} of this salt?
    (A) 3.9×10^{-5}
    (B) 3.9×10^{-6}
    (C) 2.1×10^{-7}
    (D) 1.5×10^{-8}
    (E) 1.5×10^{-9}

81. Which of the following will definitely cause the volume of a gas at STP to increase?

    I.   Decrease the pressure with the temperature held constant.
    II. Increase the pressure with a temperature decrease.
    III. Increase the temperature with a pressure increase.

    (A) I only
    (B) II only
    (C) I and III only
    (D) II and III only
    (E) I, II, and III

82. If 0.1 mol of K_2SO_4 were added to the solution in question 80, what would happen to the Ba^{2+} concentration?
    (A) It would increase.
    (B) It would decrease.
    (C) It would remain the same.
    (D) It would first increase and then decrease.
    (E) It would first decrease and then increase.

83. In a neutralization reaction of an Arrhenius acid and base, what are the products?

    I. H_2O + a sulfate
    II. H_2O + a chloride
    III. H_2O + a salt

    (A) I only
    (B) II only
    (C) I and III only
    (D) II and III only
    (E) III only

84. Which of the following particles has the least mass?
    (A) alpha particle
    (B) beta particle
    (C) proton
    (D) neutron
    (E) gamma ray

85. Which emission from a radioactive source is an electromagnetic wave and is not affected by an electric field?
    (A) alpha particle
    (B) beta particle
    (C) proton
    (D) neutron
    (E) gamma ray

STOP
IF YOU FINISH BEFORE ONE HOUR IS UP, YOU MAY GO BACK
TO CHECK YOUR WORK OR COMPLETE UNANSWERED QUESTIONS.

Answers and Explanations for Test 2

1. **(E)** The triple point of a phase diagram shows that all three states can exist at this point.

2. **(B)** The high dielectric constant of water is due to the polar nature of the water molecules. It is this property that is largely responsible for the ability of water to dissolve so many solutes.

3. **(D)** According to Graham's Law of Gaseous Diffusion (of Effusion), the rate of diffusion is inversely proportional to the square root of the molecular weight. Therefore,

$$\frac{\text{Rate}_H}{\text{Rate}_O} = \frac{\sqrt{\text{Mol. mass O}_2}}{\sqrt{\text{Mol. mass H}_2}} = \frac{\sqrt{32}}{\sqrt{2}} = \sqrt{\frac{16}{1}} = \frac{4}{1}$$

4. **(A)** The Law of Definite Composition states that when compounds form, they always form in the same ratio by mass. Water, for instance, always forms in a ratio of 1 : 8 of hydrogen to oxygen by mass. For nitrous oxide (NO) and nitrogen dioxide (NO_2) the difference in molecular mass is one atomic mass of oxygen.

5. **(B)** The first step is the ΔH for C + $\frac{1}{2}O_2$ → CO. It releases 26.4 kcal of heat. This is written as –26.46 kcal.

6. **(C)** This is the second step on the diagram.

7. **(A)** To arrive at the ΔH of this, take the total drop (–94.0 kcal) or algebraically add these reactions:

$$C + \tfrac{1}{2}O_2 \quad \rightarrow CO -26.4 \text{ kcal}$$
$$CO + \tfrac{1}{2}O_2 \rightarrow CO_2 -67.6 \text{ kcal}$$
$$\overline{C + O_2 \qquad \rightarrow CO_2 -94.0 \text{ kcal}}$$

8. **(B)** Potassium and chlorine have a large enough difference in their electronegativities to form ionic bonds. The respective positions of these two elements in the periodic chart also are indicative of the large difference in their electronegativity values.

9. **(D)** Two atoms of an element that forms a diatomic molecule always have a pure covalent bond between them since the electron attraction or electronegativity of each atom is the same.

10. **(C)** Electronegativity differences between 0.4 and 1.7 are usually indicative of polar covalent bonding. CO_2 is an interesting example of a *nonpolar molecule* with polar covalent bonds since the bonds are symmetrical in the molecule.

11. **(E)** Calcium is a metal and forms a metallic bond between atoms.

12. **(B)** The point at which all three states can exist is called the triple point.

13. **(C)** The liquid state can exist only at (C).

14. **(E)** The only point at which CO_2 can exist as both a solid and a gas under 2 atmospheres of pressure is (E).

15. **(A)** These are the alkali metals.

16. **(B)** Because of their complete outer energy levels these elements have very low boiling points.

17. **(B)** The complete outer energy level makes it difficult to remove an electron; thus the high peaks would be the noble gases.

18. **(A)** With only one electron in the outer energy level, the alkali metals lose an electron relatively easily.

19. **(E)** There is a slight increase in the stability of the *p* orbitals when each has one electron—thus the little peaks at (E).

20. **(C)** The filled *s* orbital shows a slight increase in stability—thus the little peaks at C.

21. **(D)** The dip at the D areas on the chart shows the first pairing of electrons in the *p* orbitals and indicates that this first paired electron has a lower ionization potential than in the atom, which has only one electron in each *p* orbital.

22. **(A)** The most active metals, the alkalis, have the lowest ionization potentials.

23. **(B)** The filled *p* orbitals occur in the noble gases.

101. **(T, F)** Sulfur trioxide is shown by three structural formulas because each bond is a "hybrid" of a single and a double bond. Resonance in chemistry does not mean that the bonds resonate between the structures shown in the structural drawing.

102. **(T, F)** When ΔG is negative in the Gibbs equation, the reaction is spontaneous. However, the total equation determines this, not just ΔH. The Gibbs equation is:

$$\Delta G = \Delta H - T\,\Delta S$$

103. **(T, T, CE)** One mole of each of the gases contains 6.022×10^{23} molecules, but their molecular masses are different. CO_2 is found by adding one C = 12 and two O = 32, or a total of 44 amu. H_2O is found by adding two H = 2 and one O = 16, or a total of 18 amu. So it is true that 1 mole of CO_2 is heavier than 1 mole of H_2O.

104. **(T, T, CE)** Hydrosulfuric acid is used in qualitative tests because of the distinctly colored precipitates of sulfides that it forms with many metallic ions.

105. **(T, T, CE)** Sodium chloride is an ionic crystal and its ions are hydrated by the polar water molecules.

106. **(T, T, CE)** This statement and its reason fit the rules of Le Châtelier's Principle.

107. **(T, T, CE)** The statement is true, and the reason is also true and explains the statement.

108. **(T, F)** The statement is true, but not the reason. In an equilibrium reaction, concentrations can be shown to progress like this:

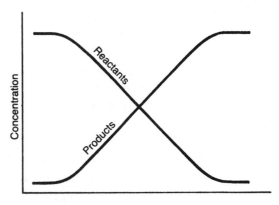

until equilibrium is reached. Then the concentrations stabilize.

109. **(F, T)** The forward and reverse reactions are occurring at equal rates when equilibrium is reached. The reactions do not stop. The concentrations remain the same at this point.

110. **(T, T, CE)** Since acetylene is known to be a linear molecule with a triple bond between the two carbons, the *sp* orbitals along the central axis with the hydrogens bonded on either end fit the experimental evidence.

111. **(F, F)** The weakest bonds between molecules are called van der Waals forces.

112. **(T, T, CE)** The terms *dilute* and *concentrated* merely indicate a relatively large amount of solvent and a small amount of solvent, respectively. You can have a dilute saturated solution if the solute is only slightly soluble.

113. **(F, T)** Cs is the most active Group I metal because it has (a) the largest atomic radius, thus making it easier to lose its outer energy level electron, and (b) the intermediate electrons help screen the positive attraction of the nucleus, also increasing the ease with which the outer electron is lost.

114. **(F, T)** The cations are positively charged ions and migrate to the cathode, while the anions are negatively charged and migrate to the anode.

115. **(T, T, CE)** There are as many electrons as there are protons in a neutral atom, and the atomic number represents the number of each.

116. (**F, T**) The first two principal energy levels fill up at 2 and 8 electrons, respectively. That leaves 7 electrons to fill the $3s$ and $3p$ orbitals like this: $3s^2$, $3p^5$. With only one electron missing in the $3p$ orbitals, the most likely oxidation number is −1.

40. (**D**) The solidification of Crisco is merely a physical change just like the formation of ice from liquid water at lower temperatures. All the others involve actual recombinations of the atoms and thus are chemical changes.

41. (**C**) Water is formed because most common fuels contain hydrogen in their structure.

42. (**E**) The other choices, in order, would be 1, $\frac{1}{10}$ or deci-, $\frac{1}{100}$ or centi-, and 100.

43. (**E**) One mole of any substance contains 6.02×10^{23} molecules. Since each water molecule is triatomic, there would be $3(6.02 \times 10^{23})$ atoms present.

44. (**E**) The noble gases are all monoatomic because of their complete outer energy level. A rule to help you remember diatomic gases is: Gases ending in *–gen* or *–ine* usually form diatomic molecules.

45. (**D**) Fractional distillation separates the hydrocarbons by taking advantage of the difference in boiling points. All other processes mentioned are used to modify the molecular structure of the hydrocarbons.

46. (**C**) The complete loss and gain of electrons is an ionic bond. All other bonds indicated are "sharing of electrons" type bonds or some form of covalent bonding.

47. (**D**) The cathode reaction releases only H_2 gas. The half-reaction is as given in reaction D.

48. (**D**) $K_a = 1.8 \times 10^{-5} = \dfrac{[H_3O^+][Ac^-]}{[HAc]} = \dfrac{(x)(x)}{0.1 - x}$. Since acetic acid is a weak acid, the 0.1-M concentration is much greater than the amount ionized (x), and so $0.1 - x$ is essentially equal to 0.1. Therefore $1.8 \times 10^{-5} = \dfrac{(x)(x)}{0.1} = \dfrac{x^2}{0.1}$ and $x = 1.3 \times 10^{-3}$.

49. (**D**) The sample went from 25% to 50% going back 100 years. If you go back another 100 years, it would be back to 100%. That is a total of 200 years.

50. (**D**) $pH = -\log[H_3O^+] = -\log[1 \times 10^{-4}] = -(-4) = 4$.

51. (**C**) Only I and II are true. Distilled water does not significantly conduct an electric current. The polarity of the water molecule is helpful in ionization and in causing substances to go into solution.

52. (**A**) SO_2 is the acid anhydride of H_2SO_3 or sulfurous acid. $H_2O + SO_2 \rightarrow H_2SO_3$.

53. (**E**) Four grams of hydrogen gas at STP represent 2 mol of hydrogen since 2 g is the gram-molecular mass of hydrogen. Each mole of a gas contains 6.02×10^{23} molecules, and 2 mol would contain $2 \times 6.02 \times 10^{23}$ or 12.04×10^{23} molecules.

54. (**B**) To solve percent composition problems, first divide the percentage given by the atomic mass:

$$12\overline{)85.7} \qquad 1\overline{)14.3}$$
$$7.14 \qquad 14.3$$

Then divide by the smallest quotient to get small whole numbers:

$$7.14\overline{)7.14} \qquad 7.14\overline{)14.3}$$
$$1.0 \qquad 2.0$$

So the empirical formula is CH_2. Since the molecular mass is 42 and the empirical formula has a molecular mass of 14, the true formula must be 3 times the empirical formula, or C_3H_6.

55. (**E**) Since the temperature (in kelvins) increases from 303 to 333 K, the volume of the gas should increase with the pressure held constant. The correct fraction is $\frac{333}{303}$.

56. **(D)** One mole of dissolved substance (which does not ionize) causes a 1.86 drop (Celsius) in the freezing point of a 1-m solution. Since 1500 g of water was used, the solution has $\frac{684}{342}$ or 2 mol in 1500 g of water. Then

$$\frac{2\text{ mol}}{1500\text{ g}} = \frac{x\text{ moles}}{1000\text{ g}}$$

$$x = \frac{2000}{1500} = 1.33\text{ mol/1000 g}$$

So the freezing point is depressed 1.33 × −1.86° = −2.48°C.

57. **(B)** The pH is −log[H⁺]. A 0.005-M solution of H_2SO_4 ionizes in a dilute solution to release two H⁺ ions per molecule of H_2SO_4. Therefore the molar concentration of H⁺ ion is 2 × 0.005 mol/L or 0.010 mol/L. Substituting this in the formula gives pH = −log(0.010) = −log(1×10^{-2}). The log of a number is the exponent to which the base 10 is raised to express that number in exponential form:

$$-\log(1 \times 10^{-2}) = -(-2) = 2$$
$$-\log(6 \times 10^{-3}) = -\log(10^7 \times 10^{-3})$$
$$= -[0.7 + (-3)] = -(-2.3) = 2.3$$
(approx.)

58. **(B)** If the solution is to be 5% sodium hydroxide, then 5% of 100 g is 5 g. Percent is always by weight unless otherwise specified.

59. **(C)** Since this equation is exothermic, higher temperatures will decrease the reaction to the right and increase the reaction to the left; so II is true. Also, statement I is true because with an increase in pressure the reaction will try to relieve that pressure by going in the direction that has the least volume. In this reaction, that would be to the right. Statement III is false because removing product in this reaction would increase the forward reaction. So I and II are true.

60. **(B)** The reaction is

$$H_2SO_4 + 2NaOH \rightarrow Na_2SO_4 + H_2O$$
$$\underset{\text{acid}}{1\text{ mol}} \quad \underset{\text{base}}{2\text{ mol}}$$

1 mol acid = $\frac{1}{2}$ mol of base
Since molarity × vol (L) = moles, then

$$M_aV_a = \tfrac{1}{2}M_bV_b$$
$$(1\ M)(x\text{ liters}) = \tfrac{1}{2}(1\ M)(0.05\text{ L})$$
$$x\text{ L} = \tfrac{1}{2}(0.05\text{ L})$$
$$x\text{ L} = 0.025\text{ L or }25\text{ mL}$$

61. **(E)** The reaction is 150 g

$$CaCO_3 + 2HCL \rightarrow CaCl_2 + H_2O + CO_2$$

The gram-molecular mass of calcium is 100 g. Then 150 g = $\frac{150}{100}$ or 1.5 mol of calcium carbonate. According to the equation, 1 mol of calcium carbonate yields 1 mol of carbon dioxide, and so 1.5 mol of calcium carbonate would yield 1.5 mol of carbon dioxide.

The gram-molecular mass of CO_2 = 44 g:

$$1.5\text{ mol of }CO_2 = 1.5 \times 44\text{ g} = 66\text{ g }CO_2$$

62. **(E)** The other choices are wrong because
(A) is lighter than air
(B) reacts with air
(C) is lighter than air
(D) needs heat to be evolved

63. **(B)** The Mg needs oxygen to form MgO; so the lid cannot be tightly sealed. All other choices are true.

64. **(C)** To ensure that all the water vapor collected in the U-tube comes from the reaction, the first drying tube is placed in the path of the hydrogen to absorb any evaporated water.

65. **(C)** Calcium chloride is deliquescent, and its weight gain of water indicates that water was formed from the reaction.

66. **(D)**

$CaCl_2$ and U-tube after	$= 57.32$ g
$CaCl_2$ and U-tube before	$= 39.32$ g
Mass of water formed	$= 18.00$ g

CuO and boat before	$= 30.23$ g
CuO and boat after	$= 14.23$ g
Mass of oxygen reacted	$= 16.00$ g

So, mass of water $= 18$ g

mass of oxygen $= 16$ g

Mass of hydrogen $= 2$ g

The ratio of mass of water to mass of hydrogen is $18 : 2$ or $9 : 1$.

67. **(D)** $1K = 39$, $1Al = 27$, $2SO_4 = 2(32 + 16 \times 4) = 192$, and $12H_2O = 12(2 + 16) = 216$. This totals 474 g.

68. **(D)**

$$\begin{array}{ccccc} x \text{ grams} & & 44.8 \text{ L} & & \\ 4Al & + & 3O_2 & \to & 2Al_2O_3 \\ 180 \text{ g} & & 67.2 \text{ L} & & \end{array}$$

$\dfrac{x}{108} = \dfrac{44.8}{67.2}$, and so $x = 72$ g.

or, using the mole method: 44.8 L $= 2$ mol

$$\begin{array}{ccccc} x \text{ mol} & & 2 \text{ mol} & & \\ 4Al & + & 3O_2 & \to & 2Al_2O_3 \\ 4 \text{ mol} & & 3 \text{ mol} & & \end{array}$$

This shows that:

$$\frac{x \text{ mol Al}}{4 \text{ mol Al}} = \frac{2 \text{ mol } O_2}{3 \text{ mol } O_2}$$

$$3x = 8$$

$$x = 8/3 \text{ mol Al}$$

$8/3$ mol Al $\times 27$ g Al/1 mol Al $= 72$ g Al

69. **(B)** Metal oxides are generally basic anhydrides.

70. **(A)** Since A is marked +, the oxidation (or loss of electrons) will occur at this pole.

71. **(C)** *Salt bridge* is the correct terminology.

72. **(A)**

$Zn \to Zn^{2+} + e^-$	$E^0 = +0.76$ V
$Cu^{2+} + 2e^- \to Cu^0$	$E^0 = +0.34$ V
Adding these	$= +1.10$ V

The voltmeter will read $+1.10$ V

73. **(D)**

$$\begin{array}{ccc} 213 \text{ g} & & x \text{ liters} \\ 2NaClO_3 & \to & 2NaCl + 3O_2 \end{array}$$

$\dfrac{213}{213} = \dfrac{x}{67.2}$, and so $x = 67.2$ L.

74. **(C)** The balanced equation has the coefficients 2, 3, 1, and 6 : $2Al(OH)_3 + 3H_2SO_4 \to Al_2(SO_4)_3 + 6H_2O$.

75. **(E)** The reaction is $NaClO_3 \to NaCl + \frac{3}{2}O_2$.

$$\Delta H_{\text{reaction}} = \Delta H_{f(\text{products})} - H_{f(\text{reactants})}$$
$$\Delta H_{\text{reaction}} = -98.2 + 0 - (-85.7)$$
$$\Delta H_{\text{reaction}} = -12.5 \text{ kcal}$$

76. **(D)** II and III are identical; isotopes differ only in the number of neutrons in the nucleus, and this affects only the atomic mass.

77. **(E)** I, II, and III, would affect this reaction.

78. **(D)** This setup depends on water displacement of an insoluble gas.

79. **(C)** The coefficients give the molar relations, so 2 mol of HCl give off 1 mol of CO_2. Given 4 mol of HCl, you have

$\dfrac{4}{2} = \dfrac{x}{1}$, then $x = \dfrac{4}{2} = 2$ mol.

80. **(E)** $K_{sp} = [B_2^{2+}] [SO_4^{2-}] = (3.9 \times 10^{-5})(3.9 \times 10^{-5}) = 1.5 \times 10^{-9}$

These two will be the same since

$$BaSO_4 \to Ba^{2+} + SO_4^{2-}$$

81. **(A)** According to the gas laws, only I will cause an increase in the volume of a confined gas.

82. **(B)** The introduction of the common ion SO_4^{2-} at 0.1 M forces the equilibrium to shift to the left and reduce the Ba^{2+} concentration.

83. **(E)** This type of neutralization always involves the formation of H_2O and a salt. I and II would be true only if the acids used were sulfuric and hydrochloric, respectively.

84. **(B)** The beta particle is a high-speed electron and has the smallest mass of the first four choices. Gamma rays are electromagnetic waves.

85. **(E)** A gamma ray is not affected by an electric field because it is an electromagnetic wave. The neutron is not an electromagnetic wave; it is a particle with no charge but with the approximate mass of a proton.

Diagnosing Your Needs

After taking Practice Test 2, check your answers against the correct ones. Then fill in the chart below.

In the space under each question number, place a check if you answered that question correctly.

Example

If your answer to question 5 was correct, place a check in the appropriate box.

Next, total the check marks for each section and insert the number in the designated block. Now do the arithmetic indicated and insert your percent for each area.

SUBJECT AREA	(✔)QUESTIONS ANSWERED CORRECTLY

I. Atomic Theory and Structure, including periodic relationships	19	20	21	23	101	115	44	76
☐ No. of checks ÷ 8 × 100 = _____%								

II. Nuclear Reactions	49	84	85
☐ No. of checks ÷ 3 × 100 = _____%			

III. Chemical Bonding and Molecular Structure	4	8	9	10	11	110	111	46
☐ No. of checks ÷ 8 × 100 = _____%								

IV. States of Matter and Kinetic Molecular Theory of Gases	3	12	13	14	43	53	55	81
☐ No. of checks ÷ 8 × 100 = _____%								

V. Solutions, including concentration units, solubility, and colligative properties	2	105	112	56	60
☐ No. of checks ÷ 5 × 100 = _____%					

VI. Acids and Bases	48	50	52	57	69	83
☐ No. of checks ÷ 6 × 100 = _____%						

VII. Oxidation-Reduction and Electrochemistry	114	116	47	51	70	71	72
☐ No. of checks ÷ 7 × 100 = _____%							

SUBJECT AREA	(✔)QUESTIONS ANSWERED CORRECTLY									
VIII. Stoichiometry	54	58	61	66	67	68	73	74	79	103
☐ No. of checks ÷ 10 × 100 = _____%										
IX. Reaction Rates									59	77
☐ No. of checks ÷ 2 × 100 = _____%										
X. Equilibrium						106	108	109	80	82
☐ No. of checks ÷ 5 × 100 = _____%										
XI. Thermodynamics: energy changes in chemical reactions, randomness, and criteria for spontaneity					5	6	7	107	75	102
☐ No. of checks ÷ 6 × 100 = _____%										
XII. Descriptive Chemistry: physical and chemical properties of elements and their familiar compounds; organic chemistry; periodic properties					1	15	16	17	18	22
					104	113	40	41	42	45
☐ No. of checks ÷ 12 × 100 = _____%										
XIII. Laboratory: equipment, procedures, observations, safety, calculations, and interpretation of results						62	63	64	65	78
☐ No. of checks ÷ 5 × 100 = _____%										

Planning Your Study

The percentages give you an idea of how you have done in the various major areas of the test. Because of the limited number of questions on some parts, these percentages may not be as reliable as the percentages for parts with larger numbers of questions. However, you should now have at least a rough idea of the areas in which you have done well and those in which you need more study.

Start your study with the areas in which you are the weakest. The corresponding chapters are indicated on page 371:

Subject Area	Chapters to Review
I. Atomic Theory and Structure, including periodic relationships	2
II. Nuclear Reactions	15
III. Chemical Bonding and Molecular Structure	3, 4
IV. States of Matter and Kinetic Molecular Theory of Gases	1
V. Solutions, including concentration units, solubility, and colligative properties	7
VI. Acids and Bases	11
VII. Oxidation-Reduction and Electrochemistry	12
VIII. Stoichiometry	5, 6
IX. Reaction Rates	9
X. Equilibrium	10
XI. Thermodynamics, including energy changes in chemical reactions, randomness, and criteria for spontaneity	8
XII. Descriptive Chemistry: physical and chemical properties of elements and their familiar compounds; organic chemistry; periodic properties	1, 2, 13, 14,
XIII. Laboratory: equipment, procedures, observations, safety, calculations, and interpretation of results.	All lab diagrams, 16

Answer Sheet for Practice Test 3

Determine the correct answer for each question. Then, using a No. 2 pencil, blacken completely the oval containing the letter of your choice

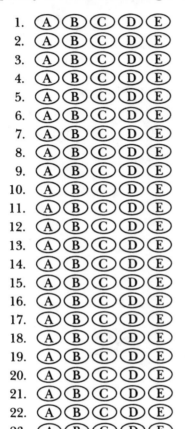

1. Ⓐ Ⓑ Ⓒ Ⓓ Ⓔ
2. Ⓐ Ⓑ Ⓒ Ⓓ Ⓔ
3. Ⓐ Ⓑ Ⓒ Ⓓ Ⓔ
4. Ⓐ Ⓑ Ⓒ Ⓓ Ⓔ
5. Ⓐ Ⓑ Ⓒ Ⓓ Ⓔ
6. Ⓐ Ⓑ Ⓒ Ⓓ Ⓔ
7. Ⓐ Ⓑ Ⓒ Ⓓ Ⓔ
8. Ⓐ Ⓑ Ⓒ Ⓓ Ⓔ
9. Ⓐ Ⓑ Ⓒ Ⓓ Ⓔ
10. Ⓐ Ⓑ Ⓒ Ⓓ Ⓔ
11. Ⓐ Ⓑ Ⓒ Ⓓ Ⓔ
12. Ⓐ Ⓑ Ⓒ Ⓓ Ⓔ
13. Ⓐ Ⓑ Ⓒ Ⓓ Ⓔ
14. Ⓐ Ⓑ Ⓒ Ⓓ Ⓔ
15. Ⓐ Ⓑ Ⓒ Ⓓ Ⓔ
16. Ⓐ Ⓑ Ⓒ Ⓓ Ⓔ
17. Ⓐ Ⓑ Ⓒ Ⓓ Ⓔ
18. Ⓐ Ⓑ Ⓒ Ⓓ Ⓔ
19. Ⓐ Ⓑ Ⓒ Ⓓ Ⓔ
20. Ⓐ Ⓑ Ⓒ Ⓓ Ⓔ
21. Ⓐ Ⓑ Ⓒ Ⓓ Ⓔ
22. Ⓐ Ⓑ Ⓒ Ⓓ Ⓔ
23. Ⓐ Ⓑ Ⓒ Ⓓ Ⓔ

ON THE ACTUAL CHEMISTRY TEST, THE RE-MAINING QUESTIONS MUST BE ANSWERED BY RETURNING TO THE SECTION OF YOUR AN-SWER SHEET YOU STARTED FOR CHEMISTRY.

ON THE ACTUAL CHEMISTRY TEST, THE FOLLOWING TYPE OF QUESTION MUST BE ANSWERED ON A SPECIAL SECTION (LABELED "CHEMISTRY") AT THE LOWER LEFT-HAND CORNER OF PAGE 2 OF YOUR ANSWER SHEET. THESE QUESTIONS WILL BE NUMBERED BEGINNING WITH 101 AND MUST BE ANSWERED ACCORDING TO THE FOLLOWING DIRECTIONS.

CHEMISTRY* Fill in oval CE only if II is a correct explanation of I.

	I	II	CE*
101.	Ⓣ Ⓕ	Ⓣ Ⓕ	◯
102.	Ⓣ Ⓕ	Ⓣ Ⓕ	◯
103.	Ⓣ Ⓕ	Ⓣ Ⓕ	◯
104.	Ⓣ Ⓕ	Ⓣ Ⓕ	◯
105.	Ⓣ Ⓕ	Ⓣ Ⓕ	◯
106.	Ⓣ Ⓕ	Ⓣ Ⓕ	◯
107.	Ⓣ Ⓕ	Ⓣ Ⓕ	◯
108.	Ⓣ Ⓕ	Ⓣ Ⓕ	◯
109.	Ⓣ Ⓕ	Ⓣ Ⓕ	◯
110.	Ⓣ Ⓕ	Ⓣ Ⓕ	◯
111.	Ⓣ Ⓕ	Ⓣ Ⓕ	◯
112.	Ⓣ Ⓕ	Ⓣ Ⓕ	◯
113.	Ⓣ Ⓕ	Ⓣ Ⓕ	◯
114.	Ⓣ Ⓕ	Ⓣ Ⓕ	◯
115.	Ⓣ Ⓕ	Ⓣ Ⓕ	◯
116.	Ⓣ Ⓕ	Ⓣ Ⓕ	◯

40. Ⓐ Ⓑ Ⓒ Ⓓ Ⓔ
41. Ⓐ Ⓑ Ⓒ Ⓓ Ⓔ
42. Ⓐ Ⓑ Ⓒ Ⓓ Ⓔ
43. Ⓐ Ⓑ Ⓒ Ⓓ Ⓔ
44. Ⓐ Ⓑ Ⓒ Ⓓ Ⓔ
45. Ⓐ Ⓑ Ⓒ Ⓓ Ⓔ
46. Ⓐ Ⓑ Ⓒ Ⓓ Ⓔ
47. Ⓐ Ⓑ Ⓒ Ⓓ Ⓔ
48. Ⓐ Ⓑ Ⓒ Ⓓ Ⓔ
49. Ⓐ Ⓑ Ⓒ Ⓓ Ⓔ
50. Ⓐ Ⓑ Ⓒ Ⓓ Ⓔ
51. Ⓐ Ⓑ Ⓒ Ⓓ Ⓔ
52. Ⓐ Ⓑ Ⓒ Ⓓ Ⓔ
53. Ⓐ Ⓑ Ⓒ Ⓓ Ⓔ
54. Ⓐ Ⓑ Ⓒ Ⓓ Ⓔ
55. Ⓐ Ⓑ Ⓒ Ⓓ Ⓔ

56. Ⓐ Ⓑ Ⓒ Ⓓ Ⓔ
57. Ⓐ Ⓑ Ⓒ Ⓓ Ⓔ
58. Ⓐ Ⓑ Ⓒ Ⓓ Ⓔ
59. Ⓐ Ⓑ Ⓒ Ⓓ Ⓔ
60. Ⓐ Ⓑ Ⓒ Ⓓ Ⓔ
61. Ⓐ Ⓑ Ⓒ Ⓓ Ⓔ
62. Ⓐ Ⓑ Ⓒ Ⓓ Ⓔ
63. Ⓐ Ⓑ Ⓒ Ⓓ Ⓔ
64. Ⓐ Ⓑ Ⓒ Ⓓ Ⓔ
65. Ⓐ Ⓑ Ⓒ Ⓓ Ⓔ
66. Ⓐ Ⓑ Ⓒ Ⓓ Ⓔ
67. Ⓐ Ⓑ Ⓒ Ⓓ Ⓔ
68. Ⓐ Ⓑ Ⓒ Ⓓ Ⓔ
69. Ⓐ Ⓑ Ⓒ Ⓓ Ⓔ
70. Ⓐ Ⓑ Ⓒ Ⓓ Ⓔ

71. Ⓐ Ⓑ Ⓒ Ⓓ Ⓔ
72. Ⓐ Ⓑ Ⓒ Ⓓ Ⓔ
73. Ⓐ Ⓑ Ⓒ Ⓓ Ⓔ
74. Ⓐ Ⓑ Ⓒ Ⓓ Ⓔ
75. Ⓐ Ⓑ Ⓒ Ⓓ Ⓔ
76. Ⓐ Ⓑ Ⓒ Ⓓ Ⓔ
77. Ⓐ Ⓑ Ⓒ Ⓓ Ⓔ
78. Ⓐ Ⓑ Ⓒ Ⓓ Ⓔ
79. Ⓐ Ⓑ Ⓒ Ⓓ Ⓔ
80. Ⓐ Ⓑ Ⓒ Ⓓ Ⓔ
81. Ⓐ Ⓑ Ⓒ Ⓓ Ⓔ
82. Ⓐ Ⓑ Ⓒ Ⓓ Ⓔ
83. Ⓐ Ⓑ Ⓒ Ⓓ Ⓔ
84. Ⓐ Ⓑ Ⓒ Ⓓ Ⓔ
85. Ⓐ Ⓑ Ⓒ Ⓓ Ⓔ

PRACTICE TEST 3

<u>Note:</u> For all questions involving solutions and/or chemical equations, assume that the system is in water unless otherwise stated.

PART A

<u>Directions:</u> Each set of lettered choices refers to the numbered statements or formulas immediately following it. Select the one lettered choice that best fits each statement or formula and then fill in the corresponding oval on the answer sheet. A choice may be used once, more than once, or not at all in each set.

<u>Questions 1–4</u> refer to the following diagram:

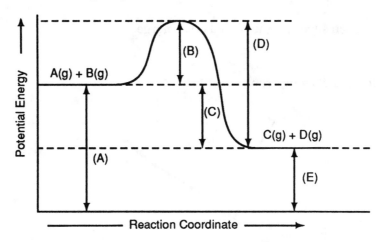

1. The activation energy of the forward reaction

2. The activation energy of the reverse reaction

3. The heat of the reaction for the forward reaction

4. The potential energy of the reactants
    (A) A
    (B) B
    (C) C
    (D) D
    (E) E

Questions 5–7 refer to the following diagram.

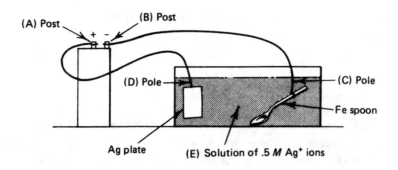

5. To plate silver on the spoon, the position to which the spoon must be connected

6. The position of the anode

7. The position from which silver that is plated out emerges

Questions 8–11 refer to the following graph:

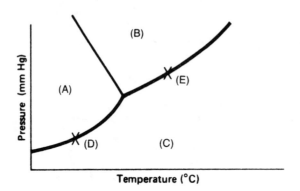

8. The solid area

9. The liquid area

10. The gaseous area

11. The area that can have both liquid and gas

Questions 12–14

    (A) 1

    (B) 2

    (C) 3

    (D) 4

    (E) 5

12. When the equation $Cu + HNO_3 \rightarrow Cu(NO_3)_2 + H_2O + NO$ is balanced, the coefficient, in the lowest whole number, of NO

13. If 6 moles of Cu react according to the above *balanced* equation, how many moles of NO will be formed?

14. If $Cu(NO_3)_2$ goes into solution as ions, it will dissociate into how many ions?

Questions 15–18

    (A) Ionic substance

    (B) Polar covalent substance

    (C) Nonpolar covalent substance

    (D) Amorphous substance

    (E) Metallic network

15. KCl

16. HCl(g)

17. CH_4

18. Li

Questions 19–23

    (A) Brownian movement

    (B) Litmus paper reaction

    (C) Phenolphthalein reaction

    (D) Dehydration

    (E) Deliquescent

19. The reason why a blue crystal of $CuSO_4 \cdot 5H_2O$ turns white when heated

20. The zigzag path of colloidal particles in light

21. The pink color in a basic solution

22. The pink color in an acid solution

23. The adsorption of water to the surface of a crystal

PART B

ON THE ACTUAL CHEMISTRY TEST, THE FOLLOWING TYPE OF QUESTION MUST BE ANSWERED ON A SPECIAL SECTION (LABELED "CHEMISTRY") AT THE LOWER LEFT-HAND CORNER OF PAGE 2 OF YOUR ANSWER SHEET. THESE QUESTIONS WILL BE NUMBERED BEGINNING WITH 101 AND MUST BE ANSWERED ACCORDING TO THE FOLLOWING DIRECTIONS.

<u>Directions:</u> Each question below consists of two statements, I in the left-hand column and II in the right-hand column. For each question, determine whether statement I is true or false <u>and</u> if statement II is true or false and fill in the corresponding T or F ovals on your answer sheet. <u>Fill in oval CE only if statement II is a correct explanation of statement I.</u>

Sample Answer Grid:

CHEMISTRY * Fill in oval CE only if
II is a correct explanation of I.

	I	II	CE*
101.	T F	T F	◯

101. Elements in the upper right of the periodic table are active nonmetals — BECAUSE — nonmetals have larger ionic radii than atomic radii.

102. A synthesis reaction that is nonspontaneous and has a negative value for the heat of formation will not occur until some heat is added — BECAUSE — nonspontaneous exothermic reactions need enough activation energy to get them started.

103. Transition elements in a particular period often have the same oxidation number — BECAUSE — they have a complete outer energy level.

104. When a crystal is added to a supersaturated solution of itself, the crystal does not change — BECAUSE — the solution cannot hold more solute.

105. Equilibrium is a static condition — BECAUSE — the forward reaction rate equals the reverse reaction rate.

106. The ionic bond is the strongest bond — BECAUSE — ionic bonds have an electrostatic attraction due to the loss and gain of electron(s).

107. Pressure applied to a gaseous system in equilibrium favors the forward reaction — BECAUSE — the product side of an equilibrium would show an increase in gas pressure.

108. If the forward reaction of an equilibrium is exothermic, adding heat to the system favors the reverse reaction BECAUSE additional heat causes a stress on the system, and the system moves in the direction that releases the stress.

109. An element that has an electron configuration of $1s^2$, $2s^2$, $3s^2$, $3p^6$, $3d^3$, $4s^2$ is a transition element BECAUSE the transition elements from Sc to Zn are filling the $3d$ orbitals.

110. The most electronegative elements in the periodic chart are found among nonmetals BECAUSE electronegativity is a measure of the ability of an atom to draw valence electrons to itself.

111. Basic anhydrides react in water to form bases BECAUSE metallic oxides react with water to form a solution that has an excess of hydroxide ions.

112. There are 3 mol of atoms in 18 g of water BECAUSE there are 6×10^{23} atoms in 1 mol.

113. Benzene is a good electrolyte BECAUSE a good electrolyte has charged ions that carry the electric current.

114. Normal butyl alcohol and 2-butanol are isomers BECAUSE isomers vary in the number of neutrons in the nucleus of the atom.

115. The reaction of $CaCO_3$ and HCl goes to completion BECAUSE reactions that form a precipitate go to completion.

116. A large number of alpha particles were deflected in the Rutherford experiment BECAUSE alpha particles that came close to the nucleus of the gold atoms were deflected.

PART C

<u>Directions:</u> Each of the questions or incomplete statements below is followed by five suggested answers or completions. Select the one that is best in each case and then fill in the corresponding oval on the answer sheet.

40. What are the simplest whole-number coefficients that balance this equation?
     _____C_4H_{10} + _____O_2 → _____CO_2 + _____H_2O
     (A) 1, 6, 4, 2
     (B) 2, 13, 8, 10
     (C) 1, 6, 1, 5
     (D) 3, 10, 16, 20
     (E) 4, 26, 16, 20

41. How many atoms are present in the formula $KAl(SO_4)_2$?
    (A) 7
    (B) 9
    (C) 11
    (D) 12
    (E) 13

42. All of the following are compounds EXCEPT
    (A) copper sulfate
    (B) carbon dioxide
    (C) washing soda
    (D) air
    (E) lime

43. What volume of gas, in liters, would 1.5 mol of hydrogen occupy at STP?
    (A) 11.2
    (B) 22.4
    (C) 33.6
    (D) 44.8
    (E) 67.2

44. What is the maximum number of electrons held in the d orbitals?
    (A) 2
    (B) 6
    (C) 8
    (D) 10
    (E) 14

45. If an element has an atomic number of 11, it will combine most readily with an element that has an atomic configuration of
    (A) $1s^2, 2s^2, 2p^6, 3s^2, 3p^1$
    (B) $1s^2, 2s^2, 2p^6, 3s^2, 3p^2$
    (C) $1s^2, 2s^2, 2p^6, 3s^2, 3p^3$
    (D) $1s^2, 2s^2, 2p^6, 3s^2, 3p^4$
    (E) $1s^2, 2s^2, 2p^6, 3s^2, 3p^5$

46. An example of a physical property is
    (A) rusting
    (B) decay
    (C) souring
    (D) low melting point
    (E) high heat of formation

47. A gas at STP which contains 6.02×10^{23} atoms and forms diatomic molecules will occupy
    (A) 11.2 L
    (B) 22.4 L
    (C) 33.6 L
    (D) 67.2 L
    (E) 1.06 qt

48. When excited electrons cascade to lower energy levels in an atom,
    (A) visible light is always emitted
    (B) the potential energy of the atom increases
    (C) the electrons always fall back to the first energy level
    (D) the electrons fall indiscriminately to all levels
    (E) the electrons fall back to the lowest unfilled energy level

49. The spectroscope uses the concept that
    (A) charged particles are evenly deflected in a magnetic field
    (B) charged particles are deflected in a magnetic field inversely to the mass of the particles
    (C) particles of heavier mass are deflected in a magnetic field to a greater degree than lighter particles
    (D) particles are evenly deflected in a magnetic field

50. The bond that includes an upper and a lower sharing of electron orbitals is called
    (A) a pi bond
    (B) a sigma bond
    (C) a hydrogen bond
    (D) a covalent bond
    (E) an ionic bond

51. What is the boiling point of water at the top of Pike's Peak?
    (A) It is 100°C.
    (B) It is >100°C since the pressure is less than at ground level.
    (C) It is <100°C since the pressure is less than at ground level.
    (D) It is >100°C since the pressure is greater than at ground level.
    (E) It is <100°C since the pressure is greater than at ground level.

52. The atomic structure of the alkane series contains the hybrid orbitals designated as
    (A) sp
    (B) sp^2
    (C) sp^3
    (D) sp^3d^2
    (E) sp^4d^3

53. Which of the following is(are) true for this reaction?

$$Cu + 4HNO_3 \rightarrow Cu(NO_3)_2 + 2H_2O + 2NO_2(g)$$

I. It is an oxidation-reduction reaction.
II. Copper is oxidized.
III. The oxidation number of nitrogen goes from +5 to +4.

(A) I, II, and III
(B) I and II only
(C) II and III only
(D) I only
(E) III only

54. Which of the following properties can be attributed to water?

I. It has a permanent dipole moment attributed to its molecular structure.
II. It is a very good conductor of electricity.
III. It has its polar covalent bonds with hydrogen on opposite sides of the oxygen atom, and thus the molecule is linear.

(A) I, II, and III
(B) I and II only
(C) II and III only
(D) I only
(E) III only

55. Which of the following define(s) an acid according to conventional acid theories?

I. It is a good proton donor.
II. It is a good electron-pair acceptor.
III. It has an excess of H_3O^+ in solution.

(A) I, II, and III
(B) I and II only
(C) II and III only
(D) I only
(E) III only

56. A nuclear reactor must include which of the following parts?

I. Electric generator
II. Fissionable fuel elements
III. Moderator

(A) I, II, and III
(B) I and II only
(C) II and III only
(D) I only
(E) III only

57. Which of the following salts will hydrolyze in water to form basic solutions?

 I. NaCl
 II. $CuSO_4$
 III. K_3PO_4

(A) I, II, and III
(B) I and II only
(C) II and III only
(D) I only
(E) III only

58. The dissolving of 1 mol of NaCl in 1000 g of water will change the boiling point of the water to
(A) 100.51°C
(B) 101.02°C
(C) 101.53°C
(D) 101.86°C
(E) 103.62°C

59. What is the structure associated with the BF_3 molecule?
(A) rectangle
(B) trigonal planar
(C) tetrahedron
(D) octahedron
(E) square

60. Which of these statements is the best expression for the sp^3 hybridization of carbon electrons?
(A) The new orbitals are one s orbital and three p orbitals.
(B) The s electron is promoted to the p orbitals.
(C) The s orbital is deformed into a p orbital.
(D) Four new and equivalent orbitals are formed.
(E) The s orbital electron loses energy to fall back into a partially filled p orbital.

61. The bonding that explains the variation of the boiling point of water from the boiling points of similarly structured molecules is
(A) hydrogen bonding
(B) van der Waals forces
(C) covalent bonding
(D) ionic bonding
(E) coordinate covalent bonding

62. If K for the reaction $H_2 + I_2 \rightleftharpoons 2HI$ is equal to 45.9 at 450°C, and 1 mol of H_2 and 1 mol of I_2 are introduced into a 1-L box at that temperature, what will be the concentration of H_2 at equilibrium?
(A) 0.114 mol/L
(B) 0.228 mol/L
(C) 0.456 mol/L
(D) 0.516 mol/L
(E) 0.772 mol/L

63. What is the gram-molecular mass of a nonionizing solid if 10 g of this solid, dissolved in 100 g of water, formed a solution that froze at −1.22°C?

    (A) 0.65 g
    (B) 6.5 g
    (C) 130 g
    (D) 154 g
    (E) 265 g

64. What is the pH of a solution with a hydroxide ion concentration of 0.00001 mol/L?

    (A) −5
    (B) −1
    (C) 5
    (D) 9
    (E) 14

65. Electrolysis of a dilute solution of sodium chloride results in the cathode product

    (A) sodium
    (B) hydrogen
    (C) chlorine
    (D) oxygen
    (E) peroxide

66. ____$C_2H_4(g)$ + ____$O_2(g)$ → ____$CO_2(g)$ + ____$H_2O(l)$

    If the equation for the above reaction is balanced with the smallest whole numbers, what is the appropriate coefficient for oxygen gas?

    (A) 1
    (B) 2
    (C) 3
    (D) 4
    (E) 5

67. Five liters of gas at STP have a mass of 12.5 g. What is the molar mass of the gas?

    (A) 12.5 g/mol
    (B) 25.0 g/mol
    (C) 47.5 g/mol
    (D) 56.0 g/mol
    (E) 125 g/mol

68. A compound whose molecular mass is 90 g contains 40.0% carbon, 6.67% hydrogen, and 53.33% oxygen. What is the true formula of the compound?

    (A) $C_2H_2O_4$
    (B) CH_2O_4
    (C) C_3H_6O
    (D) C_3HO_3
    (E) $C_3H_6O_3$

69. How many moles of CaO are needed to react with an excess of water to form 370 g of calcium hydroxide?
    (A) 1
    (B) 2
    (C) 3
    (D) 4
    (E) 5

70. To what volume, in milliliters, must 50.0 mL of 3.50 M H_2SO_4 be diluted in order to make $2 M$ H_2SO_4?
    (A) 25
    (B) 60.1
    (C) 87.5
    (D) 93.2
    (E) 101

71. If K_{eq} is small, it indicates that equilibrium occurs
    (A) at a low product concentration
    (B) at a high product concentration
    (C) after considerable time
    (D) with the help of a catalyst
    (E) with no forward reaction

72. A student measured 10.0 mL of an HCl solution into a beaker and titrated it with a standard NaOH solution that was 0.09 M. The initial NaOH buret reading was 34.7 mL while the final reading showed 49.2 mL. What is the molarity of the HCl solution?
    (A) 0.13
    (B) 0.47
    (C) 4.7
    (D) 14.5
    (E) 36.5

73. A student made the following observations in the laboratory:

    (a) Sodium metal reacted vigorously with water, while a strip of magnesium did not seem to react at all.
    (b) The magnesium strip reacted with dilute hydrochloric acid faster than an iron strip.
    (c) A copper rivet suspended in silver nitrate solution was covered with silver-colored stalactites in several days, and the resulting solution had a blue color.
    (d) Iron filings dropped into the blue solution were coated with an orange color.

    The order of *decreasing* strength reducing agents is
    (A) Na, Mg, Fe, Ag, Cu
    (B) Mg, Na, Fe, Cu, Ag
    (C) Ag, Cu, Fe, Mg, Na
    (D) Na, Fe, Mg, Cu, Ag
    (E) Na, Mg, Fe, Cu, Ag

74. A student placed water, sodium chloride, potassium dichromate, sand, chalk, and hydrogen sulfide into a distilling flask and proceeded to distill. What ingredient besides water would be found in the distillate?
(A) sodium chloride
(B) chalk
(C) sand
(D) hydrogen sulfide
(E) chrome sulfate

75. Which test would you use to prove the presence of the other substance in the distillate flask in question 74?
(A) flame test for sodium
(B) barium chloride test for sulfate
(C) silver nitrate test for chloride
(D) acid test for carbonate in chalk
(E) lead acetate test for hydrogen sulfide

76. A student used a steam-jacketed eudiometer and filled it with 32 mL of oxygen and 4 mL of hydrogen over mercury. How much of which gas would be left uncombined after sparking?
(A) none of either
(B) 3 mL H_2
(C) 24 mL O_2
(D) 28 mL O_2
(E) 30 mL O_2

77. What would be the *total* volume, in milliliters, of the gases in question 76 after sparking?
(A) 16
(B) 24
(C) 34
(D) 36
(E) 40

78. A student mixed a small amount of amyl alcohol, glacial acetic acid, and a few drops of concentrated H_2SO_4. She then gently heated the mixture until she observed an odor. What odor should she expect to be given off?
(A) banana oil
(B) oil of wintergreen
(C) pineapple
(D) apple
(E) formaldehyde

79. In which period is the most electronegative element found?
(A) 1
(B) 2
(C) 3
(D) 4
(E) 5

80. What could be the equilibrium constant for the reaction $aA + bB \rightleftharpoons cC + dD$ if A and D are solids?

(A) $\dfrac{[C]^c[D]^d}{[A]^a[B]^b}$

(B) $\dfrac{[A]^a[B]^b}{[C]^c[D]^d}$

(C) $\dfrac{[C]^c}{[B]^b}$

(D) $\dfrac{[C]^c[D]^d}{[A]^a}$

(E) $[A]^a[B]^b[C]^c[D]^d$

81. Which of the following is NOT a commonly occurring sulfur compound?
    (A) H_2S
    (B) H_2SO_4
    (C) SO_2
    (D) Ag_2S
    (E) SO_3

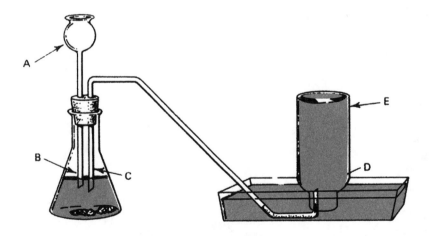

82. What letter designates an error in the above laboratory setup?
    (A) A
    (B) B
    (C) C
    (D) D
    (E) E

83. The bonding that is most significant in explaining the variation in the boiling point of water from the boiling points of similarly structured molecules is
(A) hydrogen bonding
(B) van der Waals forces
(C) convalent bonding
(D) ionic bonding
(E) coordinate covalent bonding

84. If K for the reaction $H_2 + I_2 \rightleftharpoons 2HI$ is equal to 45.9 at 450°C and 1 mole of H_2 and 1 mole of I_2 are introduced into a 1-liter box at that temperature, what will be the expression of K at equilibrium?

(A) $\dfrac{[x^2]^2}{[1-x][1-x]}$

(B) $\dfrac{[2x^2]^2}{[1-x][1-x]}$

(C) $\dfrac{[2x^2]^2}{[x][x]}$

(D) $\dfrac{[1-x][1-x]}{[2x^2]^2}$

(E) $\dfrac{[1-x][1-x]}{[x^2]^2}$

85. What is the molar mass of a nonionizing solid if 10 grams of this solid, dissolved in 200 grams of water, forms a solution that freezes at –3.72°C?
(A) 25 g/mol
(B) 50 g/mol
(C) 100 g/mol
(D) 150 g/mol
(E) 1000 g/mol

STOP
IF YOU FINISH BEFORE ONE HOUR IS UP, YOU MAY GO BACK
TO CHECK YOUR WORK OR COMPLETE UNANSWERED QUESTIONS.

Answers and Explanations for Test 3

1. **(B)** The activation energy is the energy needed to begin the reaction.
2. **(D)** For the reverse reaction to occur, the sum B + C is needed. This is shown by D.
3. **(C)** The heat of the reaction is the heat liberated between the level of potential energy of the reactants and that of the products. This is the quantity C on the diagram.
4. **(A)** The potential energy of the activated complex is the total of the original potential energy (A) and the activation energy (B).
5. **(B)** The spoon must be made the cathode to attract the Ag^+ ions.
6. **(D)** The silver plate is the anode.
7. **(E)** The solution of Ag^+ provides the silver for plating.
8. **(A)**
9. **(B)** In a phase diagram, the zones are as shown below.
10. **(C)**

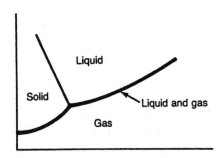

11. **(E)** The boundary of liquid and gas is shown on the diagram.
12–14. The balanced equation with half-reactions is:
12. **(B)** $3Cu^0 \rightarrow 3Cu^{2+} + 3(2e^-)$

$$\frac{2NO_3^- + 8H^+ + 2(3e^-) \rightarrow 2NO + 4H_2O}{3Cu + 8HNO_3 \rightarrow 3Cu(NO_3)_2}$$
$$+ 4H_2O + 2NO$$

13. **(D)** The coefficients show that 3 moles of Cu produces 2 moles of NO; so 6 moles of Cu produces 4 moles of NO.
14. **(C)** $Cu(NO_3)_2 \rightarrow Cu^{2+} + 2NO_3^-$ shows that the dissociation yields three ions.
15. **(A)** KCl is ionic because it is the product of a very active metal combining with a very active nonmetal.
16. **(B)** The electronegativity difference between H and Cl is between 0.5 and 1.7. This indicates an unequal sharing of electrons, which results in a polar covalent bond.
17. **(C)** Because these polar bonds are symmetrically arranged in the methane molecule, the molecule is nonpolar covalent.
18. **(E)** Lithium (Li) is a metal.
19. **(D)** When hydrated copper sulfate is heated, the crystal crumples as the water is forced out of the structure, and a white powder is the result.
20. **(A)** Brownian movement is due to molecular collisions with colloidal particles, which knock the particles about in a zigzag path noted by the reflected light from these particles.
21. **(C)** The indicator phenolphathalein turns pink in a basic solution.
22. **(B)** Litmus paper turns pink in an acid solution.
23. **(E)** A substance that is deliquescent draws water to its surface. At times it can draw enough water to form a water solution.
101. **(T, T)** The assertion is true because the nonmetals have a tendency to gain electrons readily. The reason is a true statement but does not explain the assertion.

102. (**T, T, CE**) The assertion is explained by the reason. The graphic display of this is

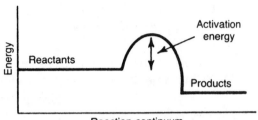

103. (**T, F**) The assertion is true, but the reasoning is false. Transition elements have incomplete inner energy levels which are being filled with the additional electrons, thus leaving the outer energy level the same in most cases. This gives these elements common oxidation numbers.

104. (**F, T**) The assertion is false, and the reason is true. A supersaturated solution is holding more than its normal solubility, and the addition of a crystal causes a crystallization to occur.

105. (**F, T**) Equilibrium is a dynamic condition because of the reason stated. The assertion is false; the reason is true.

106. (**T, T, CE**) Both statements are true, and the reason explains why ionic bonding is the strongest.

107. (**F, F**) An equilibrium system must have a gaseous reactant or product for pressure to affect the equilibrium. Then increased pressure will cause the reaction to go in the direction that reduces the concentration of gaseous substances.

108. (**T, T, CE**) The assertion is explained by the reason; both are true.

109. (**T, T, CE**) Both the assertion and the reason are true, and the assertion explains that the element's orbital designation places it in the first transition series of filling the $3d$ orbitals.

110. (**T, T, CE**) The assertion is true; the reason is true and explains why nonmetals have the highest electronegativity.

111. (**T, T, CE**) The assertion is explained by the reason; both are true.

112. (**T, T**) There are three moles of atoms in 18 g of water because 18 g is 1 mol of water molecules and each molecule has three atoms. The reason does not explain the assertion but is also true.

113. (**F, T**) Benzene is a nonionizing substance and therefore a nonelectrolyte. The reason is a true statement.

114. (**T, F**) Isomers have the same empirical formula but vary in their structural formulas.

115. (**T, T**) The reaction does go to completion, but a gaseous product is formed, not a precipitate. $2HCl + CaCO_3 \rightarrow CaCl_2 + H_2O + CO_2(g)$

116. (**F, T**) In the Rutherford experiment relatively few alpha particles were deflected, indicating a great deal of empty space in the atom. The reason is a true statement.

40. (**B**) The correct coefficients are 2, 13, 8, and 10.

41. (**D**) $1K + 1Al + 2S + 8O = 12$ total.

42. (**D**) Air is a mixture; all others are compounds. Washing soda (C) is sodium carbonate, and lime (E) is calcium oxide.

43. (**C**) One mol of a gas at STP occupies 22.4 L, and so 1.5 mol $\times$ 22.4 L = 33.6 L.

44. (**D**) The maximum number of electrons in each kind of orbital is

$s = 2$ in one orbital
$p = 6$ in three orbitals
$d = 10$ in five orbitals
$f = 14$ in seven orbitals

45. (**E**) The element with atomic number 11 is Na with 1 electron in the $3s$ orbital. It would readily combine with the element that has $3p^5$ as the outer orbital since it needs only 1 more electron to be filled.

46. **(D)** The only physical property named in the list is low melting point.

47. **(A)** If the gas is diatomic, then 6.02×10^{23} atoms will form $6.02 \times 10^{23}/2$ molecules. At STP, 6.02×10^{23} molecules occupy 22.4 L. So half that amount would occupy 11.2 L.

48. **(E)** Cascading excited electrons can fall only to the lowest energy level that is unfilled.

49. **(B)** The spectroscope uses a magnetic field to separate isotopes by bending their path. The lighter ones are bent further than the heavier ones.

50 **(A)** The pi bond is a bond between two p orbitals, like this:

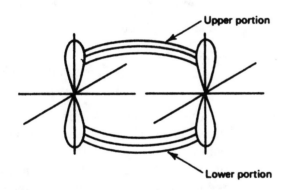

51. **(C)** At Pike's Peak (alt. approx. 14,000 ft) the pressure is lower than at ground level; therefore the vapor pressure at a lower temperature will equal the outside pressure and boiling will occur.

52. **(C)** The alkanes use the sp^3 hybrid orbitals.

53. **(A)** I, II, and III are correct.

54. **(D)** Only I is correct.

55. **(A)** I, II, and III are acid definitions.

56. **(C)** I is not necessary for the reactor, but nuclear energy is often used to operate an electric generator. The others, II and III, are necessary for fuel and neutron-speed control, respectively.

57. **(E)** III is a salt from a strong base and a weak acid, which hydrolyzes to form a basic solution with water.

58. **(B)** Since the boiling point is increased by 0.51°C for each mole of particles, 1 mol of $NaCl \rightarrow Na^+ + Cl^-$ gives 2 mol of particles. Therefore the boiling point will be 1.02° higher or 101.02°C.

59. **(B)** The sp^2 hybridized orbital in BF_3 is trigonal, planar and so is related to the triangle.

60 **(D)** When hybridization forms the sp^3 orbitals, four entirely new orbitals different from the former s and p orbitals result.

61. **(A)** Hydrogen bonding between water molecules causes the boiling point to be higher than would be expected.

62. **(B)** At the beginning of the reaction

$$[H_2] = 1 \text{ mol/L}$$
$$[I_2] = 1 \text{ mol L}$$
$$[HI] = 0$$

At equilibrium,

$$(H_2 + I_2 \rightleftharpoons 2HI)$$
(Let x = moles/liter of H_2 and I, in HI form)

$$[H_2] = (1 - x) \text{ mole/liter}$$
$$[I_2] = (1 - x) \text{ mole/liter}$$
$$[HI] = 2x \text{ mole/liter}$$

$$K = \frac{[HI]^2}{[H_2][I_2]} = \frac{(2x)^2}{(1-x)(1-x)} = 45.9$$

Taking the square root of both sides gives

$$\frac{2x}{1-x} = 6.8$$
$$x = 0.772 \text{ mol/L}$$

Therefore $[H_2] = 1 - x = 1 - 0.0772 = 0.228$ mol/L.

63. **(D)** 10 g/100 g water = 100 g/1000 g water. The freezing point depression, 1.22, is divided by 1.86, which is the depression caused by 1 mol in 1000 g of water, to find how many moles are dissolved.

1.22/1.86 = 0.65 mol
So 100 g = 0.65 mol. Then 1 mol =
$$\frac{100\ g}{0.65\ mol} = 153.8\ or\ 154\ g/mol$$

64. **(D)** The K_w of water = [H⁺] [OH⁻] = 10^{-14}. If [OH⁻] = 10^{-5} mol/L, then [H⁺] = $10^{-14}/10^{-5} = 10^{-9}$

   pH = –log [H⁺] (by definition)
   pH = –[–9]
   pH = 9

65. **(B)** When a dilute NaCl solution is electrolyzed, hydrogen is given off at the cathode, chlorine is given off at the anode, and sodium hydroxide is left in the container.

66. **(C)** The correctly balanced equation is:

   $$C_2H_4(g) + 3O_2(g) \rightarrow 2CO_2(g) + 2H_2O(l)$$

67. **(D)** One mol of a gas at STP occupies 22.4 L.

   If $\dfrac{5\ L}{12.5\ g}$, then

   $$\frac{22.4\ L}{gram\text{-}molecular\ mass}$$

   $$\frac{5}{12.5} = \frac{22.4}{x}$$
   $$x = 56\ g$$

68. **(E)** To find the simple or empirical formula, divide each percent by the element's atomic mass.

   Carbon
   40 ÷ 12 = 3.333

   Hydrogen
   6.67 ÷ 1 = 6.67

   Oxygen
   53.33 ÷ 16 = 3.33

   Next, divide each quotient by the smallest quotient in an attempt to get small whole numbers.

   3.33 ÷ 3.33 = 1 carbon
   6.67 ÷ 3.33 = 2 hydrogen
   3.33 ÷ 3.33 = 1 oxygen

The simplest formula is CH_2O, which has a molecular mass of 30.
The true molecular mass is given as 90, which is three times the simplest. Therefore the true formula is $C_3H_6O_3$.

69. **(E)** The reaction is

   $$\overset{370\ g}{CaO} + H_2O \rightarrow Ca(OH)_2$$

   $Ca(OH)_2$ molecular wt. = 74
   So 370 g ÷ 74 = 5 mol of $Ca(OH)_2$ wanted. The reaction shows 1 mol of CaO produces 1 mol of $Ca(OH)_2$, and so the answer is 5 mols.

70. **(C)** In dilution problems, this formula can be used.

   $$M_{before} \times V_{before} = M_{after} \times V_{after}$$

   Substituting gives

   $3.5 \times 50 = 2 \times (?x)$
   $x = 87.5$ mL, new volume after dilution

71. **(A)** For the K_{eq} to be small, the numerator, which is made up of the concentration(s) of the product(s) at equilibrium, must be smaller than the denominator. This generally means that equilibrium is reached rather rapidly.

72. **(A)** The amount of NaOH used is

   $$49.2 - 34.7 = 14.5\ mL$$

   Using $M_1 \times V_1 \times M_2 \times V_2$ gives
   $0.09\ M \times 14.5\ mL = M_2 \times 10\ mL$
   $M_2 = 0.13\ M$

73. **(E)** The reactions recorded indicated that the ease of losing electrons is greater in sodium than in magnesium, greater in magnesium than in iron, greater in iron than in copper, and finally greater in copper than in silver.

74. **(D)** Distillation removes only dissolved solids from the distillate. The volatile gases, like H_2S, will be carried into the distillate.

75. **(E)** The test for H_2S is to moisten lead acetate paper and look for brown-black precipitate to form, which is a positive result.

76. **(E)** H_2 to O_2 ratio by volume is 2 : 1 in the formation of water. Therefore, 4 mL H_2 will react with 2 mL of O_2 to make 4 mL of steam.
$$2H_2(g) + O_2(g) \rightarrow 2H_2O(g)$$
This leaves 30 mL of O_2 uncombined.

77. **(C)** There will be 30 mL of O_2 + 4 mL of steam = 34 mL total.

78. **(A)** She would make a small amount of amyl acetate which is an ester with a banana-like odor.

79. **(B)** The most electronegative element is fluorine (F), found in Period 2.

80. **(C)** Solids are incorporated into the K value and therefore do not appear on the right side of the equation.

81. **(E)** SO_3 is not easily formed. The commercial process uses a catalyst.

82. **(B)** The thistle tube is below the fluid level in the flask and will cause liquid to be forced up the tube when gas is evolved in the reaction.

83. **(A)** Hydrogen bonding between water molecules causes the boiling point to be higher than would be expected.

84. **(B)** At the beginning of the reaction,
$$[H_2] = 1 \text{ mol/L}$$
$$[I_2] = 1 \text{ mol/L}$$
$$[HI] = 0$$

At equilibrium,
$$H_2 + I_2 \rightleftharpoons 2HI$$
(Let x = moles/liter of H_2 and I_2 that now exist in HI form.)
Then, at equilibrium,
$$[H_2] = (1 - x) \text{ mol/L}$$
$$[I_2] = (1 - x) \text{ mol/L}$$
$$[HI] = 2x \text{ mol/L}$$

Then, substituting the above values into the equation gives
$$K = \frac{[HI]^2}{[H_2][I_2]} = \frac{(2x)^2}{(1-x)(1-x)} = 45.9$$

85. **(A)** 10 g/200 g of water = 50 g/1000 g of water (5 times as much).
The freezing point depression, 3.72°, is divided by 1.86°, which is the depression caused by 1 mol in 1000 g of water, to find how many moles are dissolved:
$$3.72°/1.86°/\text{mol} = 2 \text{ mol}$$
So if there were 50 g causing this depression and equal to 2 moles, 1 mol would be one half of 50 g, or 25 g.

Diagnosing Your Needs

After taking Practice Test 3, check your answers against the correct ones. Then fill in the chart below.

In the space under each question number, place a check if you answered that question correctly.

Example

If your answer to question 5 was correct, place a check in the appropriate box.

Next, total the check marks for each section and insert the number in the designated block. Now do the arithmetic indicated and insert your percent for each area.

SUBJECT AREA	(✔) QUESTIONS ANSWERED CORRECTLY								
I. Atomic Theory and Structure, including periodic relationships	14	110	112	116	44	48	50	59	60
☐ No. of checks ÷ 9 × 100 = _____ %									
II. Nuclear Reactions								49	56
☐ No. of checks ÷ 2 × 100 = _____ %									
III. Chemical Bonding and Molecular Structure	15	16	17	18	106	41	45	54	61	84
☐ No. of checks ÷ 10 × 100 = _____ %										
IV. States of Matter and Kinetic Molecular Theory of Gases	8	9	10	11	43	47	51	67
☐ No. of checks ÷ 8 × 100 = _____ %								
V. Solutions, including concentration units, solubility, and colligative properties	19	20	104	58	70
☐ No. of checks ÷ 5 × 100 = _____ %					
VI. Acids and Bases	21	22	113	115	55	57	64
☐ No. of checks ÷ 7 × 100 = _____ %							
VII. Oxidation-Reduction and Electrochemistry	5	6	7	53	65
☐ No. of checks ÷ 5 × 100 = _____ %					

SUBJECT AREA	(✔)QUESTIONS ANSWERED CORRECTLY								

VIII. Stoichiometry

12	13	40	63	66	68	69	76	77

☐ No. of checks ÷ 9 × 100 = _____%

IX. Reaction Rates

1	108

☐ No. of checks ÷ 2 × 100 = _____%

X. Equilibrium

105	107	62	71	80	85

☐ No. of checks ÷ 6 × 100 = _____%

XI. Thermodynamics: energy changes in chemical reactions, randomness, and criteria for spontaneity

2	3	4	102

☐ No. of checks ÷ 4 × 100 = _____%

XII. Descriptive Chemistry: physical and chemical properties of elements and their familiar compounds; organic chemistry; periodic properties

23	101	103	109	111	114	42

46	52	78	79	81

☐ No. of checks + 12 × 100 = _____%

XIII. Laboratory: equipment, procedures, observations, safety, calculations, and interpretation of results

72	73	74	75	82	83

☐ No. of checks ÷ 6 × 100 = _____%

Planning Your Study

The percentages give you an idea of how you have done in the various major areas of the test. Because of the limited number of questions on some parts, these percentages may not be as reliable as the percentages for parts with larger numbers of questions. However, you should now have at least a rough idea of the areas in which you have done well and those in which you need more study.

Start your study with the areas in which you are the weakest. The corresponding chapters are indicated on page 396.

Subject Area	Chapters to Review
I. Atomic Theory and Structure, including periodic relationships	2, 15
II. Nuclear Reactions	15
III. Chemical Bonding and Molecular Structure	3, 4
IV. States of Matter and Kinetic Molecular Theory of Gases	1
V. Solutions, including concentration units, solubility, and colligative properties	7
VI. Acids and Bases	11
VII. Oxidation-Reduction and Electrochemistry	12
VIII. Stoichiometry	5, 6
IX. Reactions Rates	9
X. Equilibrium	10
XI. Thermodynamics, including energy changes in chemical reactions, randomness, and criteria for spontaneity	8
XII. Descriptive Chemistry: physical and chemical properties of elements and their familiar compounds; organic chemistry; periodic properties	1, 2, 13, 14
XIII. Laboratory: equipment, procedures, observations, safety, calculations, and interpretation of results.	All lab diagrams, 16

Answer Sheet for Practice Test 4

Determine the correct answer for each question. Then, using a No. 2 pencil, blacken completely the oval containing the letter of your choice.

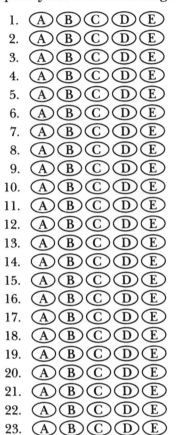

1. Ⓐ Ⓑ Ⓒ Ⓓ Ⓔ
2. Ⓐ Ⓑ Ⓒ Ⓓ Ⓔ
3. Ⓐ Ⓑ Ⓒ Ⓓ Ⓔ
4. Ⓐ Ⓑ Ⓒ Ⓓ Ⓔ
5. Ⓐ Ⓑ Ⓒ Ⓓ Ⓔ
6. Ⓐ Ⓑ Ⓒ Ⓓ Ⓔ
7. Ⓐ Ⓑ Ⓒ Ⓓ Ⓔ
8. Ⓐ Ⓑ Ⓒ Ⓓ Ⓔ
9. Ⓐ Ⓑ Ⓒ Ⓓ Ⓔ
10. Ⓐ Ⓑ Ⓒ Ⓓ Ⓔ
11. Ⓐ Ⓑ Ⓒ Ⓓ Ⓔ
12. Ⓐ Ⓑ Ⓒ Ⓓ Ⓔ
13. Ⓐ Ⓑ Ⓒ Ⓓ Ⓔ
14. Ⓐ Ⓑ Ⓒ Ⓓ Ⓔ
15. Ⓐ Ⓑ Ⓒ Ⓓ Ⓔ
16. Ⓐ Ⓑ Ⓒ Ⓓ Ⓔ
17. Ⓐ Ⓑ Ⓒ Ⓓ Ⓔ
18. Ⓐ Ⓑ Ⓒ Ⓓ Ⓔ
19. Ⓐ Ⓑ Ⓒ Ⓓ Ⓔ
20. Ⓐ Ⓑ Ⓒ Ⓓ Ⓔ
21. Ⓐ Ⓑ Ⓒ Ⓓ Ⓔ
22. Ⓐ Ⓑ Ⓒ Ⓓ Ⓔ
23. Ⓐ Ⓑ Ⓒ Ⓓ Ⓔ

ON THE ACTUAL CHEMISTRY TEST, THE RE-MAINING QUESTIONS MUST BE ANSWERED BY RETURNING TO THE SECTION OF YOUR AN-SWER SHEET YOU STARTED FOR CHEMISTRY.

ON THE ACTUAL CHEMISTRY TEST, THE FOLLOWING TYPE OF QUESTION MUST BE ANSWERED ON A SPECIAL SECTION (LABELED "CHEMISTRY") AT THE LOWER LEFT-HAND CORNER OF PAGE 2 OF YOUR ANSWER SHEET. THESE QUESTIONS WILL BE NUMBERED BEGINNING WITH 101 AND MUST BE ANSWERED ACCORDING TO THE FOLLOWING DIRECTIONS.

CHEMISTRY* Fill in oval CE only if II is a correct explanation of I.

	I	II	CE*
101.	Ⓣ Ⓕ	Ⓣ Ⓕ	◯
102.	Ⓣ Ⓕ	Ⓣ Ⓕ	◯
103.	Ⓣ Ⓕ	Ⓣ Ⓕ	◯
104.	Ⓣ Ⓕ	Ⓣ Ⓕ	◯
105.	Ⓣ Ⓕ	Ⓣ Ⓕ	◯
106.	Ⓣ Ⓕ	Ⓣ Ⓕ	◯
107.	Ⓣ Ⓕ	Ⓣ Ⓕ	◯
108.	Ⓣ Ⓕ	Ⓣ Ⓕ	◯
109.	Ⓣ Ⓕ	Ⓣ Ⓕ	◯
110.	Ⓣ Ⓕ	Ⓣ Ⓕ	◯
111.	Ⓣ Ⓕ	Ⓣ Ⓕ	◯
112.	Ⓣ Ⓕ	Ⓣ Ⓕ	◯
113.	Ⓣ Ⓕ	Ⓣ Ⓕ	◯
114.	Ⓣ Ⓕ	Ⓣ Ⓕ	◯
115.	Ⓣ Ⓕ	Ⓣ Ⓕ	◯
116.	Ⓣ Ⓕ	Ⓣ Ⓕ	◯

40. Ⓐ Ⓑ Ⓒ Ⓓ Ⓔ
41. Ⓐ Ⓑ Ⓒ Ⓓ Ⓔ
42. Ⓐ Ⓑ Ⓒ Ⓓ Ⓔ
43. Ⓐ Ⓑ Ⓒ Ⓓ Ⓔ
44. Ⓐ Ⓑ Ⓒ Ⓓ Ⓔ
45. Ⓐ Ⓑ Ⓒ Ⓓ Ⓔ
46. Ⓐ Ⓑ Ⓒ Ⓓ Ⓔ
47. Ⓐ Ⓑ Ⓒ Ⓓ Ⓔ
48. Ⓐ Ⓑ Ⓒ Ⓓ Ⓔ
49. Ⓐ Ⓑ Ⓒ Ⓓ Ⓔ
50. Ⓐ Ⓑ Ⓒ Ⓓ Ⓔ
51. Ⓐ Ⓑ Ⓒ Ⓓ Ⓔ
52. Ⓐ Ⓑ Ⓒ Ⓓ Ⓔ
53. Ⓐ Ⓑ Ⓒ Ⓓ Ⓔ
54. Ⓐ Ⓑ Ⓒ Ⓓ Ⓔ
55. Ⓐ Ⓑ Ⓒ Ⓓ Ⓔ

56. Ⓐ Ⓑ Ⓒ Ⓓ Ⓔ
57. Ⓐ Ⓑ Ⓒ Ⓓ Ⓔ
58. Ⓐ Ⓑ Ⓒ Ⓓ Ⓔ
59. Ⓐ Ⓑ Ⓒ Ⓓ Ⓔ
60. Ⓐ Ⓑ Ⓒ Ⓓ Ⓔ
61. Ⓐ Ⓑ Ⓒ Ⓓ Ⓔ
62. Ⓐ Ⓑ Ⓒ Ⓓ Ⓔ
63. Ⓐ Ⓑ Ⓒ Ⓓ Ⓔ
64. Ⓐ Ⓑ Ⓒ Ⓓ Ⓔ
65. Ⓐ Ⓑ Ⓒ Ⓓ Ⓔ
66. Ⓐ Ⓑ Ⓒ Ⓓ Ⓔ
67. Ⓐ Ⓑ Ⓒ Ⓓ Ⓔ
68. Ⓐ Ⓑ Ⓒ Ⓓ Ⓔ
69. Ⓐ Ⓑ Ⓒ Ⓓ Ⓔ
70. Ⓐ Ⓑ Ⓒ Ⓓ Ⓔ

71. Ⓐ Ⓑ Ⓒ Ⓓ Ⓔ
72. Ⓐ Ⓑ Ⓒ Ⓓ Ⓔ
73. Ⓐ Ⓑ Ⓒ Ⓓ Ⓔ
74. Ⓐ Ⓑ Ⓒ Ⓓ Ⓔ
75. Ⓐ Ⓑ Ⓒ Ⓓ Ⓔ
76. Ⓐ Ⓑ Ⓒ Ⓓ Ⓔ
77. Ⓐ Ⓑ Ⓒ Ⓓ Ⓔ
78. Ⓐ Ⓑ Ⓒ Ⓓ Ⓔ
79. Ⓐ Ⓑ Ⓒ Ⓓ Ⓔ
80. Ⓐ Ⓑ Ⓒ Ⓓ Ⓔ
81. Ⓐ Ⓑ Ⓒ Ⓓ Ⓔ
82. Ⓐ Ⓑ Ⓒ Ⓓ Ⓔ
83. Ⓐ Ⓑ Ⓒ Ⓓ Ⓔ
84. Ⓐ Ⓑ Ⓒ Ⓓ Ⓔ
85. Ⓐ Ⓑ Ⓒ Ⓓ Ⓔ

PRACTICE TEST 4

Note: For all questions involving solutions and/or chemical equations, assume that the system is in water unless otherwise stated.

PART A

Directions: Each set of lettered choices refers to the numbered statements or formulas immediately following it. Select the one lettered choice that best fits each statement or formula and then fill in the corresponding oval on the answer sheet. A choice may be used once, more than once, or not at all in each set.

Questions 1–3 refer to the following graphs:

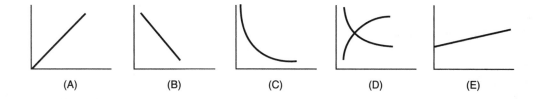

(A) (B) (C) (D) (E)

1. The graph that best shows the relationship of gas volume to temperature, with pressure held constant

2. The graph that best shows the relationship of gas volume to pressure, with temperature held constant

3. The graph that best shows the relationship of the number of grams of solute that is soluble in 100 g of H_2O at varying temperatures if the solubility begins at a small quantity and increases in slowly as there is an increase in temperature

Questions 4–7

    (A) A molecule
    (B) A mixture of compounds
    (C) An isotope
    (D) An isomer
    (E) An acid salt

4. The simplest unit of water that retains its properties

5. A commercial cake mix

6. An atom with the same number of protons as another atom of the same element but a different number of neutrons

7. Classification of $NaHCO_3$

    (A) 1
    (B) 7
    (C) 9
    (D) 10
    (E) 14

8. The atomic number of an atom with an electron dot arrangement similar to $:\ddot{\underset{..}{I}}:$

9. The number of atoms represented in the formula $Al(OH)_3$

10. The number that represents the most acid pH

Questions 11–14

    (A) Density
    (B) Equilibrium constant
    (C) Molar mass
    (D) Freezing point
    (E) Molarity

11. Can be expressed in moles per liter of solution

12. Can be expressed in grams per liter of a gas

13. Will NOT be affected by changes in temperature and pressure

14. At STP, can be used to determine the molecular mass of a pure gas

Questions 15–18

    (A) Buffer
    (B) Indicator
    (C) Arrhenius acid
    (D) Arrhenius base
    (E) Neutral condition

15. Resists a rapid change of pH

16. Exhibits different colors in acidic and basic solutions

17. At 25°C, the aqueous solution has a pH < 7.

18. At 25°C, the aqueous solution has a pH > 7.

Questions 19–23

    (A) H_2
    (B) NH_3
    (C) CO_2
    (D) HCl
    (E) O_2

19. A gas produced by the reaction of zinc with hydrochloric acid

20. A gas that is heavier than air and used in fire extinguishers

21. A gas produced by the heating of potassium chlorate

22. A gas that is slightly soluble in water and gives a weakly acid solution

23. A gas that is very soluble in water and gives a weakly basic solution

PART B

ON THE ACTUAL CHEMISTRY TEST, THE FOLLOWING TYPE OF QUESTION MUST BE ANSWERED ON A SPECIAL SECTION (LABELED "CHEMISTRY") AT THE LOWER LEFT-HAND CORNER OF PAGE 2 OF YOUR ANSWER SHEET. THESE QUESTIONS WILL BE NUMBERED BEGINNING WITH 101 AND MUST BE ANSWERED ACCORDING TO THE FOLLOWING DIRECTIONS.

Directions: Each question below consists of two statements, I in the left-hand column and II in the right-hand column. For each question, determine whether statement I is true or false *and* if statement II is true or false and fill in the corresponding T or F ovals on your answer sheet. *Fill in oval CE only if statement II is a correct explanation of statement I.*

Sample Answer Grid:

CHEMISTRY * Fill in oval CE only if
II is a correct explanation of I.

	I	II	CE*
101.	T F	T F	◯

101. According to the Kinetic Molecular Theory, gas particles are in random motion above absolute zero   BECAUSE   the degree of random motion of gas molecules varies inversely with the temperature of the gas.

102. The electron has wave properties as well as corpuscular properties   BECAUSE   the design of the experiment determines which is verified.

103. The alkanes are considered a homologous series   BECAUSE   they are made up of only hydrogen and carbon atoms.

104. When an atom of an active metal becomes an ion, the radius of the ion is less than that of the atom   BECAUSE   the nucleus of an active metallic ion has less positive charge than the electron "cloud."

105. When the heat of formation for a compound is negative, the ΔH is negative   BECAUSE   a negative heat of formation indicates that a reaction is exothermic with a negative enthalpy.

106. Water is a polar substance   BECAUSE   the sharing of the bonding electrons in water is equal.

107. A catalyst accelerates a chemical reaction   BECAUSE   a catalyst lowers the activation energy of the reaction.

108. Copper is an oxidizing agent in the reaction with silver nitrate solution   BECAUSE   copper loses electrons in a reaction with silver ions.

109. The rate of diffusion (or effusion) of hydrogen gas compared to that of helium gas is 1 : 4   BECAUSE   the rate of diffusion (or effusion) of gases varies inversely as the square root of the molecular masses.

110. A gas heated from 10°C to 100°C at constant pressure will increase in volume   BECAUSE   Charles's Law states that if the pressure remains constant, the volume varies directly as the absolute temperature varies.

111. The Gibbs free-energy equation can be used to predict the solubility of a solute   BECAUSE   the solubility of most salts increases as the temperature increases.

112. The complete electrolysis of 45 g of water will yield 40 g of H_2 and 5 g of O_2 BECAUSE   water is composed of hydrogen and oxygen in a ratio of 8 : 1 by mass.

113. 320 calories of heat will melt 4 grams of water at 0°C   BECAUSE   the heat of fusion of water is 80 calories per gram.

114. Hydrochloric acid solutions have a high level of conductivity   BECAUSE   hydrogen chloride is completely ionized in solution.

115. Water is a good solvent   BECAUSE   water shows hydrogen bonding between oxygen atoms.

116. In the periodic chart, electronegativity decreases from left to right   BECAUSE   the number of protons in the nucleus of of the atom decreases from left to right.

PART C

<u>Directions:</u> Each of the questions or incomplete statements below is followed by five suggested answers or completions. Select the one that is best in each case and then fill in the corresponding oval on the answer sheet.

40. In this graphic representation of a chemical reaction, which arrow depicts the activation energy?
    (A) A
    (B) B
    (C) C
    (D) D
    (E) E

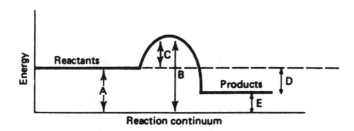

41. How many liters (STP) of O_2 can be produced by completely decomposing 2 moles of $KClO_3$?
    (A) 11.2
    (B) 22.4
    (C) 33.6
    (D) 44.8
    (E) 67.2

42. Which of the following is the correct structural representation of sodium?
    (A) 11 p   nucleus and orbital notation: $1s^2, 2s^2, 2p^6, 3s^2, 3p^6, 4s^2, 4p^3$
        11 n
    (B) 11 p   nucleus and orbital notation: $1s^2, 2s^2, 2p^6, 3s^2, 3p^6, 4s^2, 3d^1, 4p^2$
        12 n
    (C) 23 p   nucleus and orbital notation: $1s^2, 2s^2, 2p^6, 3s^1$
        23 n
    (D) 23 p   nucleus and orbital notation: $1s^2, 2s^2, 2p^6, 3s^2, 3p^6, 3d^3, 4s^2$
        23 n
    (E) 11 p   nucleus and orbital notation: $1s^2, 2s^2, 2p^6, 3s^1$
        12 n

43. Which of the following statements is true?
    (A) A catalyst cannot lower the activation energy.
    (B) A catalyst can lower the activation energy.
    (C) A catalyst affects only the activation energy of the forward reaction.
    (D) A catalyst affects only the activation energy of the reverse reaction.
    (E) A catalyst is permanently changed after the activation energy is achieved.

44. If the molecular mass of NH_3 is 17, what is the density of this compound at STP?
    (A) 0.25 g/L
    (B) 0.76 g/L
    (C) 1.52 g/L
    (D) 3.04 g/L
    (E) 9.11 g/L

45. Which bond(s) is(are) ionic?
    I. $H-Cl(g)$
    II. $S-Cl(g)$
    III. $Cs-F(s)$

    (A) I, II, and III
    (B) I and II only
    (C) II and III only
    (D) I only
    (E) III only

46. Aromatic hydrocarbons are represented by which of the following?

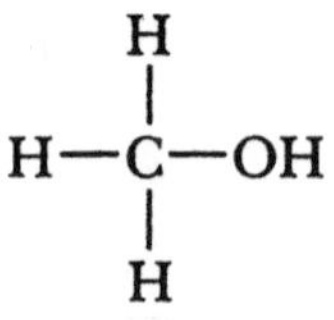

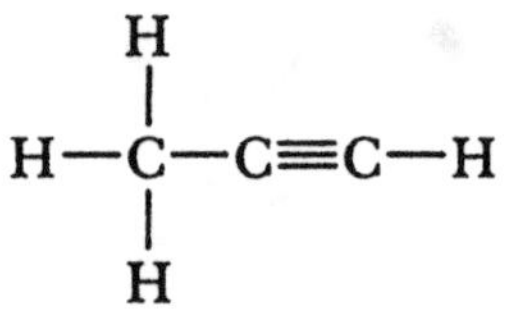

    (A) I, II, and III
    (B) I and II only
    (C) II and III only
    (D) I only
    (E) III only

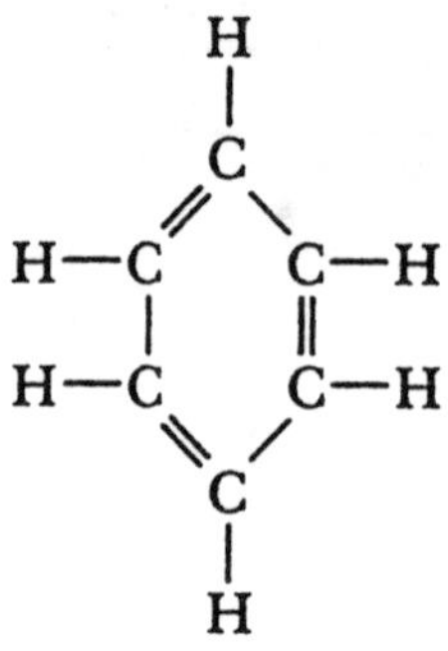

47. According to placement in the periodic table, which statement(s) regarding the first ionization energies of certain elements should be true?
    I. Li has a higher value than Na.
    II. K has a higher value than Cs.
    III. Na has a higher value than Al.

    (A) I, II, and III
    (B) I and II only
    (C) II and III only
    (D) I only
    (E) III only

48. Correctly expressed half-reactions include which of the following?
    I. $CrO_4^{2-} + 8H^+ + 6e^- \rightarrow Cr^{3+} + 4H_2O$
    II. $I^- + 6OH^- \rightarrow IO_3^- + 3H_2O + 6e^-$
    III. $MnO_4^- + 2H_2O + 3e^- \rightarrow MnO_2 + 4OH^-$

    (A) I, II, and III
    (B) I and II only
    (C) II and III only
    (D) I only
    (E) III only

What is the apparent oxidation number of the underlined element in the compound?

49. Na<u>N</u>O₃?
    (A) +1
    (B) +2
    (C) +3
    (D) +4
    (E) +5

50. <u>Ca</u>SO₄?
    (A) +1
    (B) −1
    (C) +2
    (D) −2
    (E) +3

51. <u>N</u>H₃?
    (A) +2
    (B) −2
    (C) +3
    (D) −3
    (E) +5

52. An atom with an orbital notation of $1s^2\ 2s^2\ 2p^6\ 3s^3\ 3p^4$ will probably exhibit which oxidation state?
    (A) +2
    (B) −2
    (C) +3
    (D) −3
    (E) +5

53. In this Lewis dot structure, x: what is the predictable oxidation number?
    (A) +1
    (B) −1
    (C) +2
    (D) −2
    (E) +3

Questions 54–57 refer to the following apparatus, assembled by a student:

54. The apparatus assembled is used to prepare a gas by a reaction that takes place when a solid
    (A) is heated
    (B) is exposed
    (C) reacts with a liquid
    (D) is heated in a vacuum
    (E) is decomposed by a catalyst

55. This apparatus suggests that the student plans to collect a gas that
    (A) supports combustion
    (B) is heavier than air
    (C) is not flammable
    (D) is insoluble in water
    (E) reacts with water

56. If the student is planning to prepare hydrogen, what would also be needed?
    (A) mercuric oxide
    (B) acid plus zinc
    (C) potassium chlorate
    (D) carbon disulfide
    (E) benzene

57. If you collected hydrogen gas by the displacement of water and under the conditions shown:

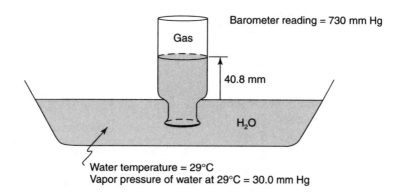

Barometer reading = 730 mm Hg

Gas

40.8 mm

H_2O

Water temperature = 29°C
Vapor pressure of water at 29°C = 30.0 mm Hg

which of the following would give you the pressure of the hydrogen in the bottle?

(A) 370 mm – 40.8 mm
(B) 730 mm – 30.0 mm
(C) 730 mm – 30.0 mm/13.6 + 40.8 mm
(D) 730 mm – 30.0 mm/13.6 – 40.8 mm
(E) 730 mm – 40.8 mm/13.6 – 30.0 mm

58. What occurs when a reaction is at equilibrium and more reactant is added to the container?
(A) The equilibrium remains unchanged.
(B) The forward reaction rate increases.
(C) The reverse reaction rate increases.
(D) The forward reaction rate decreases.
(E) The reverse reaction rate decreases.

59. How much heat energy is released when 8 g of hydrogen are burned? The thermal equation is $2H_2 + O_2 \rightarrow 2H_2O + 136.64$ kcal.
(A) 68.32 kcal
(B) 102.48 kcal
(C) 136.64 kcal
(D) 273.28 kcal
(E) 546.56 kcal

60. Would a spontaneous reaction occur between zinc *ions* and gold *atoms*?

$Zn^{2+} + 2e^- \rightleftharpoons Zn^0 \qquad E^0 = -0.76$ V
$Au^{3+} + 3e^- \rightleftharpoons Au^0 \qquad E^0 = +1.42$ V

(A) yes—reaction potential 2.18 V
(B) no—reaction potential –2.18 V
(C) yes—reaction potential 0.66 V
(D) no—reaction potential –0.66 V
(E) yes—reaction potential 0.56 V

61. Four moles of electrons ($4 \times 6.02 \times 10^{23}$ electrons) can electroplate how many grams of silver from a silver nitrate solution?
    (A) 108
    (B) 216
    (C) 324
    (D) 432
    (E) 540

62. A 5-M solution of HCl has how many moles of H^+ ion in 1 L?
    (A) 0.5
    (B) 1.0
    (C) 2.0
    (D) 2.5
    (E) 5.0

63. What is the K_{sp} for silver acetate if a saturated solution contains 2×10^{-3} mol of silver ion per liter of solution?
    (A) 2×10^{-3}
    (B) 2×10^{-6}
    (C) 4×10^{-3}
    (D) 4×10^{-6}
    (E) 4×10^{6}

64. The following data were obtained for H_2O and H_2S.

	Formula Mass	Freezing Point (°C)	Boiling Point (°C)
H_2O	18	0	100
H_2S	34	−83	−60

What is the best explanation for the variation of physical properties between these two compounds?
    (A) The H_2S has stronger bonds between molecules.
    (B) The H_2O has a great deal of hydrogen bonding.
    (C) The bond angles differ by about 15°.
    (D) The formula mass is of prime importance.
    (E) The oxygen atom has a smaller radius and thus does not bump into other molecules as often as the sulfur does.

65. What is the pH of a solution that has 0.00001 mol of H_3O^+ per liter of solution?
    (A) 2
    (B) 3
    (C) 4
    (D) 5
    (E) 9

66. How many grams of sulfur are present in 1 mol of H_2SO_4?
   (A) 2
   (B) 32
   (C) 49
   (D) 64
   (E) 98

67. What is the approximate mass, in grams, of 1 L of nitrous oxide, N_2O, at STP?
   (A) 1
   (B) 2
   (C) 11.2
   (D) 22
   (E) 44

68. If the simplest formula of a substance is CH_2 and its molecular mass is 56, what is its true formula?
   (A) CH_2
   (B) C_2H_4
   (C) C_3H_4
   (D) C_4H_8
   (E) C_5H_{10}

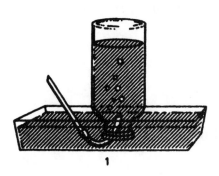

1   2

Questions 60–70 refer to the following diagrams of two methods of collecting gases.

69. Method 1 is BEST suited to collect
   (A) a gas heavier than air
   (B) a gas lighter than air
   (C) a gas that is insoluble in water
   (D) a gas that is soluble in water
   (E) a gas that has a distinct color

70. Which of these gases should be collected by method 2 because of its density and solubility?
   (A) NH_3
   (B) H_2
   (C) HCl
   (D) Br_2
   (E) He

71. What is the molar mass of $CaCO_3$?
    (A) 68 g/mol
    (B) 75 g/mol
    (C) 82 g/mol
    (D) 100 g/mol
    (E) 116 g/mol

72. What volume, in liters, will be occupied at STP by 4 g of H_2?
    (A) 11.2
    (B) 22.4
    (C) 33.6
    (D) 44.8
    (E) 56.0

73. How many moles of KOH are needed to neutralize 196 g of sulfuric acid? (H_2SO_4 = 98 amu)
    (A) 1.0
    (B) 1.5
    (C) 2.0
    (D) 4.0
    (E) 6.0

74. What volume, in liters, of $NH_3(g)$ is produced when 22.4 L of $N_2(g)$ are made to combine completely with a sufficient quantity of $H_2(g)$ under appropriate conditions?
    (A) 11.2
    (B) 22.4
    (C) 44.8
    (D) 67.0
    (E) 78.2

75. What volume, in liters, of SO_2 will result from the complete burning of 64 g of sulfur?
    (A) 2
    (B) 11.2
    (C) 44.8
    (D) 126
    (E) 158

76. The amount of energy needed to melt 5 g of ice at 0°C would also heat 1 g of water at 4°C to what condition? (Heat of fusion = 80 cal/g); (heat of vaporization = 540 cal/g.)
    (A) water at 90°C
    (B) water at 100°C
    (C) steam at 100°C
    (D) part of it would be vaporized into steam
    (E) all of it would be vaporized to steam

77. How many moles of electrons are needed to electroplate a deposit of 0.5 mol of silver from a silver nitrate solution?
    (A) 0.5
    (B) 1
    (C) 27
    (D) 54
    (E) 108

78. All of the following statements about carbon dioxide are true EXCEPT
    (A) It can be prepared by the action of acid on limestone.
    (B) It is used in fire extinguishers.
    (C) It dissolves in water at room temperature.
    (D) It sublimes rather than melts at 20°C and 1 atm pressure.
    (E) It is a product of photosynthesis in plants.

79. Three moles of H_2 and 3 mol of I_2 are introduced into a liter box at a temperature of 490°C. What will the K_{eq} expression be for this reaction? ($K_{eq} = 45.9$)

    (A) $K_{eq} = \dfrac{[H_2][I_2]}{HI}$

    (B) $K_{eq} = \dfrac{[HI]}{[H_2][I_2]}$

    (C) $K_{eq} = \dfrac{2x}{(x)(x)}$

    (D) $K_{eq} = \dfrac{(2x)^2}{(3-x)^2}$

    (E) $K_{eq} = \dfrac{(3-x)^2}{(2x)^2}$

80. If the following reaction has achieved equilibrium in a closed system,

$$N_2O_4(g) \rightleftharpoons 2NO_2(g)$$

which item(s) is(are) increased by decreasing the size of the container?
    I. the value of K_{eq}
    II. the concentration of $N_2O_4(g)$
    III. the rate of the reverse reaction

    (A) I, II, and III
    (B) I and II only
    (C) II and III only
    (D) I only
    (E) III only

81. Which of the following correctly completes the nuclear reaction $^{14}_{7}N + ^{4}_{2}He \rightarrow$ ___ $+ ^{1}_{1}H$?
    (A) $^{17}_{8}O$
    (B) $^{16}_{9}O$
    (C) $^{17}_{8}N$
    (D) $^{17}_{7}N$
    (E) $^{16}_{8}O$

82. How many grams of NaCl will be needed to make 100 mL of a 2-M solution?
    (A) 5.85
    (B) 11.7
    (C) 29.2
    (D) 58.5
    (E) 117

83. How many grams of H_2SO_4 are in 1000 g of a 10% solution? (1 mol of H_2SO_4 = 98 g.)
    (A) 1.0
    (B) 9.8
    (C) 10
    (D) 98
    (E) 100

84. If 1 mol of ethyl alcohol in 1000 g of water depresses the freezing point by 1.86°C, what will be the freezing point of a solution of 1 mol of ethyl alcohol in 500 g of water?
    (A) –0.93°C
    (B) –1.86°C
    (C) –2.79°C
    (D) –3.72°C
    (E) –5.58°C

85. Which of the nuclear reactions below shows the release of a beta particle?
    (A) $^{235}_{92}U + ^{1}_{0}n \rightarrow ^{93}_{36}Kr + ^{140}_{56}Ba + 3^{1}_{0}n$
    (B) $^{210}_{84}Po \rightarrow ^{206}_{82}Pb + ^{4}_{2}He$
    (C) $^{14}_{6}C \rightarrow ^{14}_{7}N + ^{0}_{-1}\beta$
    (D) $^{106}_{47}Ag + ^{0}_{-1}e \rightarrow ^{106}_{46}Pd$
    (E) $^{38}_{19}K \rightarrow ^{38}_{18}Ar + ^{0}_{+1}\beta$

STOP
IF YOU FINISH BEFORE ONE HOUR IS UP, YOU MAY GO BACK
TO CHECK YOUR WORK OR COMPLETE UNANSWERED QUESTIONS.

Answers and Explanations for Test 4

1. **(A)** The volume of a gas increases as the temperature increases provided that the pressure remains constant. This is a direct proportion. Heating a balloon is a good example of this.

2. **(B)** The volume of a gas decreases as the pressure is increased provided that the temperature is held constant. This is shown by the inversely proportional curve in (B). Pressure increase on a closed cylinder is a good example.

3. **(E)** The chart shows that there is a starting quantity in solution, and as you move to the right a slight positive slope indicates a directly proportional change.

4. **(A)** This is the definition of any molecule.

5. **(B)** A commercial cake mix is a mixture of ingredients.

6. **(C)** This is the definition of an isotope.

7. **(E)** An acid salt contains one or more H atoms in the salt formula separating a positive ion and the hydrogen-bearing negative ion. For example, Na_2SO_4 is a *normal* salt and $NaHSO_4$ is an *acid* salt because of the presence of H in the hydrogen sulfate ion.

8. **(C)** An atom with atomic number 9 would have a 2,7 electron configuration, which matches the outer energy level of iodine.

9. **(B)** Three $(OH)^-$ each have two atoms = 6 atoms plus one A1 = 7.

10. **(A)** pH from 1 to 6 is acid, 7 neutral, 8 to 14 basic. Most acid is 1.

11. **(E)** Molarity is defined as moles of solute per liter of solution.

12. **(A)** Gas densities can be expressed in grams per liter.

13. **(C)** Molecular mass is not affected by pressure and temperature.

14. **(A)** If the density of a gas is known, the mass of 1 L can be multiplied by 22.4 to find the molecular mass because 1 mol occupies 22.4 L at STP.

15. **(A)** Buffers resist changes in pH.

16. **(B)** Color change is the function of indicators.

17. **(C)** On the pH scale, from 1 to 6 is acid, and 7 is neutral.

18. **(D)** On the pH scale, from 8 to 14 is basic.

19. **(A)** $Zn + 2HCl \rightarrow ZnCl_2 + H_2(g)$ is the reaction that occurs.

20. **(C)** Only CO_2, with a molecular mass of 44, is heavier than air, for which the molecular mass is 29.

21. **(E)** $2KClO_3 \rightarrow 2KCl + 3O_2(g)$ is the reaction that occurs.

22. **(C)** CO_2 is slightly soluble in water, forming carbonic acid, H_2CO_3, which is a weak acid.

23. **(B)** NH_3 is very soluble in water and forms a solution of the weak ammonium hydroxide base.

101. **(T, F)** The assertion is true, but the degree of motion is directly related to the temperature.

102. **(T, T, CE)** Assertion and reason are true; the electron can be treated as either an electromagnetic wave or a bundle of negative charge.

103. **(T, T)** A homologous series increases each member by a constant number of carbons and hydrogens. Examples are the alkane, alkene, and alkyne series, which each increase the chain by a CH_2 group. The reason is true but does not explain the assertion.

104. **(T, F)** The nuclear charge is greater than the electron cloud. The reason is false.

105. **(T, T, CE)** A negative heat of formation indicates that the reaction is exothermic and the enthalpy is negative.

106. **(T, F)** Water is a polar molecule because there is unequal, not equal, sharing of bonding electrons.

107. **(T, T, CE)** This is a function of a catalyst—to either speed up or slow down a reaction without permanent change to itself. Assertion and reason are true.

108. (**F, T**) The Cu is losing electrons and thus being oxidized; the assertion is false. It is furnishing electrons and thus is a reducing agent; the reason is true.

109. (**F, T**) $H_2 = 2$, $He = 4$ (molecular mass); then inversely, $\sqrt{4} : \sqrt{2} = 2 : \sqrt{2}$ is the rate of diffusion of hydrogen to helium. The assertion is false, the reason, true.

110. (**T, T, CE**) Since the volume is being heated at constant pressure, it expands. The temperatures are converted to kelvins (K) by adding 273° to the Celsius readings. The fraction must be $\frac{373}{273}$, and this will increase the volume.

111. (**F, T**) Gibbs free energy is useful in indicating the conditions under which a chemical reaction will occur. It is not related to solubility. It is true that, generally speaking, solubility of a solute increases with an increase in the temperature of the solvent.

112. (**F, F**) Water is $\frac{1}{9}$ hydrogen and $\frac{8}{9}$ oxygen by weight. Both assertion and reason are false.

113. (**T, T, CE**) Four grams of ice would require 4×80 cal/g or 320 cal to melt.

114. (**T, T, CE**) HCl in solution ionizes virtually completely, forming a very conductive solution.

115. (**T, F**) The reason why water is a good solvent is false.

116. (**F, F**) Both assertion and reason are false. The opposite is true for both statements.

40. (**C**) The energy necessary to get the reaction started, which is the activation energy, is shown at C.

41. (**E**) $2KClO_3 \rightarrow 2KCl + 3O_2 \uparrow$ shows that 2 mol of $KClO_3$ yields 3 mol of O_2.

$$3 \text{ mol} \times \frac{22.4 \text{ L}}{1 \text{ mol}} = 67.2 \text{ L}$$

42. (**E**) The atomic number gives the number of protons in the nucleus and the total number of electrons. The atomic mass indicates the total number of protons and neutrons in the nucleus—for Na, 23 (11 protons plus 12 neutrons).

43. (**B**) A catalyst can speed up a reaction by lowering the activation energy needed to start the reaction and then keep it going.

44. (**B**) $\text{Density} = \dfrac{\text{Mass}}{\text{Volume}}$. For gases this is expressed as grams per liter. Since 1 gram-molecular mass of a gas occupies 22.4 L, 17 g/22.4 L = 0.76 g/L.

45. (**E**) Choice III is made up of elements from extreme sides of the periodic table and will therefore form ionic bonds.

46. (**E**) Only III is a ring hydrocarbon of the aromatic series.

47. (**B**) Since Li is higher in Group 1A than Na, and K is higher than Cs, they have smaller radii and hence higher ionization energies. Al is to the right of Na and therefore has a higher ionization energy.

48. (**C**) Only II and III are correctly balanced. To be correct, I should have only $3e^-$.

49. (**E**) ⎫ These answers are based on the fact that the total charge using the assigned oxidation numbers for all the atoms in a compound is zero.

50. (**C**) ⎬

51. (**C**) ⎭

52. (**B**) This orbital notation shows 6 electrons in the third energy level. The atom would like to gain $2e^-$ to fill the $3p$ and thereby gain a –2 charge.

53. (**C**) With this structure, the atom would tend to lose these electrons and acquire a 2+ charge.

54. (**C**) Assembled, the apparatus would look like this:

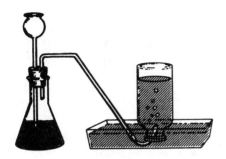

and could be used to prepare a gas by reacting a solid with a liquid.

55. (**D**) The setup depends on the property of insolubility of the gas collected over water.

56. (**B**) This apparatus would need an acid to cause the reaction.

57. (**E**) The pressure in the bottle will be less than atmospheric pressure by the Hg equivalent height of the 30 mm of water above the level in the collecting pan. This is calculated as 40.8 mm water/(13.6 mm water/1 mm Hg) and must be subtracted from atmospheric pressure. The other adjustment is to subtract the vapor pressure of water in the collected gas since it was collected over water. The amount is given as 30.0 mm Hg. So each of these must be subtracted from 730 mm Hg, the given atmospheric pressure.

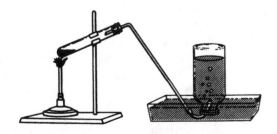

58. (**B**) The equilibrium shifts in the direction that tends to relieve the stress and thus regain equilibrium.

59. (**D**) The thermal reaction shows 2 mol of hydrogen reacting or 4 g. Therefore 8 g would release twice the amount of energy; 2×136.64 kcal = 273.28 kcal.

60. (**B**) The reaction potential calculation would be

$$Zn^{2+} + 2e^- \rightarrow Zn^0 \qquad E^0 = -0.76 \text{ V}$$
$$\underline{Au^0 \rightarrow Au^{3+} + 3e^- \quad E^0 = -1.42 \text{ V}}$$
$$Zn^{2+} + Au^0 \rightarrow Zn^0 + Au^{3+} \quad E^0 = -2.18 \text{ V}$$

61. (**D**) Since $Ag^+ + 1e^- \rightarrow Ag^0$, 1 mol of electrons yields 1 mol of silver;

1 mol silver = 6.02×10^{23} atoms
4×108 g/mol = 432 g

62. (**E**) $5\ M = \dfrac{5 \text{ mol}}{\text{liter}}$, and since HCl ionizes completely, there would be 5 mol of H^+ and 5 mol of Cl^- ions.

63. (**D**) $K_{sp} = [Ag^+] [C_2H_3O_2^-]$
$\qquad = [2 \times 10^{-3}] [2 \times 10^{-3}]$
(Since $AgC_2H_3O_2 \rightleftharpoons Ag^+ + C_2H_3O_2^-$, the silver ion and acetate ion concentrations are equal.)

$$K_{sp} = 4 \times 10^{-6}$$

64. (**B**) It is the explanation for the observed high boiling point and high freezing point of water compared to hydrogen sulfide.

65. (**D**) pH = $-\log [H^+]$
$\qquad = -\log [10^{-5}] = -[-5] = 5$

66. (**B**) 1 mol H_2SO_4 contains 1 molar mass of sulfur, that is, 32 g.

67. (**B**) N_2O = 44 g/mol
$\qquad\qquad (2 \times 14 + 16 = 44)$
1 mol of a gas occupies 22.4 L. So 44 g/22.4 L = 1.99 g/L.

68. (**D**) CH_2 = 14
$\qquad (12 + 2 = 14$ molecular mass)
$56 \div 14 = 4$
So $4 \times CH_2 = C_4H_3$

69. (**C**) Only insoluble gases can be collected this way.

70. (**C**) HCl is very soluble in water and heavier than air, and so it is suited to method 2.

71. (**D**) Ca = 40
$\quad$ C $\ = 12$
$\quad 30 = \dfrac{48}{100 \text{ g}}$

72. **(D)** The gram-molecular mass of H_2 is 2 g. 4 g is 2 mol, and each mole occupies 22.4 L. $2 \times 22.4 = 44.8$ L.

73. **(D)**
$$
\begin{array}{cc}
 & x \text{ mol} \quad 196 \text{ g} \\
2KOH + H_2SO_4 \rightarrow K_2SO_4 + 2H_2O \\
2 \text{ mol} \quad 98 \text{ g}
\end{array}
$$

So $\dfrac{x \text{ mol}}{2 \text{ mol}} = \dfrac{196 \text{ g}}{98 \text{ g}}$

Then $x = 4$ mol

74. **(C)**
$$
\begin{array}{cc}
22.4 \text{ L} \qquad x \text{ liters} \\
N_2 + 3H_2 \rightarrow 2NH_3 \\
1 \text{ L} \qquad\quad 2 \text{ L}
\end{array}
$$

So $\dfrac{22.4 \text{ L}}{1 \text{ L}} = \dfrac{x \text{ liters}}{2 \text{ L}}$.

Then $x = 44.8$ L.

75. **(C)**
$$
\begin{array}{cc}
64 \text{ g} \qquad x \text{ liters} \\
S + O_2 \rightarrow SO_2 \\
32 \text{ g} \qquad 22.4 \text{ L}
\end{array}
$$

So $\dfrac{64 \text{ g}}{32 \text{ g}} = \dfrac{x \text{ liters}}{22.4 \text{ L}}$ or

$64 \, \cancel{g} \times \dfrac{1 \, \cancel{\text{mol S}}}{32 \, \cancel{g}} \times \dfrac{1 \text{ mol } SO_2}{1 \, \cancel{\text{mol S}}}$

$= 2$ mol SO_2

$2 \, \cancel{\text{mol } SO_2} \times \dfrac{22.4 \text{ L}}{1 \, \cancel{\text{mol}}} = 44.8$ L

Then $x = 44.8$ L.

76. **(D)** 5 g ice to water $= 5 \times 80$ cal $= 400$ cal. 1 g at 4°C can go to 100°C as water and absorb 1 cal/°C. So 400 cal $- (100° - 4°)$ $= 400 - 96 = 304$ cal, and 304 cal can change $\dfrac{304 \text{ cal}}{540 \text{ cal/g}}$ or 0.56 g of water to steam. So 0.56 g steam and 0.44 g water would remain.

77. **(A)** Since Ag^+ gains 1 e^- to become Ag^0, 0.5 mol would require 0.5 mol of electrons.

78. **(E)** CO_2 is a reactant in photosynthesis, not a product. The reaction is $6CO_2 + 6H_2O \rightarrow C_6H_{12}O_6 + 6O_2(g)$ simple sugar

or

$6CO_2 + 5H_2O \rightarrow C_6H_{12}O_5 + 6O_2(g)$
cellulose

79. **(D)** $K_{eq} = \dfrac{[HI]^2}{[H_2][I_2]}$

Let $x =$ moles of H_2 and also I_2 that combine to form HI.

Then at equilibrium $[H_2] = 3 - x$, $[I_2] = 3 - x$, $[HI] = 2x$.

So $K_{eq} = \dfrac{(2x)^2}{(3-x)(3-x)}$.

80. **(C)** In a closed system, decreasing the size of the container will cause the pressure to increase. When pressure is applied to an equilibrium involving gases, the reaction that lowers the pressure by decreasing the number of molecules will increase in rate. In this reaction, the reverse reaction, changing two molecules to one, increases in rate, thus reducing pressure while also increasing the concentration of N_2O_4.

81. **(A)** This is Rutherford's famous artifical transmutation experiment done in 1919.

82. **(B)** $2 M = \dfrac{2 \text{ mol}}{1000 \text{ mol}}$

2 mol of NaCl $= 2 \times 58.5$ g
$\qquad\qquad\qquad = 117.0$ g
$2 M = 117$ g/1000 mL,

so $\dfrac{117 \text{ g}}{1000 \text{ mol}} = \dfrac{x \text{ grams}}{100 \text{ mL}}$,

$x = 11.7$ g

83. **(E)** Percent is by mass. So 10% is 0.1×1000 g or 100 g.

84. **(D)** First find the molality. 1 mol in 500 g $= 2$ mol in 1000 g. Then $2 \times 1.86°C = 3.72°C$ drop from 0°C or $-3.72°C$.

85. **(C)** The nuclear reactions shown release (A) neutron, (B) alpha particle, (C) beta particle, (D) none, (E) positron.

Diagnosing Your Needs

After taking Practice Test 4, check your answers against the correct ones. Then fill in the chart below.

In the space under each question number, place a check if you answered that question correctly.

Example

If your answer to question 5 was correct, place a check in the appropriate box.

Next, total the check marks for each section and insert the number in the designated block. Now do the arithmetic indicated and insert your percent for each area.

SUBJECT AREA　　　　　　　　　　　　　　(✔) QUESTIONS ANSWERED CORRECTLY

I. **Atomic Theory and Structure,** including periodic relationships	6	8	102	107	42	47	52	53
No. of checks ÷ 8 × 100 = _____%								

II. **Nuclear Reactions**							81	85
No. of checks ÷ 2 × 100 = _____%								

III. **Chemical Bonding and Molecular Structure**	4	9	106	45	49	50	51	62	64
No. of checks ÷ 9 × 100 = _____%									

IV. **States of Matter and Kinetic Molecular Theory of Gases**	1	2	14	101	109	110	113	44
No. of checks ÷ 8 × 100 = _____%								

V. **Solutions,** including concentration units, solubility, and colligative properties				3	11	82	83	84
No. of checks ÷ 5 × 100 = _____%								

VI. **Acids and Bases**		7	10	15	16	17	18	114
No. of checks ÷ 7 × 100 = _____%								

VII. **Oxidation-Reduction and Electrochemistry**		108	112	48	60	61	77
No. of checks ÷ 6 × 100 = _____%							

SUBJECT AREA	(✔) QUESTIONS ANSWERED CORRECTLY								
VIII. Stoichiometry	41	66	67	68	71	72	73	74	75
▢ No. of checks ÷ 9 × 100 = _____%									

IX. Reaction Rates	107	43
▢ No. of checks ÷ 2 × 100 = _____%		

X. Equilibrium	58	63	65	79	80
▢ No. of checks ÷ 5 × 100 = _____%					

XI. Thermodynamics: energy changes in chemical reactions, randomness, and criteria for spontaneity	105	111	40	59	76
▢ No. of checks ÷ 5 × 100 = _____%					

XII. Descriptive Chemistry: physical and chemical properties of elements and their familiar compounds; organic chemistry; periodic properties	5	12	13	19	20	21	22
	23	103	115	116	46	78	
▢ No. of checks ÷ 13 × 100 = _____%							

XIII. Laboratory: equipment, procedures, observations, safety, calculations, and interpretation of results	54	55	56	57	69	70
▢ No. of checks ÷ 6 × 100 = _____%						

Planning Your Study

The percentages give you an idea of how you have done in the various major areas of the test. Because of the limited number of questions on some parts, these percentages may not be as reliable as the percentages for parts with larger numbers of questions. However, you should now have at least a rough idea of the areas in which you have done well and those in which you need more study.

Start your study with the areas in which you are the weakest. The corresponding chapters are indicated on page 419.

Subject Area	Chapters to Review
I. Atomic theory and Structure, including periodic relationships	2
II. Nuclear Reactions	15
III. Chemical Bonding and Molecular Structure	3, 4
IV. States of Matter and Kinetic Molecular Theory of Gases	1
V. Solutions, including concentration units, solubility, and colligative properties	7
VI. Acids and Bases	11
VII. Oxidation-Reduction and Electrochemistry	12
VIII. Stoichiometry	5, 6
IX. Reaction Rates	9
X. Equilibrium	10
XI. Thermodynamics, including energy changes in chemical reactions, randomness, and criteria for spontaneity	8
XII. Descriptive Chemistry: physical and chemical properties of elements and their familiar compounds; organic chemistry; periodic properties	1, 2, 13, 14
XIII. Laboratory: equipment, procedures, observations, safety, calculations, and interpretation of results	All lab diagrams, 16

Answer Sheet for Practice Test 5

Determine the correct answer for each question. Then, using a No. 2 pencil, blacken completely the oval containing the letter of your choice.

1. Ⓐ Ⓑ Ⓒ Ⓓ Ⓔ
2. Ⓐ Ⓑ Ⓒ Ⓓ Ⓔ
3. Ⓐ Ⓑ Ⓒ Ⓓ Ⓔ
4. Ⓐ Ⓑ Ⓒ Ⓓ Ⓔ
5. Ⓐ Ⓑ Ⓒ Ⓓ Ⓔ
6. Ⓐ Ⓑ Ⓒ Ⓓ Ⓔ
7. Ⓐ Ⓑ Ⓒ Ⓓ Ⓔ
8. Ⓐ Ⓑ Ⓒ Ⓓ Ⓔ
9. Ⓐ Ⓑ Ⓒ Ⓓ Ⓔ
10. Ⓐ Ⓑ Ⓒ Ⓓ Ⓔ
11. Ⓐ Ⓑ Ⓒ Ⓓ Ⓔ
12. Ⓐ Ⓑ Ⓒ Ⓓ Ⓔ
13. Ⓐ Ⓑ Ⓒ Ⓓ Ⓔ
14. Ⓐ Ⓑ Ⓒ Ⓓ Ⓔ
15. Ⓐ Ⓑ Ⓒ Ⓓ Ⓔ
16. Ⓐ Ⓑ Ⓒ Ⓓ Ⓔ
17. Ⓐ Ⓑ Ⓒ Ⓓ Ⓔ
18. Ⓐ Ⓑ Ⓒ Ⓓ Ⓔ
19. Ⓐ Ⓑ Ⓒ Ⓓ Ⓔ
20. Ⓐ Ⓑ Ⓒ Ⓓ Ⓔ
21. Ⓐ Ⓑ Ⓒ Ⓓ Ⓔ
22. Ⓐ Ⓑ Ⓒ Ⓓ Ⓔ
23. Ⓐ Ⓑ Ⓒ Ⓓ Ⓔ

ON THE ACTUAL CHEMISTRY TEST, THE FOLLOWING TYPE OF QUESTION MUST BE ANSWERED ON A SPECIAL SECTION (LABELED "CHEMISTRY") AT THE LOWER LEFT-HAND CORNER OF PAGE 2 OF YOUR ANSWER SHEET. THESE QUESTIONS WILL BE NUMBERED BEGINNING WITH 101 AND MUST BE ANSWERED ACCORDING TO THE FOLLOWING DIRECTIONS.

CHEMISTRY* Fill in oval CE only if II is a correct explanation of I.

	I	II	CE*
101.	Ⓣ Ⓕ	Ⓣ Ⓕ	◯
102.	Ⓣ Ⓕ	Ⓣ Ⓕ	◯
103.	Ⓣ Ⓕ	Ⓣ Ⓕ	◯
104.	Ⓣ Ⓕ	Ⓣ Ⓕ	◯
105.	Ⓣ Ⓕ	Ⓣ Ⓕ	◯
106.	Ⓣ Ⓕ	Ⓣ Ⓕ	◯
107.	Ⓣ Ⓕ	Ⓣ Ⓕ	◯
108.	Ⓣ Ⓕ	Ⓣ Ⓕ	◯
109.	Ⓣ Ⓕ	Ⓣ Ⓕ	◯
110.	Ⓣ Ⓕ	Ⓣ Ⓕ	◯
111.	Ⓣ Ⓕ	Ⓣ Ⓕ	◯
112.	Ⓣ Ⓕ	Ⓣ Ⓕ	◯
113.	Ⓣ Ⓕ	Ⓣ Ⓕ	◯
114.	Ⓣ Ⓕ	Ⓣ Ⓕ	◯
115.	Ⓣ Ⓕ	Ⓣ Ⓕ	◯
116.	Ⓣ Ⓕ	Ⓣ Ⓕ	◯

ON THE ACTUAL CHEMISTRY TEST, THE REMAINING QUESTIONS MUST BE ANSWERED BY RETURNING TO THE SECTION OF YOUR ANSWER SHEET YOU STARTED FOR CHEMISTRY.

40. Ⓐ Ⓑ Ⓒ Ⓓ Ⓔ
41. Ⓐ Ⓑ Ⓒ Ⓓ Ⓔ
42. Ⓐ Ⓑ Ⓒ Ⓓ Ⓔ
43. Ⓐ Ⓑ Ⓒ Ⓓ Ⓔ
44. Ⓐ Ⓑ Ⓒ Ⓓ Ⓔ
45. Ⓐ Ⓑ Ⓒ Ⓓ Ⓔ
46. Ⓐ Ⓑ Ⓒ Ⓓ Ⓔ
47. Ⓐ Ⓑ Ⓒ Ⓓ Ⓔ
48. Ⓐ Ⓑ Ⓒ Ⓓ Ⓔ
49. Ⓐ Ⓑ Ⓒ Ⓓ Ⓔ
50. Ⓐ Ⓑ Ⓒ Ⓓ Ⓔ
51. Ⓐ Ⓑ Ⓒ Ⓓ Ⓔ
52. Ⓐ Ⓑ Ⓒ Ⓓ Ⓔ
53. Ⓐ Ⓑ Ⓒ Ⓓ Ⓔ
54. Ⓐ Ⓑ Ⓒ Ⓓ Ⓔ
55. Ⓐ Ⓑ Ⓒ Ⓓ Ⓔ

56. Ⓐ Ⓑ Ⓒ Ⓓ Ⓔ
57. Ⓐ Ⓑ Ⓒ Ⓓ Ⓔ
58. Ⓐ Ⓑ Ⓒ Ⓓ Ⓔ
59. Ⓐ Ⓑ Ⓒ Ⓓ Ⓔ
60. Ⓐ Ⓑ Ⓒ Ⓓ Ⓔ
61. Ⓐ Ⓑ Ⓒ Ⓓ Ⓔ
62. Ⓐ Ⓑ Ⓒ Ⓓ Ⓔ
63. Ⓐ Ⓑ Ⓒ Ⓓ Ⓔ
64. Ⓐ Ⓑ Ⓒ Ⓓ Ⓔ
65. Ⓐ Ⓑ Ⓒ Ⓓ Ⓔ
66. Ⓐ Ⓑ Ⓒ Ⓓ Ⓔ
67. Ⓐ Ⓑ Ⓒ Ⓓ Ⓔ
68. Ⓐ Ⓑ Ⓒ Ⓓ Ⓔ
69. Ⓐ Ⓑ Ⓒ Ⓓ Ⓔ
70. Ⓐ Ⓑ Ⓒ Ⓓ Ⓔ

71. Ⓐ Ⓑ Ⓒ Ⓓ Ⓔ
72. Ⓐ Ⓑ Ⓒ Ⓓ Ⓔ
73. Ⓐ Ⓑ Ⓒ Ⓓ Ⓔ
74. Ⓐ Ⓑ Ⓒ Ⓓ Ⓔ
75. Ⓐ Ⓑ Ⓒ Ⓓ Ⓔ
76. Ⓐ Ⓑ Ⓒ Ⓓ Ⓔ
77. Ⓐ Ⓑ Ⓒ Ⓓ Ⓔ
78. Ⓐ Ⓑ Ⓒ Ⓓ Ⓔ
79. Ⓐ Ⓑ Ⓒ Ⓓ Ⓔ
80. Ⓐ Ⓑ Ⓒ Ⓓ Ⓔ
81. Ⓐ Ⓑ Ⓒ Ⓓ Ⓔ
82. Ⓐ Ⓑ Ⓒ Ⓓ Ⓔ
83. Ⓐ Ⓑ Ⓒ Ⓓ Ⓔ
84. Ⓐ Ⓑ Ⓒ Ⓓ Ⓔ
85. Ⓐ Ⓑ Ⓒ Ⓓ Ⓔ

PRACTICE TEST 5

<u>Note</u>: For all questions involving solutions and/or chemical equations, assume that the system is in water unless otherwise stated.

PART A

<u>Directions</u>: Each set of lettered choices refers to the numbered statements or formulas immediately following it. Select the one lettered choice that best fits each statement or formula and then fill in the corresponding oval on the answer sheet. A choice may be used once, more than once, or not at all in each set.

<u>Questions 1–4</u> refer to the periodic group shown as they are aligned in the periodic table with letter choices substituted for the respective elements.

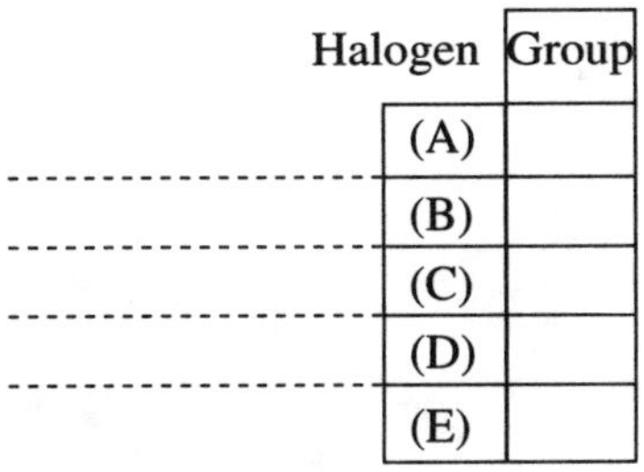

Halogen	Group
(A)	
(B)	
(C)	
(D)	
(E)	

1. The element that is most active chemically

2. The element with the smallest ionic radius

3. The element with the lowest first ionization potential

4. The element that first shows some metallic properties at room temperature

<u>Questions 5–7</u>

(A) X^+
(B) X^{2+}
(C) X^{3+}
(D) XO_3^{2-}
(E) XO_4^{2-}

5. A type of ion found in sodium acetate

6. A type of ion found in aluminum oxide

7. A type of ion found in potassium phosphate

Questions 8–12

    (A) Avogadro's number
    (B) $P_1V_1 = P_2V_2$
    (C) $V_1T_2 = V_2T_1$
    (D) Dalton's Theory
    (E) Gay-Lussac's Law

8. Proposes basic postulates concerning elements and atoms

9. Proposes a relationship between the combining volumes of gases with respect to the reactants and gaseous products

10. Proposes a temperature-volume relationship for gases

11. Proposes a concept regarding the number of particles in a mole

12. Proposes a volume-pressure relationship for gases

Questions 13–16

    (A) R$-$OH

    (B) R$-$O$-$R*

    (C) 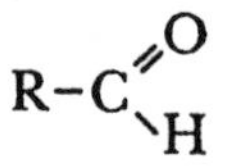 $R-C\begin{smallmatrix} \nearrow O \\ \searrow H \end{smallmatrix}$

    (D) 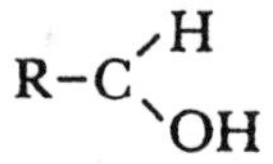 $R-C\begin{smallmatrix} \nearrow H \\ \searrow OH \end{smallmatrix}$

    (E) 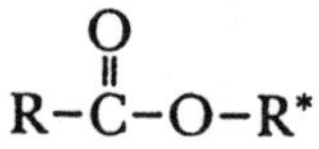 $R-\overset{\overset{\textstyle O}{\|}}{C}-O-R^*$

    (*Alkyl group that is not necessarily the same as R.)

13. The organic structure designation that includes the functional group of an aldehyde

14. The organic structure designation that includes the functional group of an acid

15. The organic structure designation that includes the functional group of an ester

16. The organic structural designation that includes the functional group of an ether

<u>Questions 17–21</u>

    (A) $H_2(g)$
    (B) $CO_2(g)$
    (C) $2N_2O(g)$
    (D) $2NaCl(aq)$
    (E) $H_2SO_4(aq)$

17. The expression that can be used to designate a linear nonpolar molecule

18. The expression that can be used to designate 2 mol of atoms

19. The expression that can be used to designate 3 mol of atoms

20. The expression that can be used to designate 3 mol of ions

21. The expression that can be used to designate 6 mol of atoms

<u>Questions 22–25</u>

    (A) Ionic substance
    (B) Metallic substance
    (C) Polar covalent molecule
    (D) Nonpolar covalent molecule
    (E) Aromatic organic structure

22. Carbon tetrachloride

23. Cesium chloride

24. Hydrogen chloride

25. Benzene

PART B

ON THE ACTUAL CHEMISTRY TEST, THE FOLLOWING TYPE OF QUESTION MUST BE ANSWERED ON A SPECIAL SECTION (LABELED "CHEMISTRY") AT THE LOWER LEFT-HAND CORNER OF PAGE 2 OF YOUR ANSWER SHEET. THESE QUESTIONS WILL BE NUMBERED BEGINNING WITH 101 AND MUST BE ANSWERED ACCORDING TO THE FOLLOWING DIRECTIONS.

<u>Directions</u>: Each question below consists of two statements, I in the left-hand column and II in the right-hand column. For each question, determine whether statement I is true or false <u>and</u> if statement II is true or false and fill in the corresponding T or F ovals on your answer sheet. <u>Fill in oval CE only if statement II is a correct explanation of statement I.</u>

Sample Answer Grid:

CHEMISTRY * Fill in oval CE only if
II is a correct explanation of I.

	I	II	CE*
101.	(T) (F)	(T) (F)	◯

101. A catalyst can accelerate a chemical reaction   BECAUSE   a catalyst can decrease the activation energy required for the reaction to occur.

102. Molten sodium chloride is a good electrical conductor   BECAUSE   sodium chloride in the molten state allows ions to move freely.

103. Ice is less dense than liquid water   BECAUSE   water molecules are nonpolar.

104. Two isotopes of the same element have the same mass number   BECAUSE   isotopes of an element have the same number of protons.

105. A saturated solution can be classified as dilute   BECAUSE   a solute can have a very low solubility in a solvent.

106. Nitrogen is an element   BECAUSE   the gas exists in the form of diatomic molecules at room temperature and atmospheric pressure.

107. A reaction is at equilibrium when it reaches completion   BECAUSE   the concentrations of the reactants in a state of equilibrium equal the concentrations of the products.

108. The anions in an electrochemical cell migrate to the cathode   BECAUSE   positively charged ions are attracted to the negatively charged cathode in an electrochemical cell.

109. A solution with pH = 5 has a higher concentration of hydronium ions than a solution with pH = 3   BECAUSE   pH is defined as $-\log[H^+]$.

110. An endothermic reaction can be spontaneous   BECAUSE   both the enthalpy and entropy changes affect the reaction's Gibbs free energy change.

111. Weak acids have a small value for the equilibrium constant, K_a BECAUSE   the concentration of the hydronium ion is in the numerator of the K_a expression.

112. As you proceed across the periodic chart, the electronegativity increases   BECAUSE   the ionic radius of a nonmetallic ion is usually greater than the atomic radius of the atom.

113. A pi bond is formed between the lobes of adjacent p orbitals in the same plane of two atoms that contain only one electron each

BECAUSE   both lobes of a single p orbital can hold two electrons of opposite spin.

114. H_2S and H_2O have a significant difference in their boiling points

BECAUSE   hydrogen sulfide has a higher degree of hydrogen bonding than water.

PART C

Directions: Each of the questions or incomplete statements below is followed by five suggested answers or completions. Select the one that is best in each case and then fill in the corresponding oval on the answer sheet.

40. Two immiscible liquids, when shaken together vigorously, may form
    (A) a solution
    (B) a tincture
    (C) a sediment
    (D) a hydrated solution
    (E) a colloidal dispersion

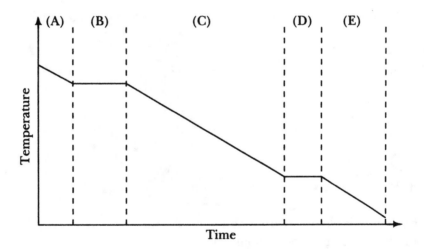

41. A thermometer is used to record the cooling of a confined pure substance over a period of time. During which interval on the cooling graph above is the system undergoing a change of state from a liquid to a solid?

42. If a principal energy level of an atom contains 18 electrons, they will be arranged in orbitals according to the pattern
    (A) $s^6p^6d^6$
    (B) $s^2p^6d^{10}$
    (C) $s^2d^6f^{10}$
    (D) $s^2p^6f^{10}$
    (E) $s^6p^2f^{14}$

43. For a particular organic compound, which of the following pairs can represent the empirical and the molecular formulas, respectively?
   (A) CH and CH_4
   (B) CH and C_6H_6
   (C) CH_2 and C_2H_2
   (D) CH_2 and C_2H_3
   (E) CH_3 and C_3H_6

44. A liter of hydrogen is at 5°C temperature and under 640 torr pressure. If the temperature is raised to 60°C and the pressure decreased to 320 torr, how will the liter volume be modified?
   (A) $1\,L \times \dfrac{5}{60} \times \dfrac{640}{32}$
   (B) $1\,L \times \dfrac{60}{5} \times \dfrac{320}{640}$
   (C) $1\,L \times \dfrac{278}{333} \times \dfrac{640}{320}$
   (D) $1\,L \times \dfrac{333}{278} \times \dfrac{640}{320}$
   (E) $1\,L \times \dfrac{333}{278} \times \dfrac{320}{640}$

45. Of the following statements about the number of subatomic particles in an ion of $^{32}_{16}S^{2-}$, which is (are) true?
   I. 16 protons
   II. 14 neutrons
   III. 18 electrons

   (A) II only
   (B) III only
   (C) I and II only
   (D) I and III only
   (E) I, II, and III

46. The most active metallic elements are found in
   (A) the upper right corner of the periodic table
   (B) the lower right corner of the periodic table
   (C) the upper left corner of the periodic table
   (D) the lower left corner of the periodic table
   (E) the middle of the table, just beyond the transition elements

47. If 1 mol of each of these substances is dissolved in 1000 g of water, which solution will have the highest boiling point?
   (A) NaCl
   (B) KCl
   (C) $CaCl_2$
   (D) $C_6H_{10}O_5$
   (E) $C_{12}H_{22}O_{11}$

48. A tetrahedral molecule, XY_4, would be formed if X uses the orbital configuration
    (A) p^2
    (B) s^2
    (C) sp
    (D) sp^2
    (E) sp^3

49. In the following reaction, how many liters of SO_2 at STP will result from the complete burning of pure sulfur in 8 L of oxygen?

$$S(s) + O_2(g) \rightarrow SO_2(g)$$

    (A) 1
    (B) 4
    (C) 8
    (D) 16
    (E) 32

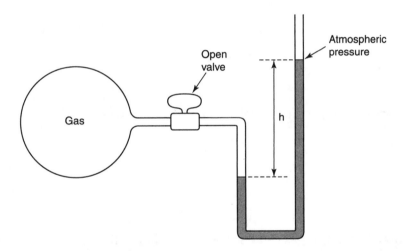

50. In the above laboratory setup to measure the pressure of a confined gas, what will be true concerning the calculated pressure on the confined gas?
    (A) The gas pressure will be the same as the atmospheric pressure.
    (B) The gas pressure will be less than the atmospheric pressure.
    (C) The gas pressure will be greater than the atmospheric pressure.
    (D) The difference in the height (h) of mercury levels is equal to the pressure of the gas.
    (E) The height (h) of mercury has no effect on the pressure calculation since it is only used to enclose the gas volume.

51. Which of the following changes in the experiment shown in question 50 would cause the pressure in the glass container to vary from that shown?
    (A) Using a U-tube of a greater diameter and maintaining the height of the mercury.
    (B) Increasing the temperature of gas in the flask.
    (C) Increasing the length of the upper portion of the right side of the tubing.
    (D) Using a U-tube of a smaller diameter and maintaining the height of the mercury.
    (E) Replacing the flask with one that has the same volume but has a flat bottom.

52. Which of the following can be classified as amphoteric?
    (A) Na_3PO_4
    (B) HCl
    (C) $NaOH$
    (D) HSO_4^-
    (E) $C_2O_4^{2-}$

53. Standard conditions (STP) are
    (A) 0°C and 2 atm
    (B) 32°F and 76 torr
    (C) 273 K and 760 mm Hg
    (D) 1°C and 7.6 cm Hg
    (E) 0 K and 760 mm Hg

54. Laboratory results showed the composition of a compound to be 58.81% barium, 13.73% sulfur, and 27.46% oxygen. What is the empirical formula of the compound?
    (A) $BaSO_4$
    (B) BaS_2O
    (C) Ba_2SO_3
    (D) BaS_2O_4
    (E) Ba_2SO_4

55. What is the percentage composition of calcium in calcium hydroxide, $Ca(OH)_2$ (1 mol = 74 g)?
    (A) 40%
    (B) 43%
    (C) 54%
    (D) 69%
    (E) 74%

56. How many grams of hydrogen gas can be produced from the following reaction if 65 g of zinc and 65 g of HCl are present in the reaction?

    $$Zn(s) + 2HCl(aq) \rightarrow ZnCl_2(aq) + H_2(g)$$

    (A) 1
    (B) 1.8
    (C) 3.6
    (D) 7
    (E) 58

57. The following statements were recorded while preparing carbon dioxide gas in the laboratory. Which one involves an interpretation of the data rather than an observation?
    (A) No liquid was transferred from the reaction bottle to the beaker.
    (B) The quantity of solid minerals decreased.
    (C) The cloudiness in the last bottle of limewater was caused by the product of the reaction of the colorless gas and the limewater.
    (D) The bubbles of gas rising from the mineral remained colorless throughout the experiment.
    (E) There was a 4°C rise in temperature in the reaction vessel during the experiment.

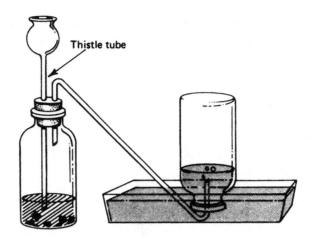

58. The above laboratory setup can be used to prepare

   I. $CO_2(g)$
  II. $H_2(g)$
 III. any soluble gas that is heavier than air

(A) I only
(B) I and II only
(C) I and III only
(D) II and III only
(E) I, II, and III

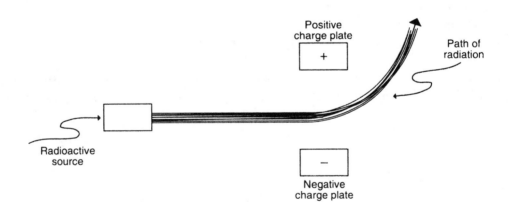

59. In the experiment diagrammed above, the radiation emitted by the source is probably

(A) $_{-1}^{0}e$
(B) $_{2}^{4}He$
(C) $_{2}^{1}H$
(D) a gamma ray
(E) $_{1}^{0}e$

60. Which of the following is true regarding the aqueous dissociation of HCN, $K_a = 4.9 \times 10^{-10}$, at 25°C?

    I. At equilibrium, $[H^+] = [CN^-]$
    II. At equilibrium, $[H^+] > [CN^-]$
    III. HCN(aq) is a strong acid.
    (A) I only
    (B) II only
    (C) I and II only
    (D) I and III only
    (E) I, II, and III

61. If 0.8 mol of NO is converted to NO_2 in the following reaction, what amount of heat will be evolved?

$$2NO(g) + O_2(g) \rightleftharpoons 2NO_2(g) + 150 \text{ kJ}$$

    (A) 30 kJ
    (B) 60 kJ
    (C) 80 kJ
    (D) 130 kJ
    (E) 150 kJ

62. How does a Brønsted-Lowry acid differ from its conjugate base?
    (A) The acid has one more proton.
    (B) The acid has one less proton.
    (C) The acid has one more electron.
    (D) The acid has one less electron.
    (E) The acid and base have the same number of protons and electrons.

63. Two containers having 1 mol of hydrogen gas and 1 mol of oxygen gas, respectively, are opened. What will be the ratio of the rate of effusion of the hydrogen to that of the oxygen?
    (A) $\sqrt{2} : 1$
    (B) $4 : 1$
    (C) $8 : 1$
    (D) $16 : 1$
    (E) $\sqrt{32} : 1$

64. A molecule in which the electron configuration is a resonance hybrid is
    (A) SO_2
    (B) C_2H_6
    (C) Cl_2
    (D) HBr
    (E) NaCl

65. What is the pH of a solution in which the $[OH^-]$ is 1.0×10^{-4}?
    (A) −4
    (B) +4
    (C) +7
    (D) −10
    (E) +10

66. If 0.365 g of hydrogen chloride is dissolved to make 1 L of solution (Cl = 35.5 and H = 1), the pH concentration of the solution is
    (A) 0.001
    (B) 0.01
    (C) 1
    (D) 2
    (E) 12

67. In the laboratory, a sample of hydrated salt was heated at 110°C for 30 min until all the water was driven off. The data were as follows:

    Mass of the hydrate before heating = 250 g
    Mass of the hydrate after heating   = 160 g

    From these data, what was the percent of water by mass in the original sample?
    (A) 26.5
    (B) 36
    (C) 47
    (D) 56
    (E) 90

68. Which of the following oxides dissolves in water to form an acidic solution?
    (A) Na_2O
    (B) CaO
    (C) Al_2O_3
    (D) ZnO
    (E) SO_3

69. In the laboratory, 20.0 mL of an aqueous solution of calcium hydroxide, $Ca(OH)_2$, was used in a titration. A drop of phenolphthalein was added to it to indicate the endpoint. The solution turned colorless after 20.0 mL of a standard solution of 0.050 M HCl solution was added. What was the molarity of the $Ca(OH)_2$?
    (A) 0.01 M
    (B) 0.025 M
    (C) 0.50 M
    (D) 0.75 M
    (E) 1.00 M

70. Which of the following reactions will NOT go to completion?
    (A) $Zn(s) + HCl(aq) \rightarrow ZnCl_2(aq) + H_2(g)$
    (B) $CaCO_3 + 2HCl \rightarrow CaCl_2 + H_2O + CO_2(g)$
    (C) $Ag^+ + HCl \rightarrow AgCl(s) + H^+$
    (D) $Cu + 2H^+ \rightarrow Cu^{2+} + H_2$; $E^0 = -0.34$ V
    (E) $H_2SO_4 + 2NaOH \rightarrow 2Na_2SO_4 + 2H_2O$

71. For a laboratory experiment, a student placed sodium hydroxide crystals on a watch glass, assembled the titration equipment, and prepared a solution of 0.10 M sulfuric acid. Then he weighed 4 g of sodium hydroxide and added it to enough water to make 1 L of solution. What might be a source of error in the results of the titration?
    (A) Some sulfuric acid evaporated.
    (B) The sulfuric acid became more concentrated.
    (C) The NaOH solution gained weight, thus increasing its molarity.
    (D) The NaOH crystals gained H_2O weight, thus making the solution less than 0.1 M.
    (E) The evaporation of sulfuric acid solution countered the absorption of H_2O by the NaOH solution.

72. If 60 g of NO is reacted with sufficient O_2 to form NO_2 that is removed during the reaction, how many grams of NO_2 can be produced? (Molar mass: NO = 30 g, NO_2 = 46 g)
    (A) 46
    (B) 60
    (C) 92
    (D) 120
    (E) 180

73. Based on the information shown, each of the following equations represents a reaction in which the entropy, ΔS is positive except:
    (A) $CaCO_3(s) \rightarrow CaO(s) \rightarrow CO_2(g)$
    (B) $Zn(s) + 2H^+(aq) \rightarrow H_2(g) + Zn^{2+}(aq)$
    (C) $2C_2H_6(g) + 7O_2(g) \rightarrow 4CO_2(g) + 6H_2O(g)$
    (D) $NaCl(s) \rightarrow Na^+(aq) + Cl^-(aq)$
    (E) $N_2(g) + 3H_2(g) \rightarrow 2NH_3(g)$

74. $Cl_2(g) + 2Br^-$ (excess) $\rightarrow$?
    When 1 mol of chlorine gas reacts completely with excess KBr solution, as shown above, the products obtained are
    (A) 1 mol of Cl^- ions and 1 mol of Br^-
    (B) 1 mol of Cl^- ions and 2 mol of Br^-
    (C) 1 mol of Cl^- ions and 1 mol of Br_2
    (D) 2 mol of Cl^- ions and 1 mol of Br_2
    (E) 2 mol of Cl^- ions and 2 mol of Br_2

75. Two water solutions are made in the laboratory, one of glucose (molar mass = 180 g/mol), the other of sucrose (molar mass = 342 g/mol). If the glucose solution had 180 grams in 1000 g of water and the sucrose has 342 g in 1000 g of water, which statement about the freezing points of the solutions is the most accurate?
    (A) The glucose solution would have the lower freezing point.
    (B) The sucrose solution would have the lower freezing point.
    (C) The freezing point of the sucrose solution would be lowered twice as much as that of the glucose solution.
    (D) Both solutions would have the same freezing point.
    (E) The freezing point of the solutions would not be affected because the solutes are both nonpolar.

76. Which K_a value indicates the strongest acid?
    (A) 1.3×10^{-2}
    (B) 6.7×10^{-5}
    (C) 5.7×10^{-10}
    (D) 4.4×10^{-7}
    (E) 1.8×10^{-16}

77. What mass of $CaCO_3$ is needed to produce 11.2 L of CO_2 when the calcium carbonate is reacted with an excess amount of hydrochloric acid? (Molar mass of $CaCO_3 = 100$ g, $HCl = 36.5$ g, $CO_2 = 44$ g.)
    (A) 25 g
    (B) 44 g
    (C) 50 g
    (D) 100 g
    (E) none of the above

78. By experimentation it is found that a saturated solution of $BaSO_4$ at 25°C contains 3.9×10^{-5} mol/L of Ba^{2+} ions. What is the K_{sp} of the $BaSO_4$?
    (A) 1.5×10^{-4}
    (B) 1.5×10^{-9}
    (C) 1.5×10^{-10}
    (D) 3.9×10^{-10}
    (E) 39×10^{-9}

79. What is the ΔH^0 value for the decomposition of sodium chlorate, given the following information?

$$NaClO_3(s) \rightarrow NaCl(s) + \tfrac{3}{2}O_2(g)$$

ΔH_f^0 values for the substances are $NaClO_3(s) = -85.7$ kcal/mol, $NaCl(s) = -98.2$ kcal/mol $O_2(g) = 0$ kcal/mol.
    (A) 12.5 kcal
    (B) –12.5 kcal
    (C) 173.9 kcal
    (D) –173.9 kcal

    (E) $\tfrac{3}{2}(173.9\,\text{kcal})$

$$AgCl(s) \rightleftharpoons Ag^+(aq) + Cl^-(aq)$$

80. To the silver chloride salt equilibrium reaction shown above, a beaker of concentrated HCl (12 M) is added. Which is the best description of what will occur?
    (A) More salt will go into solution, and K_{sp} will remain the same.
    (B) More salt will go into solution, and K_{sp} will increase.
    (C) Salt will come out of the solution, and K_{sp} will remain the same.
    (D) Salt will come out of the solution, and K_{sp} will decrease.
    (E) No change in concentration will occur, and K_{sp} will increase.

81. When this equation is balanced and all coefficients are reduced to lowest whole-number terms, what is the coefficient of H_2O?

$$HCl + KMnO_4 \rightarrow H_2O + KCl + MnCl_2 + Cl_2$$

    (A) 1
    (B) 2
    (C) 5
    (D) 8
    (E) 16

82. Each of the following systems is at equilibrium in a closed container. A decrease in the total volume of each container increases the number of moles of product(s) for which system?
    (A) $2NH_3(g) \rightleftharpoons N_2(g) + 3H_2(g)$
    (B) $H_2(g) + Cl_2(g) \rightleftharpoons 2HCl(g)$
    (C) $2NO(g) + O_2(g) \rightleftharpoons 2NO_2(g)$
    (D) $CO(g) + H_2O(g) \rightleftharpoons CO_2(g) + H_2(g)$
    (E) $Fe_3O_4(s) + 4H_2(g) \rightleftharpoons 3Fe(s) + 4H_2O(g)$

83. Which element(s), when forming an ionic bond, has(have) the following orbital notation?

$$1s^2, 2s^2, 2p^6, 3s^2, 3p^6$$

    I. Potassium
    II. Sulfur
    III. Chlorine

    (A) I only
    (B) III only
    (C) I and III only
    (D) II and III only
    (E) I, II, and III

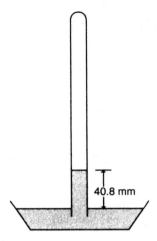

84. Hydrogen gas is collected in a eudiometer tube over water as shown above. The water level inside the tube is 40.8 mm higher than that outside. The barometric pressure is 730 mm Hg. The water vapor pressure at the room temperature of 29°C is found in a handbook to be 30.0 mm Hg. What is the pressure of the dry hydrogen?
    (A) 659.2 mm Hg
    (B) 689.2 mm Hg
    (C) 697.0 mm Hg
    (D) 740.8 mm Hg
    (E) 800.8 mm Hg

85. How many moles of electrons are required to reduce 2.93 g of nickel ions from melted $NiCl_2$? (molar mass = 58.7 g/mol)
    (A) 0.05
    (B) 0.10
    (C) 1.0
    (D) 1.5
    (E) 2.0

STOP
IF YOU FINISH BEFORE ONE HOUR IS UP, YOU MAY GO BACK
TO CHECK YOUR WORK OR COMPLETE UNANSWERED QUESTIONS.

Answers and Explanations for Test 5

1. **(A)** In the halogen family, the most active nonmetal would be the top element, fluorine, because it has the highest electronegativity.

2. **(A)** As you proceed down a group, the ionic radius increases as additional energy levels are filled further from the nucleus. Therefore fluorine, the top element, has the smallest ionic radius.

3. **(E)** Since astatine has the largest atomic radius and its outer electrons are shielded from the protons by a large number of interior electrons, it has the lowest first ionization potential.

4. **(D)** Some physical characteristics of metal are found in iodine, the fourth halogen down in the group.

5. **(A)** Sodium acetate has Na^+ and $C_2H_3O_2^-$ ions.

6. **(C)** Aluminum oxide has Al^{3+} and O^{2-} ions.

7. **(A)** Potassium phosphate has a K^+ and a PO_4^{3-} ion.

8. **(D)** John Dalton is credited with the basic postulates of atomic theory.

9. **(E)** Gay-Lussac is credited with the idea that when gases combine, they do so in ratios of small whole numbers that are in relationship to the volumes of the reactants and the volumes of the products under the same conditions.

10. **(C)** Charles is credited with this temperature-volume relationship of gases: $V_1T_2 = V_2T_1$

11. **(A)** Avogadro is credited with the concept regarding the number of particles in a mole, 6.02×10^{23}, and this number bears his name.

12. **(B)** Boyle is credited with the $P_1V_1 = P_2V_2$ relationship of gases.

13. **(C)** This includes the functional group of an aldehyde.

14. **(D)** This includes the functional group of an organic acid.

15. **(E)** An ester is the equivalent of an organic salt since it is usually formed from an organic alcohol, R—OH, plus an organic acid, R^*-C(=O)-OH. The bonding is R-O[HHO]-C(=O)-R^*, which gives R-O-C(=O)-R^* + H_2O. (R^* indicates that this hydrogen branch need not be the same as R.)

16. **(B)** This includes the functional group of an ether. It can be formed by the dehydration of two alcohol molecules. The reaction is

$$R-O[H + HO]-R^* \rightarrow R-O-R^* + H_2O.$$

17. **(B)** The molecular structure of carbon dioxide is $O = C = O$, where the oxygens are 180° apart and, although the bonding is polar to the carbon, counteract each other to constitute a nonpolar molecule.

18. **(A)** A hydrogen gas molecule is diatomic; it has 2 mol of atoms in each mole of molecules, represented by H_2.

19. **(B)** Each CO_2 has three atoms per molecule; hence the expression can represent 3 mol of atoms in 1 mol of molecules.

20. **(E)** With complete ionization $H_2SO_4 \rightarrow 2H^+ + SO_4^{2-}$, or 3 mol of ions per mole of H_2SO_4.

21. **(C)** The expression $2N_2O$ represents two triatomic molecules or 6 mol of atoms.

22. **(D)** Carbon tetrachloride has four hydrogens, each bonded to an sp^3 hybridized orbital. The four sp^3 orbitals of carbon are in a tetrahedral orientation so that the resulting configuration balances the polarity of each bond, forming a nonpolar covalent molecule.

23. **(A)** Because cesium is a very active metal, the chloride formed has an ionic bond and is an ionic substance.

24. **(C)** Chlorine's stronger electronegativity draws the hydrogen electron closer to itself more often, and so the charge distribution in the molecule is uneven, causing a negative and positive polarity. Therefore hydrogen chloride is a polar covalent molecule.

25. **(E)** Benzene is an aromatic (ring) structure classified as an organic substance.

101. **(T, T, CE)** A catalyst can accelerate a chemical reaction by lowering the activation energy required for the reaction to occur.

102. **(T, T, CE)** Sodium chloride is an ionic substance and when molten is a good electrical conductor. The reason is that in the molten state the ions are free to migrate to the anode and cathode.

103. **(T, F)** Ice is less dense than liquid water. Water molecules, however, are polar, not nonpolar, and water expands as these molecules arrange themselves into a crystal lattice.

104. **(F, T)** It is false that isotopes of the same element have the same mass number. Isotopes of the same element have the same number of protons but vary in the number of neutrons in the nucleus. This means they will have the same atomic number but different mass numbers.

105. **(T, T, CE)** A solute that has a very low solubility in a particular solvent can reach the saturated point in a solution with that solvent, but the solution can still be classified as dilute. The term *dilute* merely refers to a small amount of solute in a solvent.

106. **(T, T)** It is true that nitrogen is an element, but the reason is that it is composed of only one kind of atom. That it is diatomic at room temperature does not explain its classification as an element.

107. **(F, F)** A reaction at equilibrium has reached a point where the forward and reverse reactions are occurring at an equal rate. The concentrations of the reactants and products are not necessarily equal and are described by the K_{eq} value at the temperature of the reaction.

108. **(F, T)** In an electrochemical reaction, anions migrate to the anode. It is true that positively charged ions (cations) are attracted to the negatively charged cathode.

109. **(F, T)** The value pH = 5 can be expressed as $[H_3O^+] = 1 \times 10^{-5}$ mol/L and pH = 3 as $[H_3O^+] = 1 \times 10^{-3}$ mol/L. Thus pH = 3 represents a larger concentration of hydronium ions, $[H_3O^+]$.

110. **(T, T, CE)** The first statement is true. The change in Gibbs free energy, ΔG, depends on enthalpy change, ΔH, and entropy change, ΔS, from the equation $\Delta G = \Delta H - T\,\Delta S$. So statement II is also true and explains the first.

111. **(T, T, CE)** In the expression for the equilibrium constant of an acid, $[H_3O^+]$ is in the numerator:
$$K_a = \frac{[H_3O^+][Y^-]}{[HY]}.$$
As $[H_3O^+]$ decreases in the weak acids, the numerator becomes smaller as the denominator gets larger, therefore giving smaller K_a values.

112. **(T, T)** Electronegativity values do increase across periods in the periodic chart. This fact has no relation, however, to the increase in the radius of a nonmetallic atom as it becomes an ion.

113. **(T, F)** A pi bond is formed between *p* lobes of adjacent atoms and in the same plane:

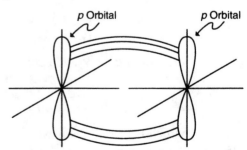

However, each *p* orbital consisting of two lobes can hold a total of two electrons, and so the reason is false.

114. **(T, F)** The assertion that the boiling points of H_2S and H_2O are significantly different is true, but the reason is false. Water has the higher degree of hydrogen bonding.

40. **(E)** Two immiscible liquids, when shaken together vigorously, may each have small, colloid-size particles dispersed in the other liquid.

41. **(D)** In the chart, the first plateau must represent the condensation from gas to liquid because there is a second, lower plateau, which represents the second change of state, from liquid to solid.

42. **(B)** If the electrons have the same principal energy level, they will fill the s^2, p^6, and then the d^{10} level. This progression is from the lowest energy sublevel to the highest, to accommodate 18 electrons.

43. **(B)** The empirical formula identifies the atoms present and the lowest whole-number ratio of their occurrence, as CH. The molecular formula, which gives the actual composition of the compound, must then be a multiple of the empirical formula C_6H_6.

44. **(D)** Considering the concepts behind Charles's Law and Boyle's Law, you can arrive at the fraction used in kelvins and torrs. The volume must increase with an increase in temperature and also increase with a decrease in pressure. Therefore the fractions are $\dfrac{333\ K}{278\ K}$ and $\dfrac{640\ torr}{320\ torr}$. Using the combined gas law equation,

$$\frac{V_1 P_1}{T_1} = \frac{V_2 P_2}{T_2}$$

and solving for V_2, you get

$$V_2 = \frac{T_2}{T_1} \times \frac{P_1}{P_2}$$

which is the same answer.

45. **(D)** The atomic mass, 32, is the total number of neutrons and protons.

Since the atomic number, 16, gives the number of protons, $32 - 16 = 16$, or the number of neutrons, statement II is false. Since this ion has a charge of -2, it has two more electrons than protons, or 18 electrons. Statements I and III are true.

46. **(D)** The most active metallic elements are found in the lower left corner of the periodic table.

47. **(C)** The rise in the boiling point depends on the number of particles in solution. One mole of $CaCl_2$ gives 3 mol of ions, more than any other substance listed:

$$CaCl_2 \rightarrow Ca^{2+} + 2Cl^-$$

The number of moles of ions given by the other substances are as follows: (A) = 2, (B) = 2, (D) = 1, (E) = 1.

48. **(E)** The sp^3 hybrid has the tetrahedron configuration. The sp^2 (D) is trigonal planar. The sp (C) is linear. The s (B) and p (A) are the usual orbital structure.

49. **(C)** The reaction is

$$\begin{array}{cc} 8\ L & x\ liters \\ S(s) + O_2(g) \rightarrow SO_2(g) \end{array}$$

The given (8 L) and the unknown (x liters) are shown above. Since the equation, according to Gay-Lussac's Law, shows that 1 vol of oxygen yields 1 vol of sulfur dioxide, then

$$8\ \cancel{L\,O_2} \times \frac{1\ L\ SO_2}{1\ \cancel{L\,O_2}} = 8\ L\ SO_2$$

50. **(C)** Since the mercury level in the U-tube is higher on the left side, it indicates that the pressure in the flask is higher than the atmospheric pressure exerted on the open end of the tube on the right side. If the pressure inside the flask were the same as the atmospheric pressure, the height of mercury in both sides of the U-tube would be the same.

51. (**B**) The only change listed that will change the pressure of the gas inside the flask is to increase the temperature. This will cause the pressure to increase.

52. (**D**) An amphoteric substance must be able to be a proton, H^+, donor and a proton receiver. The bisulfate ion, HSO_4^-, is the only choice that can either accept a proton to become H_2SO_4 or lose a proton and become the sulfate ion, SO_4^{2-}.

53. (**C**) Standard conditions are 273 K and 760 mm Hg.

54. (**A**) Dividing the percentage of each element by the atomic mass of that element gives the basic ratio of atoms but not necessarily in whole numbers. Thus, (Ba) $58.8 \div 137 = 0.43$, (S) $13.7 \div 32 = 0.43$, (O) $27.5 \div 16 = 1.72$.
Because atoms occur in whole numbers, you now must manipulate these numbers mathematically to get whole numbers. Usually dividing each number by the smallest one helps to accomplish this: (Ba) $0.43 \div 0.43 = 1$, (S) $0.43 \div 0.43 = 1$, (O) $1.72 \div 0.43 = 4$. The empirical formula is $BaSO_4$.

55. (**C**) The percentage composition can be found by dividing each total atomic mass in the formula by the molar mass of the compound.

$$Ca = 40 \quad 40 \div 74 \times 100\% = 54\%\ Ca$$
$$2O = 32 \quad 32 \div 74 \times 100\% = 43\%\ O$$
$$2H = \underline{\ 2\ } \quad 2 \div 74 \times 100\% = 2.7\%\ H$$
$$74$$

56. (**B**) The solution setup is

$$\underset{65\,g}{\overset{65\,g}{Zn}} + \underset{73\,g}{\overset{65\,g}{2HCl}} \rightarrow ZnCl_2 + \underset{2\,g}{\overset{x\,grams}{H_2(g)}}$$

Note that the equation mass is calculated under the substances that have mass units above them. According to the calculated equation masses, 73 g of HCl would be needed to react with 65 g of zinc. Since there is only 65 g of HCl, we use this and disregard the 65 g of Zn.

$$65\ g\ \text{HCl} \times \frac{2\ g\ H_2}{73\ g\ \text{HCl}}$$
$$= 1.78\ \text{or}\ 1.8\ g\ H_2$$

57. (**C**) All the other statements represent observations because they merely record what was seen.

58. (**B**) III, cannot be used because of solubility.

59. (**A**) Because the paths of the particles from the radiocactive source bend toward the positive plate, the source must be negatively charged. The only radioactive particle that fits this criterion is the alpha particle, ^4_2He.

60. (**A**) A small K_a indicates that this is a weak acid. When HCN ionizes, it can be shown that $HCN \rightleftharpoons H^+ + CN^-$. This is a molar ratio of $1:1$; so that statement I must be true.

61. (**B**) Since the equation shows that 2 mol of NO react to release 150 kJ, the solution is

$$0.8\ \text{mol NO} \times \frac{150\ kJ}{2\ \text{mol NO}} = 60\ kJ$$

This problem would be solved in the same manner if the heat had been expressed in kilocalories. To convert one unit to the other, use $4.18 \times 10^3\ J = $ kcal.

62. (**A**) In the Brønsted-Lowry acid-base theory:

$$HCl + NH_3 \rightleftharpoons NH_4^+ + Cl^-$$
acid   base   conjugate   conjugate
                acid   base

63. (**B**) According to Graham's Law, the rates of effusion of two gases are inversely proportional to the square roots of their molar masses. Since 1 mol $O_2 = 32\ g$ and 1 mol $H_2 = 2\ g$,

$$\frac{H_2\ \text{rate}}{O_2\ \text{rate}} = \frac{\sqrt{32}}{\sqrt{2}} = \sqrt{\frac{16}{1}} = \frac{4}{1}$$

Therefore, the effusion rate of hydrogen is four times faster than that of oxygen.

64. **(A)** The structure of SO_2 is a resonance hybrid, shown as follows:

$$\overset{\displaystyle S}{\underset{O}{\diagup}}\overset{O}{} \longleftrightarrow \overset{\displaystyle S}{\underset{O}{\diagup}}\overset{O}{}$$

65. **(E)** The K_w of water is

$$K_w = [H^+][OH^-] = 10^{-14}$$

If $[OH^-] = 1.0 \times 10^{-4}$, then

$$[H^+] = \frac{10^{-14}}{10^{-4}} = 10^{-10}$$

and

$$pH = -\log[H^+] = 10$$

66. **(D)** $0.365 \text{ g } HCl \times \dfrac{1 \text{ mol HCl}}{36.5 \text{ g } HCl}$

$= 0.01 \text{ mol HCl}$

$0.01 \text{ mol HCl} \rightarrow 0.01 \text{ mol H}^+ +$
0.01 mol Cl^-

If $[H^+] = 0.01 = 1 \times 10^{-2} \text{ mol/L}$, then $pH = 2$.

67. **(B)**

$$\begin{array}{r} \text{Hydrate mass before heating} = 250 \text{ g} \\ -\text{Hydrate mass after heating} = 160 \text{ g} \\ \hline \text{Water loss} = 90 \text{ g} \end{array}$$

$$\dfrac{90 \text{ g mass loss}}{250 \text{ g original mass}} \times 100\%$$

$= 36\%$ by mass

68. **(E)** Since metallic oxides are basic anhydrides, the only nonmetallic oxide is SO_3. The reaction is

$$SO_3 + H_2O \rightarrow H_2SO_4 \text{ (sulfuric acid)}$$

69. **(B)** In the titration, the reaction is

$$2HCl + Ca(OH)_2 \rightarrow CaCl_2 + 2H_2O$$

The acid-to-base ratio is $2:1$, or (moles acid used) = 2(moles base used), and so

$M_aV_a = 2M_bV_b$, where M is the molarity and V is the volume expressed in liters, then

$$M_b = \frac{M_aV_a}{2V_b}$$

$$M_b = \frac{0.05 \ M \times 0.02 \text{ L}}{2(0.02 \text{ L})} = 0.025 \ M$$

70. **(D)** All the reactions will go to completion except (D), which will not occur spontaneously at all. If E^0 had been positive, however, this redox reaction would occur.

71. **(D)** Sodium hydroxide is hygroscopic and will attract water to its surface area. This water will influence its mass; consequently there will be less sodium hydroxide in the mass used.

72. **(C)** $\overset{60 \text{ g}}{2NO} + O_2 \rightarrow \overset{x \text{ grams}}{2NO_2}$
$\underset{60 \text{ g}}{}\underset{92 \text{ g}}{}$

This is a mass stoichiometry problem. The equation masses are placed beneath the substances that have quantities above them. Using the equation relationship gives

$$60 \text{ g } NO \times \frac{92 \text{ g NO}_2}{60 \text{ g } NO} = 92 \text{ g NO}_2$$

Using the mole method gives

$$60 \text{ g } NO \times \frac{1 \text{ mol NO}}{30 \text{ g } NO} = 2 \text{ mol NO}$$

Using the equation coefficients gives

$$2 \text{ mol } NO \times \frac{2 \text{ mol NO}_2}{2 \text{ mol } NO} = 2 \text{ mol NO}_2$$

Then

$$2 \text{ mol } NO_2 \times \frac{46 \text{ g NO}_2}{1 \text{ mol } NO_2} = 92 \text{ g NO}_2$$

73. **(E)** Since in this reaction 4 vol of gases form 2 vol of product, the randomness of the system is decreasing. Therefore, entropy is decreasing and ΔS is negative. In all the other reactions randomness is increasing.

74. **(D)** The reaction is

$$Cl_2(g) + 2Br^- \text{ (excess)} \rightarrow 2Cl^- + Br_2(g)$$

Here 2 mol of chloride ions and 1 mol of Br_2 molecules are produced.

75. **(D)** Since both solutions were 1 m and neither compound ionized into more particles, the freezing points are the same.

76. **(A)** $HY \rightleftharpoons H^+ + Y^-$
Since $[H^+]$ is in the numerator of K_a,

$$K_a = \frac{[H^+][Y^-]}{[HY]}$$

The stronger the acid, the greater the $[H^+]$ and K_a values. Of the choices given, 1.3×10^{-2} is the largest.

77. **(C)** $\overset{x \text{ grams}}{\underset{100 \text{ g}}{CaCO_3}} + 2HCl \rightarrow CaCl_2 + H_2O + \overset{11.2 \text{ L}}{\underset{22.4 \text{ L}}{CO_2}}$

This is a mass-volume problem with the equation mass and volume indicated below the equation. Using the factor-label method gives

$$11.2 \text{ L } CO_2 \times \frac{100 \text{ g } CaCO_3}{22.4 \text{ L } CO_2}$$
$$= 50 \text{ g } CaCO_3$$

Using the mole method gives

$$11.2 \text{ L } CO_2 \times \frac{1 \text{ mol } CO_2}{22.4 \text{ L } CO_2} = 0.5 \text{ mol } CO_2$$

$$0.5 \text{ mol } CO_2 \times \frac{1 \text{ mol } CaCO_3}{1 \text{ mol } CO_2}$$
$$= 0.5 \text{ mol } CaCO_3$$

$$0.5 \text{ mol } CaCO_3 \times \frac{100 \text{ g } CaCO_3}{1 \text{ mol } CaCO_3}$$
$$= 50 \text{ g } CaCO_3$$

78. **(B)** $K_{sp} = [Ba^{2+}][SO_4^{2-}]$
If $[Ba^{2+}] = 3.9 \times 10^{-5}$, then $[SO_4^{2-}]$ must also equal the same amount, and so

$$K_{sp} = [3.9 \times 10^{-5}][3.9 \times 10^{-5}]$$
$$= 15.2 \times 10^{-10} \text{ or } 1.5 \times 10^{-9}.$$

79. **(B)** $\Delta H^0_{reaction} = \Delta H^0_f \text{ (products)} - \Delta H^0_f$
(reactants)

$$\Delta H^0_{reaction} = -98.2 \text{ kcal} - (-85.7 \text{ kcal})$$
$$\Delta H^0_{reaction} = -12.5 \text{ kcal}$$

80. **(C)** Since the HCl solution will add to the chloride ion concentration, according to Le Châtelier's Principle the equilibrium will shift in the direction to reduce this disturbance, and so K_{sp} will remain the same but salt will come out of the solution. This process:

$$AgCl \rightleftharpoons Ag^+ + Cl^-$$

will continue until K_{sp} is reestablished. This phenomenon is called the "common ion effect."

81. **(D)** $HCl + KMnO_4 \rightarrow H_2O + KCl + MnCl_2 + Cl_2$
oxidation: $2Cl^- \rightarrow Cl_2 + 2e^-$
reduction: $8H^+ + MnO_4^- + 5e^- \rightarrow Mn^{2+} + 4H_2O$
5 (oxidation reaction) and 2 (reduction reaction) will balance the e^- gain and loss, giving

$$10Cl^- \rightarrow 5Cl_2 + 10e^-$$
$$16H^+ + 2MnO_4^- + 10e^- \rightarrow 2Mn^{2+}$$
$$+ 8H_2O$$

Therefore

$$16HCl + 2KMnO_4 \rightarrow 8H_2O + 2KCl$$
$$+ 2MnCl_2 + 5Cl_2$$

82. **(C)** A decrease in volume will cause the equilibrium to shift in the direction that has less volume(s) of gas(es). In every case except (C) this is the reverse reaction, which decreases the product. The coefficients give the volume relationships.

83. **(E)** Each of the elements when becoming ions has this configuration of 18 electrons.
K with 19 electrons $- e^- \rightarrow K^+$ with 18 electrons
S with 16 electrons $+ 2e^- \rightarrow S^{2-}$ with 18 electrons
Cl with 17 electrons $+ e^- \rightarrow Cl^-$ with 18 electrons

84. **(C)** Change the water height to the equivalent Hg height:

$$40.8 \; \text{mm H}_2\text{O} \times \frac{1 \; \text{mm Hg}}{13.6 \; \text{mm H}_2\text{O}}$$

$$= 3 \; \text{mm Hg}$$

Adjust for the difference in height to get the gas pressure. Pressure on the gas is

730 mm Hg – 3 mm Hg = 727 mm Hg

Vapor pressure of H_2O at 29°C accounts for 30 mm Hg pressure. Therefore

727 mm Hg – 30 mm Hg = 697 mm Hg

85. **(B)** The reaction is

$$\text{Ni}^{2+} + 2e^- \rightarrow \text{Ni(s)}$$

Since 2 mol of electrons are required to form 1 mol of nickel, and

$$2.9 \; \text{g Ni} \times \frac{1 \; \text{mol Ni}}{58.7 \; \text{g Ni}} = 0.05 \; \text{mol Ni}$$

then

$$0.05 \; \text{mol Ni} \times \frac{2 \; \text{mol electrons}}{1 \; \text{mol Ni}}$$

$$= 0.1 \; \text{mol of electrons}$$

Diagnosing Your Needs

This section will help you to diagnose your need to review the various categories tested by SAT II: Chemistry.

After taking the diagnostic test, check your answers against the correct ones. Then fill in the chart below.

In the space under each question number, place a check (✓) if you answered that question correctly.

Example

If your answer to question 5 was correct, place a check in the appropriate box.

Next, total the check marks for each section and insert the number in the designated block. Then do the arithmetic indicated and insert your percentage for each area.

	SUBJECT AREA	(✔) QUESTIONS ANSWERED CORRECTLY								
I.	**Atomic Theory and Structure,** including periodic relationships	1	2	3	5	6	7	8	42	
	☐ No. of checks ÷ 8 × 100 = _____%									
II.	**Nuclear Reactions**						104	45	59	
	☐ No. of checks ÷ 3 × 100 = _____%									
III.	**Chemical Bonding and Molecular Structure**	17	20	22	23	24	113	114	48	83
	☐ No. of checks ÷ 9 × 100 = _____%									
IV.	**States of Matter and Kinetic Molecular Theory of Gases**	9	10	12	41	44	49	53	63	
	☐ No. of checks ÷ 8 × 100 = _____%									
V.	**Solutions,** including concentration units, solubility, and colligative properties				105	40	47	69	75	
	☐ No. of checks ÷ 5 × 100 = _____%									
VI.	**Acids and Bases**	109	111	52	62	65	66	68	76	
	☐ No. of checks ÷ 8 × 100 = _____%									
VII.	**Oxidation-Reduction and Electrochemistry**		102	108	70	74	81	85		
	☐ No. of checks ÷ 6 × 100 = _____%									

SUBJECT AREA   (✔) QUESTIONS ANSWERED CORRECTLY

VIII. Stoichiometry	11	18	19	106	112	43	54	55	56
	72	77							

☐ No. of checks ÷ 11 × 100 = _____%

IX. Reaction Rates								101	107

☐ No. of checks ÷ 2 × 100 = _____%

X. Equilibrium						60	78	80	82

☐ No. of checks ÷ 4 × 100 = _____%

XI. Thermodynamics: energy changes in chemical reactions, randomness, and criteria for spontaneity						110	61	73	79

☐ No. of checks ÷ 4 × 100 = _____%

XII. Descriptive Chemistry: physical and chemical properties of elements and their familiar compounds; organic chemistry; periodic properties	4	13	14	15	16	21	25	103
					46	51	64	

☐ No. of checks ÷ 11 × 100 = _____%

XIII. Laboratory: equipment, procedures, observations, safety, calculations, and interpretation of results	50	57	58	67	71	84

☐ No. of checks ÷ 6 × 100 = _____%

Planning Your Study

The percentages give you an idea of how you have done in the various major areas of the test. Because of the limited number of questions on some parts, these percentages may not be as reliable as the percentages for parts with larger numbers of questions. However, you should now have at least a rough idea of the areas in which you have done well and those in which you need more study. (There are four more practice tests in the back of this book that may be used in a diagnostic manner as well.)

Start your study with the areas in which you are the weakest. The corresponding chapters are indicated on page 447.

Subject Area	Chapters to Review
I. Atomic Theory and Structure, including periodic relationships	2
II. Nuclear Reactions	15
III. Chemical Bonding and Molecular Structure	3, 4
IV. States of Matter and Kinetic Molecular Theory of Gases	1
V. Solutions, including concentration units, solubility, and colligative properties	7
VI. Acids and Bases	11
VII. Oxidation-Reduction and Electrochemistry	12
VIII. Stoichiometry	5, 6
IX. Reaction Rates	9
X. Equilibrium	10
XI. Thermodynamics, including energy changes in chemical reactions, randomness, and criteria for spontaneity	8
XII. Descriptive Chemistry: physical and chemical properties of elements and their familiar compounds; organic chemistry; periodic properties	1, 2, 13, 14
XIII. Laboratory: equipment, procedures, observations, safety, calculations, and interpretation of results.	All lab diagrams, 16

After you have spent some time in reviewing your weaker areas, plan a schedule of work that spans the 6 weeks before the test. Unless you set up a regular study pattern and goals, you probably will not prepare sufficiently.

The following schedule provides such a plan. Note that weekends are left free, and the time spans are held to 1- or 2-hour blocks. This will be time well spent!

	Monday	Tuesday	Wednesday	Thursday	Friday
First Week:	Ch. 1: 1 hr	Ch. 2: 2 hr	Ch 2: 1 hr.	Ch. 3: 2 hr	Ch. 3: 1 hr
Second Week:	Ch. 4: 1 hr	Ch. 4: 1 hr	Ch. 5: 2 hr	Ch. 5: 1 hr	Ch. 6: 1 hr
Third Week:	Ch. 6: 2 hr	Ch. 7: 2 hr	Ch. 8: 1 hr	Ch. 9: 2 hr	Ch. 10: 1 hr
Fourth Week:	Ch. 10: 1 hr	Ch. 11: 2 hr	Ch. 12: 2 hr	Ch. 13: 1 hr	Ch. 14: 2 hr
Fifth Week:	Chs. 15, 16: 1 hr	Take Practice Test 1: 1 hr Study ans.: 1 hr	Review weakest areas: 1 hr	Take Practice Test 2: 1 hr Study ans.: 1 hr	Review weakest areas: 1 hr
Sixth Week:	Take Practice Test 3: 1 hr Study ans.: 1 hr	Review weakest areas: 1 hr	Take Practice Test 4: 1 hr Study ans.: 1 hr	Review weakest areas: 1 hr	Review a practice test already taken: 1 hr *Go to bed early.*

Final Preparation—The Day Before the Test

The day before the test, review one of the practice tests you have already taken. Study again the directions for each type of question. Long hours of study at this point will probably only heighten your anxiety, so just look over the answer section of the practice test and refer to any chapter in the book if you need more information. This type of limited, relaxed review will probably make you feel more comfortable and better prepared.

Get together the materials that you will need. They are

- Your admission ticket.
- Your identification. (You will not be admitted without some type of positive identification such as a student I.D. card with picture or a driver's license.)
- Two No. 2 pencils with erasers.
- Calculator use is *not* allowed during the Chemistry Test.

You should also go over this checklist:

A. Plan your activities so that you will have time for a good night's sleep.
B. Lay out comfortable clothes for the next day.
C. Review the following helpful tips about taking the test:
   1. Read the directions carefully. You have one hour to complete the test. You can go back to questions you skipped.
   2. In each group of questions, first answer those that you know. Temporarily skip difficult questions, but mark them in the margin so you can go back if you have time. Keep in mind that an easy question answered correctly counts as much as a difficult one.
   3. Avoid haphazard guessing since this will probably lower your score. Guess smart! If you can eliminate one or more of the choices to a question, it will generally be to your advantage to guess which of the remaining answers is correct. Your score will be based on the number right minus a fraction of the number answered incorrectly.
   4. Omit questions when you have no idea how to answer them. Remember that you neither gain nor lose credit for questions you do not answer.
   5. Pace your time throughout the hour.
   6. Mark the answer grid clearly and correctly. Be sure each answer is placed in the proper space and within the oval. *Erase all stray marks completely*.
   7. You may write as much as you like in the test booklet. Use it as a scratch pad. Only the answers on the answer sheet are scored for credit.
D. Set your alarm clock so as to allow plenty of time to dress, eat your usual (or even a better) breakfast, and reach the test center without haste or anxiety.

EQUATIONS AND TABLES FOR REFERENCE

Some Important Formulas and Equations

Density (d)

$$\text{Density} = \frac{\text{mass}}{\text{volume}} \quad d = m/v$$

Percent Error

Percent error =

$$\frac{\text{measured value} - \text{accepted value}}{\text{accepted value}} \times 100\%$$

Percent Yield

$$\text{Percent yield} = \frac{\text{actual yield}}{\text{expected yield}} \times 100\%$$

Percentage Composition

Percentage composition by mass

$$= \frac{\text{mass of element}}{\text{mass of compound}} \times 100\%$$

Boyle's Law

$$P_1V_1 = P_2V_2$$

Charles's Law

$$V_1T_2 = V_2T_1$$

Dalton's Law of Partial Pressures

$$P_T = p_a + p_b + p_c + \cdots$$

Ideal Gas Law

$$PV = nRT$$

Molarity (M)

$$\text{molarity} = \frac{\text{moles of solute}}{\text{liters of solution}}$$

Molality (m)

$$\text{Molality} = \frac{\text{moles of solute}}{\text{kilograms of solvent}}$$

Boiling Point Elevation

$$\Delta T_b = K_b m$$

where K_b is the molal boiling point elevation constant

Freezing Point Depression

$$\Delta T_f = K_f m$$

where K_f is the molal boiling point elevation constant

Rate of Reaction

$$\text{Rate} = k[\text{A}]^x[\text{B}]^y$$

where [A] and [B] are molar concentrations of reactants and k is a rate constant

Hess's Law

$$\Delta H_{\text{net}} = \Delta H_1 + \Delta H_2$$

Entropy Change

$$\Delta S = S_{\text{products}} - S_{\text{reactants}}$$

Gibbs Free Energy

$$\Delta G = \Delta H - T\,\Delta S$$

TABLES FOR REFERENCE

Ⓐ

DENSITY AND BOILING POINTS OF SOME COMMON GASES			
Name		*Density* grams/liter at STP*	*Boiling Point* (at 1 atm) K
Air	—	1.29	—
Ammonia	NH_3	0.771	240
Carbon dioxide	CO_2	1.98	195
Carbon monoxide	CO	1.25	82
Chlorine	Cl_2	3.21	238
Hydrogen	H_2	0.0899	20
Hydrogen chloride	HCl	1.64	188
Hydrogen sulfide	H_2S	1.54	212
Methane	CH_4	0.716	109
Nitrogen	N_2	1.25	77
Nitrogen (II) oxide	NO	1.34	121
Oxygen	O_2	1.43	90
Sulfur dioxide	SO_2	2.92	263

*STP is defined as 273 K and 1 atm

Ⓑ

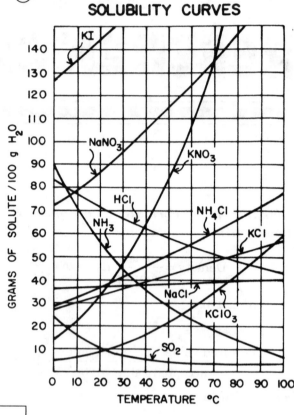

SOLUBILITY CURVES

Ⓒ

TABLE OF SOLUBILITIES IN WATER											
i—nearly insoluble ss—slightly soluble s—soluble d—decomposes n—not isolated	acetate	bromide	carbonate	chloride	chromate	hydroxide	iodide	nitrate	phosphate	sulfate	sulfide
Aluminum	ss	s	n	s	n	i	s	s	i	s	d
Ammonium	s	s	s	s	s	s	s	s	s	s	s
Barium	s	s	i	s	i	s	s	s	i	i	d
Calcium	s	s	i	s	s	ss	s	s	i	ss	d
Copper II	s	s	i	s	i	i	n	s	i	s	i
Iron II	s	s	i	s	n	s	s	s	i	s	i
Iron III	s	s	n	s	i	i	n	s	i	ss	d
Lead	s	ss	i	ss	i	i	ss	s	i	i	i
Magnesium	s	s	i	s	s	i	s	s	i	s	d
Mercury I	ss	i	i	i	ss	n	i	s	i	ss	i
Mercury II	s	ss	i	s	ss	i	i	s	i	d	i
Potassium	s	s	s	s	s	s	s	s	s	s	s
Silver	ss	i	i	i	ss	n	i	s	i	ss	i
Sodium	s	s	s	s	s	s	s	s	s	s	s
Zinc	s	s	i	s	s	i	s	s	i	s	i

Ⓓ

SELEFCTED POLYATOMIC IONS			
Hg_2^{2+}	dimercury (I)	CrO_4^{2-}	chromate
NH_4^+	ammonium	$Cr_2O_7^{2-}$	dichromate
$C_2H_3O_2^-$	acetate	MnO_4^-	permanganate
CH_3COO^-		MnO_4^{2-}	manganate
CN^-	cyanide	NO_2^-	nitrite
CO_3^{2-}	carbonate	NO_3^-	nitrate
HCO_3^-	hydrogen carbonate	OH^-	hydroxide
		PO_4^{3-}	phosphate
$C_2O_4^{2-}$	oxalate	SCN^-	thiocyanate
ClO^-	hypochlorite	SO_3^{2-}	sulfite
ClO_2^-	chlorite	SO_4^{2-}	sulfate
ClO_3^-	chlorate	HSO_4^-	hydrogen sulfate
ClO_4^-	perchlorate	$S_2O_3^{2-}$	thiosulfate

Ⓔ

STANDARD ENERGIES OF FORMATION OF COMPOUNDS AT 1 ATM AND 298 K		
Compound	**Heat (Enthalpy) of Formation*** kcal/mol (ΔH_f^0)	**Free Energy of Formation** kcal/mol (ΔG_f^0)
Aluminum oxide Al_2O_3(s)	−400.5	−378.2
Ammonia NH_3(g)	−11.0	−3.9
Barium sulfate $BaSO_4$(s)	−352.1	−325.6
Calcium hydroxide $Ca(OH)_2$(s)	−235.7	−214.8
Carbon dioxide CO_2(g)	−94.1	−94.3
Carbon monoxide CO(g)	−26.4	−32.8
Copper (II) sulfate $CuSO_4$(s)	−184.4	−158.2
Ethane C_2H_6(g)	−20.2	−7.9
Ethene (ethylene) C_2H_4(g)	12.5	16.3
Ethyne (acetylene) C_2H_2(g)	54.2	50.0
Hydrogen fluoride HF(g)	−64.8	−65.3
Hydrogen iodide HI(g)	6.3	0.4
Iodine chloride ICl(g)	4.3	−1.3
Lead (II) oxide PbO(s)	−51.5	−45.0
Magnesium oxide MgO(s)	−143.8	−136.1
Nitrogen (II) oxide NO(g)	21.6	20.7
Nitrogen (IV) oxide NO_2(g)	7.9	12.3
Potassium chloride KCl(s)	−104.4	−97.8
Sodium chloride $NaCl$(s)	−98.3	−91.8
Sulfur dioxide SO_2(g)	−70.9	−71.7
Water H_2O(g)	−57.8	−54.6
Water $H_2O(\ell)$	−68.3	−56.7

*Minus sign indicates an exothermic reaction.
Sample equations:

$$2Al(s) + \tfrac{3}{2}O_2(g) \rightarrow Al_2O_3(s) + 400.5 \text{ kcal}$$

$$2Al(s) + \tfrac{3}{2}O_2(g) \rightarrow Al_2O_3(s) \quad \Delta H = -400.5 \text{ kcal/mol}$$

Ⓖ

HEATS OF REACTION AT 1 ATM AND 298 K	
Reaction	**$\Delta H_{reaction}$ (kcal)**
$CH_4(g) + 2O_2(g) \rightarrow CO_2(g) + 2H_2O(\ell)$	−212.8
$C_3H_8(g) + 5O_2(g) \rightarrow 3CO_2(g) + 4H_2O(\ell)$	−530.6
$CH_3OH(\ell) + \tfrac{3}{2}O_2(g) \rightarrow CO_2(g) + 2H_2O(\ell)$	−173.6
$C_6H_{12}O_6(s) + 6O_2(g) \rightarrow 6CO_2(g) + 6H_2O(\ell)$	−669.9
$CO(g) + \tfrac{1}{2}O_2(g) \rightarrow CO_2(g)$	−67.7
$C_8H_{18}(\ell) + \tfrac{25}{2}O_2(g) \rightarrow 8CO_2(g) + 9H_2O(\ell)$	−1302.7
$KNO_3(s) \xrightarrow{H_2O} K^+(aq) + NO_3^-(aq)$	+8.3
$NaOH(s) \xrightarrow{H_2O} Na^+(aq) + OH^-(aq)$	−10.6
$NH_4Cl(s) \xrightarrow{H_2O} NH_4^+(aq) + Cl^-(aq)$	+3.5
$NH_4NO_3(s) \xrightarrow{H_2O} NH_4^+(aq) + NO_3^-(aq)$	+6.1
$NaCl(s) \xrightarrow{H_2O} Na^+(aq) + Cl^-(aq)$	+0.9
$KClO_3(s) \xrightarrow{H_2O} K^+(aq) + ClO_3^-(aq)$	+9.9
$LiBr(s) \xrightarrow{H_2O} Li^+(aq) + Br^-(aq)$	−11.7
$H^+(aq) + OH^-(aq) \rightarrow H_2O(\ell)$	−13.8

Ⓕ

SELECTED RADIOISOTOPES		
Nuclide	**Half-Life**	**Decay Mode**
^{198}Au	2.69 d	β^-
^{14}C	5730 y	β^-
^{60}Co	5.26 y	β^-
^{137}Cs	30.23 y	β^-
^{220}Fr	27.5 s	α
^{3}H	12.26 y	β^-
^{131}I	8.07 d	β^-
^{37}K	1.23 s	β^+
^{42}K	12.4 h	β^-
^{85}Kr	10.76 y	β^-
^{85m}Kr*	4.39 h	γ
^{16}N	7.2 s	β^-
^{32}P	14.3 d	β^-
^{239}Pu	2.44×10^4 y	α
^{226}Ra	1600 y	α
^{222}Rn	3.82 d	α
^{90}Sr	28.1 y	β^-
^{99}Tc	2.13×10^5 y	β^-
^{99m}Tc*	6.01 h	γ
^{232}Th	1.4×10^{10} y	α
^{233}U	1.62×10^5 y	α
^{235}U	7.1×10^8 y	α
^{238}U	4.51×10^9 y	α

y = years; d = days; h = hours; s = seconds
*m = meta stable or excited state of the same nucleus. Gamma decay from such a state is called an isomeric transition (IT).

Nuclear isomers are different energy states of the same nucleus, each having a different measurable lifetime.

Ⓗ

SYMBOLS USED IN NUCLEAR CHEMISTRY		
alpha particle	$^4_2$He	α
beta particle (electron)	$^0_{-1}$e	β^-
gamma radiation		γ
neutron	$^1_0$n	n
proton	$^1_1$H	p
deuteron	$^2_1$H	
triton	$^3_1$H	
positron	$^0_{+1}$e	β^+

Ⓘ

RELATIVE STRENGTHS OF ACIDS IN AQUEOUS SOLUTION AT 1 ATM AND 298 K

Conjugate Pairs ACID BASE	K_a
$HI = H^+ + I^-$	very large
$HBr = H^+ + Br^-$	very large
$HCl = H^+ + Cl^-$	very large
$HNO_3 = H^+ + NO_3^-$	very large
$H_2SO_4 = H^+ + HSO_4^-$	large
$H_2O + SO_2 = H^+ + HSO_3^-$	1.5×10^{-2}
$HSO_4^- = H^+ + SO_4^{2-}$	1.2×10^{-2}
$H_3PO_4 = H^+ + H_2PO_4^-$	7.5×10^{-3}
$Fe(H_2O)_6^{3+} = H^+ + Fe(H_2O)_5(OH)^{2+}$	8.9×10^{-4}
$HNO_2 = H^+ + NO_2^-$	4.6×10^{-4}
$HF = H^+ + F^-$	3.5×10^{-4}
$Cr(H_2O)_6^{3+} = H^+ + Cr(H_2O)_5(OH)^{2+}$	1.0×10^{-4}
$CH_3COOH = H^+ + CH_3COO^-$	1.8×10^{-5}
$Al(H_2O)_6^{3+} = H^+ + Al(H_2O)_5(OH)^{2+}$	1.1×10^{-5}
$H_2O + CO_2 = H^+ + HCO_3^-$	4.3×10^{-7}
$HSO_3^- = H^+ + SO_3^{2-}$	1.1×10^{-7}
$H_2S = H^+ + HS^-$	9.5×10^{-8}
$H_2PO_4^- = H^+ + HPO_4^{2-}$	6.2×10^{-8}
$NH_4^+ = H^+ + NH_3$	5.7×10^{-10}
$HCO_3^- = H^+ + CO_3^{2-}$	5.6×10^{-11}
$HPO_4^{2-} = H^+ + PO_4^{3-}$	2.2×10^{-13}
$HS^- = H^+ + S^{2-}$	1.3×10^{-14}
$H_2O = H^+ + OH^-$	1.0×10^{-14}
$OH^- = H^+ + O^{2-}$	$< 10^{-36}$
$NH_3 = H^+ + NH_2^-$	very small

Note: $H^+(aq) = H_3O^+$

Sample equation: $HI + H_2O = H_3O^+ + I^-$

Ⓙ

CONSTANTS FOR VARIOUS EQUILBRIA AT 1 ATM AND 298 K

$H_2O(\ell) = H^+(aq) + OH^-(aq)$	$K_w = 1.0 \times 10^{-14}$
$H_2O(\ell) + H_2O(\ell) = H_3O^+(aq) + OH^-(aq)$	$K_w = 1.0 \times 10^{-14}$
$CH_3COO^-(aq) + H_2O(\ell) = CH_3COOH(aq) + OH^-(aq)$	$K_b = 5.6 \times 10^{-10}$
$Na^+F^-(aq) + H_2O(\ell) = Na^+(OH)^- + HF(aq)$	$K_b = 1.5 \times 10^{-11}$
$NH_3(aq) + H_2O(\ell) = NH_4^+(aq) + OH^-(aq)$	$K_b = 1.8 \times 10^{-5}$
$CO_3^{2+}(aq) + H_2O(\ell) = HCO_3^-(aq) + OH^-(aq)$	$K_b = 1.8 \times 10^{-4}$
$Ag(NH_3)_2(aq) = Ag^+(ag) + 2NH_3(aq)$	$K_{eq} = 8.9 \times 10^{-8}$
$N_2(g) + 3H_2(g) = 2NH_3(g)$	$K_{eq} = 6.7 \times 10^5$
$H_2(g) + I_2(g) = 2HI(g)$	$K_{eq} = 3.5 \times 10^{-1}$

Compound	K_{sp}	Compound	K_{sp}
AgBr	5.0×10^{-13}	Li_2CO_3	2.5×10^{-2}
AgCl	1.8×10^{-10}	$PbCl_2$	1.6×10^{-5}
Ag_2CrO_4	1.1×10^{-12}	$PbCO_3$	7.4×10^{-14}
AgI	8.3×10^{-17}	$PbCrO_4$	2.8×10^{-13}
$BaSO_4$	1.1×10^{-10}	PbI_2	7.1×10^{-9}
$CaSO_4$	9.1×10^{-6}	$ZnCO_3$	1.4×10^{-11}

STANDARD ELECTRODE POTENTIALS

Ionic Concentrations 1 M Water At 298 K, 1 atm

Half-Reaction	E^0 (volts)
$F_2(g) + 2e^- \rightarrow 2F^-$	+2.87
$8H^+ + MnO_4^- + 5e^- \rightarrow Mn^{2+} + 4H_2O$	+1.51
$Au^{3+} + 3e^- \rightarrow Au(s)$	+1.50
$Cl_2(g) + 2e^- \rightarrow 2Cl^-$	+1.36
$14H^+ + Cr_2O_7^{2-} + 6e^- \rightarrow 2Cr^{3+} + 7H_2O$	+1.23
$4H^+ + O_2(g) + 4e^- \rightarrow 2H_2O$	+1.23
$4H^+ + MnO_2(s) + 2e^- \rightarrow Mn^{2+} + 2H_2O$	+1.22
$Br_2(\ell) + 2e^- \rightarrow 2Br^-$	+1.09
$Hg^{2+} + 2e^- \rightarrow Hg(\ell)$	+0.85
$Ag^+ + e^- \rightarrow Ag(s)$	+0.80
$Hg_2^{2+} + 2e^- \rightarrow 2Hg(\ell)$	+0.80
$Fe^{3+} + e^- \rightarrow Fe^{2+}$	+0.77
$I_2(s) + 2e^- \rightarrow 2I^-$	+0.54
$Cu^+ + e^- \rightarrow Cu(s)$	+0.52
$Cu^{2+} + 2e^- \rightarrow Cu(s)$	+0.34
$4H^+ + SO_4^{2-} + 2e^- \rightarrow SO_2(aq) + 2H_2O$	+0.17
$Sn^{4+} + 2e^- \rightarrow Sn^{2+}$	+0.15
$2H^+ + 2e^- \rightarrow H_2(g)$	0.00
$Pb^{2+} + 2e^- \rightarrow Pb(s)$	−0.13
$Sn^{2+} + 2e^- \rightarrow Sn(s)$	−0.14
$Ni^{2+} + 2e^- \rightarrow Ni(s)$	−0.26
$Co^{2+} + 2e^- \rightarrow Co(s)$	−0.28
$Fe^{2+} + 2e^- \rightarrow Fe(s)$	−0.45
$Cr^{3+} + 3e^- \rightarrow Cr(s)$	−0.74
$Zn^{2+} + 2e^- \rightarrow Zn(s)$	−0.76
$2H_2O + 2e^- \rightarrow 2OH^- + H_2(g)$	−0.83
$Mn^{2+} + 2e^- \rightarrow Mn(s)$	−1.19
$Al^{3+} + 3e^- \rightarrow Al(s)$	−1.66
$Mg^{2+} + 2e^- \rightarrow Mg(s)$	−2.37
$Na^+ + e^- \rightarrow Na(s)$	−2.71
$Ca^{2+} + 2e^- \rightarrow Ca(s)$	−2.87
$Sr^{2+} + 2e^- \rightarrow Sr(s)$	−2.89
$Ba^{2+} + 2e^- \rightarrow Ba(s)$	−2.91
$Cs^+ + e^- \rightarrow Cs(s)$	−2.92
$K^+ + e^- \rightarrow K(s)$	−2.93
$Rb^+ + e^- \rightarrow Rb(s)$	−2.98
$Li^+ + e^- \rightarrow Li(s)$	−3.04

VAPOR PRESSURE OF WATER

°C	torr (mm Hg)	°C	torr (mm Hg)
0	4.6	26	25.2
5	6.5	27	26.7
10	9.2	28	28.3
15	12.8	29	30.0
16	13.6	30	31.8
17	14.5	40	55.3
18	15.5	50	92.5
19	16.5	60	149.4
20	17.5	70	233.7
21	18.7	80	355.1
22	19.8	90	525.8
23	21.1	100	760.0
24	22.4	105	906.1
25	23.8	110	1074.6

PHYSICAL CONSTANTS AND CONVERSION FACTORS

Name	Symbol	Value(s)	Units
Angstrom unit	Å	1×10^{-10} m	meter
Avogadro number	N_A	6.02×10^{23} per mol	
Charge of electron	e	1.60×10^{-19} C	coulomb
electron volt	eV	1.60×10^{-19} J	joule
Speed of light	c	3.00×10^8 m/s	meters/second
Planck's constant	h	6.63×10^{-34} J·s	joule-second
		1.58×10^{-37} kcal·s	kilocalorie-second
Universal gas constant	R	0.0821 L·atm/mol·K	liter-atmosphere/ mole-kelvin
		1.98 cal/mol·K	calories/mole-kelvin
		8.31 J/mol·K	joules/mole-kelvin
Atomic mass unit	μ(amu)	1.66×10^{-24} g	gram
Volume standard, liter	L	1×10^3 cm^3 = 1 dm^3	cubic centimeters, cubic decimeter
Standard pressure, atmosphere	atm	101.3 kPa	kilopascals
		760 mm Hg	millimeters of mercury
		760 torr	torr
Heat equivalent, kilocalorie	kcal	4.18×10^3 J	joules

Physical Constants for H₂O

Molal freezing point depression . . . . . . 1.86°C
Molal boiling point elevation. . . . . 0.52°C
Heat of fusion . . . . 79.72 cal/g
Heat of vaporization . . . . 539.4 cal/g

STANDARD UNITS

Symbol	Name	Quantity
m	meter	length
kg	kilogram	mass
Pa	pascal	pressure
K	kelvin	thermodynamic temperature
mol	mole	amount of substance
J	joule	energy, work, quantity of heat
s	second	time
C	coulomb	quantity of electricity
V	volt	electric potential, potential difference
L	liter	volume

PREFIXES USED WITH SI UNITS

See Table on page 10, "Prefixes Used with SI Units."

RADII OF SOME ATOMS AND IONS

	1* IA	2 IIA	13 IIIA	14 IVA	15 VA	16 VIA	17 VIIA
1	0.037 H 1+ ?						0.037 H 0.208
2	Li 0.152 1+ 0.060	Be 0.111 2+ 0.031	B 0.088 3+ 0.020	C 0.077 4+ 0.015	N 0.070 3− 0.071	O 0.066 2− 0.140	F 0.064 1− 0.136
3	Na 0.154 1+ 0.095	Mg 0.160 2+ 0.065	Al 0.143 3+ 0.050	Si 0.117 4+ 0.041	P 0.110 3− 0.212	S 0.104 2− 0.184	Cl 0.099 1− 0.181
4	K 0.227 1+ 0.133	Ca 0.197 2+ 0.099	Ga 0.122 3+ 0.062	Ge 0.122 4+ 0.053	As 0.121 3− 0.222	Se 0.116 2− 0.198	Br 0.110 1− 0.195
5	Rb 0.244 1+ 0.148	Sr 0.215 2+ 0.113	In 0.163 3+ 0.081	Sn 0.141 4+ 0.071	Sb 0.141 5+ 0.062	Te 0.137 2− 0.221	I 0.133 1− 0.216
6	Cs 0.265 1+ 0.169	Ba 0.217 2+ 0.135	Tl 0.170 3+ 0.095	Pb 0.175 4+ 0.084	Bi 0.155 5+ 0.074	Po 0.167	At 0.140
7	Fr 0.270	Ra 0.220 2+ 0.152					

*Preferred IUPAC designation

THE CHEMICAL ELEMENTS

(Atomic masses in this table are based on the atomic mass of carbon-12 being exactly 12.)

Name	Symbol	Atomic Number	Atomic Mass	Name	Symbol	Atomic Number	Atomic Mass
Actinium	Ac	89	(227)	Mercury	Hg	80	200.59
Aluminum	Al	13	26.98	Molybdenum	Mo	42	95.94
Americium	Am	95	(243)	Neilsbohrium	Ns	107	(262)
Antimony	Sb	51	121.75	Neodymium	Nd	60	144.24
Argon	Ar	18	39.95	Neon	Ne	10	20.18
Arsenic	As	33	74.92	Neptunium	Np	93	237.05
Astatine	At	85	(210)	Nickel	Ni	28	58.71
Barium	Ba	56	137.34	Niobium	Nb	41	92.90
Berkelium	Bk	97	(247)	Nitrogen	N	7	14.01
Beryllium	Be	4	9.01	Nobelium	No	102	(259)
Bismuth	Bi	83	208.98	Osmium	Os	76	190.2
Boron	B	5	10.81	Oxygen	O	8	16.00
Bromine	Br	35	79.90	Palladium	Pd	46	106.4
Cadmium	Cd	48	112.40	Phosphorus	P	15	30.97
Cesium	Cs	55	132.91	Platinum	Pt	78	195.09
Calcium	Ca	20	40.08	Plutonium	Pu	94	(244)
Californium	Cf	98	(251)	Polonium	Po	84	(210)
Carbon	C	6	12.01	Potassium	K	19	39.10
Cerium	Ce	58	140.12	Praseodymium	Pr	59	140.90
Chlorine	Cl	17	35.45	Promethium	Pm	61	(145)
Chromium	Cr	24	52.00	Protactinium	Pa	91	231.04
Cobalt	Co	27	58.93	Radium	Ra	88	(226)
Copper	Cu	29	63.55	Radon	Rn	86	(222)
Curium	Cm	96	(247)	Rhenium	Re	75	186.2
Dysprosium	Dy	66	162.50	Rhodium	Rh	45	102.91
Einsteinium	Es	99	(254)	Rubidium	Rb	37	85.47
Erbium	Er	68	167.26	Ruthenium	Ru	44	101.07
Europium	Eu	63	151.96	Rutherfordium	Rf	104	(261)
Fermium	Fm	100	(257)	Samarium	Sm	62	150.35
Fluorine	F	9	19.00	Scandium	Sc	21	44.95
Francium	Fr	87	(223)	Selenium	Se	34	78.96
Gadolinium	Gd	64	157.25	Silicon	Si	14	28.09
Gallium	Ga	31	69.72	Silver	Ag	47	107.89
Germanium	Ge	32	72.59	Sodium	Na	11	22.99
Gold	Au	79	196.97	Strontium	Sr	38	87.62
Hafnium	Hf	72	178.49	Sulfur	S	16	32.06
Hahnium	Ha	105	(262)	Tantalum	Ta	73	180.95
Hassia	Hs	108	(265)	Technetium	Tc	43	(99)
Helium	He	2	4.00	Tellurium	Te	52	127.60
Holmium	Ho	67	164.93	Terbium	Tb	65	158.92
Hydrogen	H	1	1.008	Thallium	Tl	81	204.37
Indium	In	49	114.82	Thorium	Th	90	232.03
Iodine	I	53	126.90	Thulium	Tm	69	168.93
Iridium	Ir	77	192.2	Tin	Sn	50	118.69
Iron	Fe	26	55.85	Titanium	Ti	22	47.90
Krypton	Kr	36	83.80	Tungsten	W	74	183.85
Lanthanum	La	57	138.91	Unnilhexium	Unh	106	(263)
Lawrencium	Lr	103	(261)	Uranium	U	92	238.03
Lead	Pb	82	207.19	Vanadium	V	23	50.94
Lithium	Li	3	6.94	Xenon	Xe	54	131.30
Lutetium	Lu	71	174.97	Ytterbium	Yb	70	173.04
Magnesium	Mg	12	24.31	Yttrium	Y	39	88.91
Manganese	Mn	25	54.94	Zinc	Zn	30	65.37
Meitnerium	Mt	109	(266)	Zirconium	Zr	40	91.22
Mendelevium	Md	101	(258)				

*A number in parentheses is the mass number of the most stable isotope.

Periodic Table of

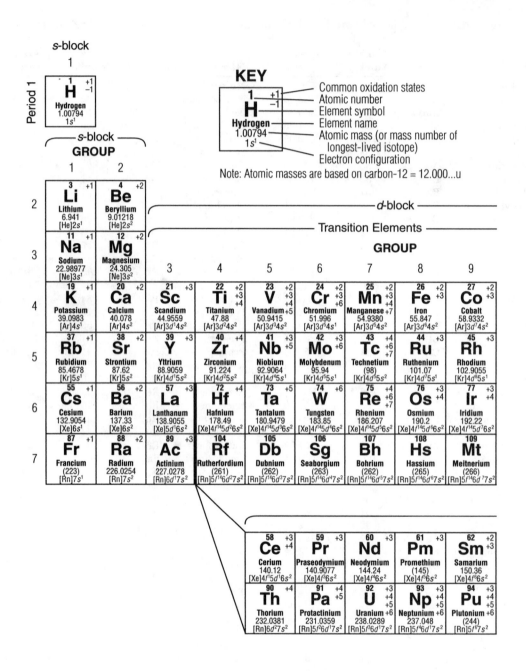

KEY

Common oxidation states
Atomic number
Element symbol
Element name
Atomic mass (or mass number of longest-lived isotope)
Electron configuration

1 +1
−1

H

Hydrogen
1.00794
$1s^1$

Note: Atomic masses are based on carbon-12 = 12.000...u

s-block

Period 1

1

1 +1
−1

H

Hydrogen
1.00794
$1s^1$

— *s*-block —
GROUP

1   2

— *d*-block —

— Transition Elements —

GROUP

3 4 5 6 7 8 9

	1	2	3	4	5	6	7	8	9
2	3 +1 **Li** Lithium 6.941 [He]$2s^1$	4 +2 **Be** Beryllium 9.01218 [He]$2s^2$							
3	11 +1 **Na** Sodium 22.98977 [Ne]$3s^1$	12 +2 **Mg** Magnesium 24.305 [Ne]$3s^2$							
4	19 +1 **K** Potassium 39.0983 [Ar]$4s^1$	20 +2 **Ca** Calcium 40.078 [Ar]$4s^2$	21 +3 **Sc** Scandium 44.9559 [Ar]$3d^14s^2$	22 +2 +3 +4 **Ti** Titanium 47.88 [Ar]$3d^24s^2$	23 +2 +3 +4 +5 **V** Vanadium 50.9415 [Ar]$3d^34s^2$	24 +2 +3 +6 **Cr** Chromium 51.996 [Ar]$3d^54s^1$	25 +2 +3 +4 +7 **Mn** Manganese 54.9380 [Ar]$3d^54s^2$	26 +2 +3 **Fe** Iron 55.847 [Ar]$3d^64s^2$	27 +2 +3 **Co** Cobalt 58.9332 [Ar]$3d^74s^2$
5	37 +1 **Rb** Rubidium 85.4678 [Kr]$5s^1$	38 +2 **Sr** Strontium 87.62 [Kr]$5s^2$	39 +3 **Y** Yttrium 88.9059 [Kr]$4d^15s^2$	40 +4 **Zr** Zirconium 91.224 [Kr]$4d^25s^2$	41 +3 +5 **Nb** Niobium 92.9064 [Kr]$4d^45s^1$	42 +3 +6 **Mo** Molybdenum 95.94 [Kr]$4d^55s^1$	43 +4 +6 +7 **Tc** Technetium (98) [Kr]$4d^55s^2$	44 +3 **Ru** Ruthenium 101.07 [Kr]$4d^75s^1$	45 +3 **Rh** Rhodium 102.9055 [Kr]$4d^85s^1$
6	55 +1 **Cs** Cesium 132.9054 [Xe]$6s^1$	56 +2 **Ba** Barium 137.33 [Xe]$6s^2$	57 +3 **La** Lanthanum 138.9055 [Xe]$5d^16s^2$	72 +4 **Hf** Hafnium 178.49 [Xe]$4f^{14}5d^26s^2$	73 +5 **Ta** Tantalum 180.9479 [Xe]$4f^{14}5d^36s^2$	74 +6 **W** Tungsten 183.85 [Xe]$4f^{14}5d^46s^2$	75 +6 +7 **Re** Rhenium 186.207 [Xe]$4f^{14}5d^56s^2$	76 +3 +4 **Os** Osmium 190.2 [Xe]$4f^{14}5d^66s^2$	77 +3 +4 **Ir** Iridium 192.22 [Xe]$4f^{14}5d^76s^2$
7	87 +1 **Fr** Francium (223) [Rn]$7s^1$	88 +2 **Ra** Radium 226.0254 [Rn]$7s^2$	89 +3 **Ac** Actinium 227.0278 [Rn]$6d^17s^2$	104 **Rf** Rutherfordium (261) [Rn]$5f^{14}6d^27s^2$	105 **Db** Dubnium (262) [Rn]$5f^{14}6d^37s^2$	106 **Sg** Seaborgium (263) [Rn]$5f^{14}6d^47s^2$	107 **Bh** Bohrium (262) [Rn]$5f^{14}6d^57s^2$	108 **Hs** Hassium (265) [Rn]$5f^{14}6d^67s^2$	109 **Mt** Meitnerium (266) [Rn]$5f^{14}6d^77s^2$

58 +3 +4 **Ce** Cerium 140.12 [Xe]$4f^15d^16s^2$	59 +3 **Pr** Praseodymium 140.9077 [Xe]$4f^36s^2$	60 +3 **Nd** Neodymium 144.24 [Xe]$4f^46s^2$	61 +3 **Pm** Promethium (145) [Xe]$4f^56s^2$	62 +2 +3 **Sm** Samarium 150.36 [Xe]$4f^66s^2$
90 +4 **Th** Thorium 232.0381 [Rn]$6d^27s^2$	91 +4 +5 **Pa** Protactinium 231.0359 [Rn]$5f^26d^17s^2$	92 +3 +4 +5 +6 **U** Uranium 238.0289 [Rn]$5f^36d^17s^2$	93 +3 +4 +5 +6 **Np** Neptunium 237.048 [Rn]$5f^46d^17s^2$	94 +3 +4 +5 +6 **Pu** Plutonium (244) [Rn]$5f^67s^2$

the Elements

s-block
18

2	0
He	
Helium	
4.00260	
$1s^2$	

p-block
GROUP

13	14	15	16	17	18

5 +3	6 −4, +2, +4	7 −3, −2, −1, +1, +2, +3, +4, +5	8 −2	9 −1	10 0
B	**C**	**N**	**O**	**F**	**Ne**
Boron 10.81	Carbon 12.011	Nitrogen 14.0067	Oxygen 15.9994	Fluorine 18.998403	Neon 20.1797
$[He]2s^22p^1$	$[He]2s^22p^2$	$[He]2s^22p^3$	$[He]2s^22p^4$	$[He]2s^22p^5$	$[He]2s^22p^6$

13 +3	14 −4, +2, +4	15 −3, +3, +5	16 −2, +4, +6	17 −1, +1, +3, +5, +7	18 0
Al	**Si**	**P**	**S**	**Cl**	**Ar**
Aluminum 26.98154	Silicon 28.0855	Phosphorus 30.97376	Sulfur 32.066	Chlorine 35.453	Argon 39.948
$[Ne]3s^23p^1$	$[Ne]3s^23p^2$	$[Ne]3s^23p^3$	$[Ne]3s^23p^4$	$[Ne]3s^23p^5$	$[Ne]3s^23p^6$

10	11	12

28 +2, +3	29 +1, +2	30 +2	31 +3	32 −4, +2, +4	33 −3, +3, +5	34 −2, +4, +6	35 −1, +1, +5	36 0, +2
Ni	**Cu**	**Zn**	**Ga**	**Ge**	**As**	**Se**	**Br**	**Kr**
Nickel 58.69	Copper 63.546	Zinc 65.39	Gallium 69.72	Germanium 72.61	Arsenic 74.9216	Selenium 78.96	Bromine 79.904	Krypton 83.80
$[Ar]3d^84s^2$	$[Ar]3d^{10}4s^1$	$[Ar]3d^{10}4s^2$	$[Ar]3d^{10}4s^24p^1$	$[Ar]3d^{10}4s^24p^2$	$[Ar]3d^{10}4s^24p^3$	$[Ar]3d^{10}4s^24p^4$	$[Ar]3d^{10}4s^24p^5$	$[Ar]3d^{10}4s^24p^6$

46 +2, +4	47 +1	48 +2	49 +3	50 +2, +4	51 −3, +3, +5	52 −2, +4, +6	53 −1, +1, +5, +7	54 0, +2, +4, +6
Pd	**Ag**	**Cd**	**In**	**Sn**	**Sb**	**Te**	**I**	**Xe**
Palladium 106.42	Silver 107.8682	Cadmium 112.41	Indium 114.82	Tin 118.710	Antimony 121.757	Tellurium 127.60	Iodine 126.9045	Xenon 131.29
$[Kr]4d^{10}$	$[Kr]4d^{10}5s^1$	$[Kr]4d^{10}5s^2$	$[Kr]4d^{10}5s^25p^1$	$[Kr]4d^{10}5s^25p^2$	$[Kr]4d^{10}5s^25p^3$	$[Kr]4d^{10}5s^25p^4$	$[Kr]4d^{10}5s^25p^5$	$[Kr]4d^{10}5s^25p^6$

78 +2, +4	79 +1, +3	80 +1, +2	81 +1, +3	82 +2, +4	83 +3, +5	84 +2, +4	85	86 0
Pt	**Au**	**Hg**	**Tl**	**Pb**	**Bi**	**Po**	**At**	**Rn**
Platinum 195.08	Gold 196.9665	Mercury 200.59	Thallium 204.383	Lead 207.2	Bismuth 208.9804	Polonium (209)	Astatine (210)	Radon (222)
$[Xe]4f^{14}5d^96s^1$	$[Xe]4f^{14}5d^{10}6s^1$	$[Xe]4f^{14}5d^{10}6s^2$	$[Xe]4f^{14}5d^{10}6s^26p^1$	$[Xe]4f^{14}5d^{10}6s^26p^2$	$[Xe]4f^{14}5d^{10}6s^26p^3$	$[Xe]4f^{14}5d^{10}6s^26p^4$	$[Xe]4f^{14}5d^{10}6s^26p^5$	$[Xe]4f^{14}5d^{10}6s^26p^6$

110	111	112
Uun	**Uuu**	**Uub**
Ununnilium (269)	Unununium (272)	Ununbium (277)
$[Rn]5f^{14}6d^87s^2$	$[Rn]5f^{14}6d^97s^2$	$[Rn]5f^{14}6d^{10}7s^2$

f-block

63 +2, +3	64 +3	65 +3	66 +3	67 +3	68 +3	69 +3	70 +2, +3	71 +3
Eu	**Gd**	**Tb**	**Dy**	**Ho**	**Er**	**Tm**	**Yb**	**Lu**
Europium 151.96	Gadolinium 157.25	Terbium 158.9254	Dysprosium 162.50	Holmium 164.9304	Erbium 167.26	Thulium 168.9342	Ytterbium 173.04	Lutetium 174.967
$[Xe]4f^76s^2$	$[Xe]4f^75d^16s^2$	$[Xe]4f^96s^2$	$[Xe]4f^{10}6s^2$	$[Xe]4f^{11}6s^2$	$[Xe]4f^{12}6s^2$	$[Xe]4f^{13}6s^2$	$[Xe]4f^{14}6s^2$	$[Xe]4f^{14}5d^16s^2$

95 +3, +4, +5, +6	96 +3	97 +3, +4	98 +3	99	100	101	102	103
Am	**Cm**	**Bk**	**Cf**	**Es**	**Fm**	**Md**	**No**	**Lr**
Americium (243)	Curium (247)	Berkelium (247)	Californium (251)	Einsteinium (252)	Fermium (257)	Mendelevium (258)	Nobelium (259)	Lawrencium (262)
$[Rn]5f^77s^2$	$[Rn]5f^76d^17s^2$	$[Rn]5f^97s^2$	$[Rn]5f^{10}7s^2$	$[Rn]5f^{11}7s^2$	$[Rn]5f^{12}7s^2$	$[Rn]5f^{13}7s^2$	$[Rn]5f^{14}7s^2$	$[Rn]5f^{14}6d^17s^2$

GLOSSARY OF COMMON TERMS

(See Index for additional references)

absolute temperature — Temperature measured on the absolute scale, which has its origin at absolute zero. *See also* **Kelvin scale**.

absorption — The process of taking up by capillary, osmotic, chemical, or solvent action, as a sponge absorbs water.

acid — A water solution that has an excess of hydrogen ions; an acid turns litmus paper pink or red, has a sour taste, and neutralizes bases to form salts.

acid anhydride — A nonmetallic oxide that, when placed in water, reacts to form an acid solution.

acid salt — A salt formed by replacing part of the hydrogen ions of a dibasic or tribasic acid with metallic ions.
   Examples: $NaHSO_4$, NaH_2PO_4.

actinide series — The series of radioactive elements starting with actinium, no. 89, and ending with lawrencium, no. 103.

activated charcoal — A specially treated and finely divided form of carbon, which possesses a high degree of adsorption.

activation energy — The minimum energy necessary to start a reaction.

adsorption — The adhesion (in an extremely thin layer) of the molecules of gases, of dissolved substances, or of liquids to the surfaces of solid or liquid bodies with which they come into contact.

alcohol — An organic hydroxyl compound formed by replacing one or more hydrogen atoms of a hydrocarbon with an equal number of hydroxyl (OH) groups.

aldehyde — An organic compound formed by dehydrating oxidized alcohol; contains the characteristic –CHO group.

alkali — Usually, a strong base, such as sodium hydroxide or potassium hydroxide.

alkaline — Referring to any substance that has basic properties.

alkyl — A substituent obtained from a saturated hydrocarbon by removing one hydrogen atom.
   Examples: methyl (CH_3^-, ethyl $C_2H_5^-$).

alkylation — The combining of a saturated hydrocarbon with an unsaturated one.

allotropic forms — Forms of the same element that differ in their crystalline structures.

alloy — A substance composed of two or more metals, which are intimately mixed; usually made by melting the metals together.

alpha particles — Positively charged helium nuclei.

alum — A double sulfate of a monovalent metal and a trivalent metal, such as K_2SO_4 $Al_2(SO_4)_3$ $24H_2O$; also, a common name for commercial aluminum sulfate.

amalgam — An alloy of mercury and another metal.

amine — A compound such as CH_3NH_2, derived from ammonia by substituting one or more hydrocarbon substituents for hydrogen atoms.

amino acid — One of the "building blocks" of proteins; contains one or more NH_2^- groups that have replaced the same number of hydrogen atoms in an organic acid.

amorphous — Having no definite crystalline structure.

amphoteric — Referring to a hydroxide that may have either acidic or basic properties, depending on the substance with which it reacts.

analysis — The breaking down of a compound into two or more simpler substances.

anhydride — A compound derived from another compound by the removal of water; it will combine with water to form an acid (acid anhydride) or a base (basic anhydride).

anhydrous — containing no water.

anion — An ion or particle that has a negative charge and thus is attracted to a positively charged anode.

anode — The electrode in an electrolytic cell that has a positive charge and attracts negative ions.

aromatic compounds — A compound whose basic structure contains the benzene ring; it usually has an odor.

aryl — A radical obtained from a ring hydrocarbon by removing one hydrogen atom from the ring.

atmosphere — The layer of gases surrounding the earth; also, a unit of pressure (1 atm = approx. 760 mm Hg or 1 torr).

atom — The smallest particle of an element that retains the properties of that element and can enter into a chemical reaction.

atomic energy — *See* **nuclear energy**, a more accurate term.

atomic mass — The average mean value of the isotopic masses of the atoms of an element. It indicates the relative mass of the element as compared with that of carbon-12, which is assigned a mass of exactly 12 atomic mass units.

atomic mass unit — One twelfth of the mass of a carbon-12 atom; equivalent to 1.660531×10^{-27} kg (abbreviation: amu or μ).

atomic number — The number that indicates the order of an element in the periodic system; numerically equal to the number of protons in the nucleus of the atom, or the number of negative electrons located outside the nucleus of the atom.

atomic radius — One-half the distance between adjacent nuclei in the crystalline or solid phase of an element; the distance from the atomic nucleus to the valence electrons.

Aufbau Principle — A principle stating that an electron occupies the lowest energy orbital that can receive it.

Avogadro's hypothesis — *See under* **laws.**

Avogadro's number — The number of molecules in 1 gram-molecular volume of a substance, or the number of atoms in 1 gram-atomic mass of an element; equal to 6.022169×10^{23}. *See also* **mole.**

barometer — An instrument, invented by Torricelli in 1643, used for measuring atmospheric pressure.

base — A water solution that contains an excess of hydroxide ions; a proton acceptor; a base turns litmus paper blue and neutralizes acids to form salts.

basic anhydride — A metallic oxide that forms a base when placed in water.

beta particles — High-speed, negatively charged electrons emitted in radiation.

binary — Referring to a compound composed of two elements, such as H_2O.

boiling point — The temperature at which the vapor pressure of a liquid equals the atmospheric pressure.

bond energy — The energy needed to break a chemical bond and form a neutral atom.

bonding — The union of atoms to form compounds or molecules by filling their outer shells of electrons. This can be done through giving and taking electrons (ionic) or by sharing electrons (covalent).

Boyle's Law — *See under* **laws.**

brass — An alloy of copper and zinc.

breeder reactor — A nuclear reactor in which more fissionable material is produced than is used up during operation.

Brownian movement — Continuous zigzagging movement of colloidal particles in a dispersing medium, as viewed through an ultramicroscope.

buffer — A substance that when added to a solution, makes changing the pH of the solution more difficult.

calorie — A unit of heat; the amount of heat needed to raise the temperature of 1 gram of water 1 degree on the Celsius scale.

calorimeter — An instrument used to measure the amount of heat liberated or absorbed during a change.

carbohydrate — A compound of carbon, hydrogen, and oxygen, with hydrogen and oxygen usually present in the ratio of 2 : 1, as in H_2O.

carbonated water — Water containing dissolved carbon dioxide.

carbon dating — The use of radioactive carbon-14 to estimate the age of ancient materials, such as archeological or paleontological specimens.

catalyst — A substance that speeds up or slows down a reaction without being permanently changed itself.

cathode — The electrode in an electrolytic cell that is negatively charged and attracts positive ions.

cathode rays — Streams of electrons given off by the cathode of a vacuum tube.

cation — An ion that has a positive charge.

Celsius scale — A temperature scale divided into 100 equal divisions and based on water freezing at 0° and boiling at 100°. Synonymous with centigrade.

chain reaction — A reaction produced during nuclear fission when at least one neutron from each fission produces another fission, so that the process becomes self-sustaining without additional external energy.

Charles's Law — *See under* **laws.**

chemical change — A change that alters the atomic structures of the substances involved and results in different properties.

chemical property — A property that determines how a substance will behave in a chemical reaction.

chemistry — The science concerned with the compositions of substances and the changes that they undergo.

colligative properties — Properties of a solution that depend primarily on the concentration, not the type, of particles present.

colloids — Particles larger than those found in a solution but smaller than those in a suspension.

Combining volumes — *See Gay-Lussac's, under* **laws.**

combustion — A chemical action in which both heat and light are given off.

compound — A substance composed of elements chemically united in definite proportions by weight.

condensation — (a) A change from gaseous to liquid state; (b) the union of like or unlike molecules with the elimination of water, hydrogen chloride, or alcohol.

Conservation of energy — *See under* **laws**.

Conservation of matter — *See under* **laws**.

control rod — In a nuclear reactor, a rod of a certain metal such as cadmium, which controls the speed of the chain reaction by absorbing neutrons.

coordinate covalence — Covalence in which both electrons in a pair come from the same atom.

covalent bonding — Bonding accomplished through the sharing of electrons so that atoms can fill their outer shells.

critical mass — The smallest amount of fissionable material that will sustain a chain reaction.

critical temperature — The temperature above which no gas can be liquefied, regardless of the pressure applied.

crystalline — Having a definite molecular or ionic structure.

crystallization — The process of forming definitely shaped crystals when water is evaporated from a solution of the substance.

cyclotron — A device used to accelerate charged particles to high energies for bombarding the nuclei of atoms.

Dalton's law of partial pressures — *See under* **laws**.

decomposition — The breaking down of a compound into simpler substances or into its constituent elements.

definite composition — *See under* **laws**.

dehydrate — To take water from a substance.

dehydrating agent — A substance able to withdraw water from another substance, thereby drying it.

deliquescence — The absorption by a substance of water from the air so that the substance becomes wet.

denatured alcohol — Ethyl alcohol that has been "poisoned" in order to produce (by avoiding federal tax) a cheaper alcohol for industrial purposes.

density — The mass per unit volume of a substance; the mathematical formula is $D = m/V$, where D = density, m = mass, and V = volume.

destructive distillation — The process of heating an organic substance, such as coal, in the absence of air to break it down into solid and volatile products.

deuterium — An isotope of hydrogen, sometimes called heavy hydrogen, with an atomic mass of 2.

dew point — The highest temperature at which water vapor condenses out of the air.

dialysis — The process of separation of a solution by diffusion through a semipermeable membrane.

diffusion — The process whereby gases or liquids intermingle freely of their own accord.

dipole-dipole attraction — A relatively weak force of attraction between polar molecules; a component of van der Waals forces.

disaccharide — A sugar, like $C_{12}H_{22}O_{11}$, that hydrolyzes to form two molecules of simpler sugars.

displacement — A change by which an element takes the place of another element in a compound.

dissociation (ionic) — The separation of the ions of an electrovalent compound as a result of the action of a solvent.

distillation — The process of first vaporizing a liquid and then condensing the vapor back into a liquid, leaving behind the nonvolatile impurities.

double bond — A bond between atoms involving two electron pairs. In organic chemistry: unsaturated.

double displacement — A reaction in which two chemical substances exchange ions with the formation of two new compounds.

dry ice — Solid carbon dioxide.

ductile — Capable of being drawn into thin wire.

effervescence — The rapid escape of excess gas that has been dissolved in a liquid.

efflorescence — The loss by a substance of its water of hydration on exposure to air at ordinary temperatures.

effusion — The flow of a gas through a small aperture.

Einstein equation — The equation $E = mc^2$, which relates mass to energy; E is energy in ergs, m is mass in grams, and c is velocity of light, 3×10^{10} cm/s.

electrode — A terminal of an electrolytic cell.

electrode potential — The difference in potential between an electrode and the solution in which it is immersed.

electrolysis — The process of separating the ions in a compound by means of electrically charged poles.

electrolysis (electrolytic) cell — A cell in which electrolysis is carried out.

electrolyte — A liquid that will conduct an electric current.

electron — A negatively charged particle found outside the nucleus of the atom; it has a mass of 9.109×10^{-28} g.

electron dot symbol — *See* **Lewis dot symbol**.

electronegativity — The numerical expression of the relative strength with which the atoms of an element attract valence electrons to themselves; the higher the number, the greater the attraction.

electron volt — A unit for expressing the kinetic energy of subatomic particles; the energy acquired by an electron when it is accelerated by a potential difference of 1 V; equals 1.6×10^{-12} erg or 23.1 kcal/mol (abbreviation: eV).

electroplating — Depositing a thin layer of (usually) a metallic element on the surface of another metal by electrolysis.

electrovalence — The bond of attraction between negatively and positively charged ions.

element — One of the more than 100 "building blocks" of which all matter is composed. An element consists of atoms of only one kind and cannot be decomposed further by ordinary chemical means.

empirical formula — A formula that shows only the simplest ratio of the numbers and kinds of atoms, such as CH_4.

emulsifying agent — A colloidal substance that forms a film about the particles of two immiscible liquids so that one liquid remains suspended in the other.

emulsion — A suspension of fine particles or droplets of one liquid in another, the two liquids being immiscible in each other; the droplets are surrounded by a colloidal (emulsifying) agent.

endothermic — Referring to a chemical reaction that results in an overall absorption of heat from its surroundings.

energy — The capacity to do work. In every chemical change energy is either given off or taken in. Forms of energy are heat, light, motion, sound, and electrical, chemical, and nuclear energy.

enthalpy — The heat content of a chemical system.

entropy — The measure of the randomness that exists in a system.

equation — A shorthand method of showing the changes that take place in a chemical reaction.

equilibrium — The point in a reversible reaction at which the forward reaction is occurring at the same rate as the opposing reaction.

erg — A unit of energy or work done by a force of 1 dyne (1/980 g of force) acting through a distance of 1 cm; equals 2.4×10^{-11} kcal.

ester — An organic salt formed by the reaction of an alcohol with an organic (or inorganic) acid.

esterification — A chemical reaction between an alcohol and an acid in which an ester is formed.

ether — An organic compound containing the —O— group.

eudiometer — A graduated glass tube into which gases are placed and subjected to an electric spark; used to measure the individual volumes of combining gases.

evaporation — The process in which molecules of a liquid (or a solid) leave the surface in the form of vapor.

exothermic — Referring to a chemical reaction that results in the giving off of heat to its surroundings.

Fahrenheit scale — The temperature scale that has 32° as the freezing point of water and 212° as the boiling point.

fallout — The residual radioactivity from an atmospheric nuclear test, which eventually settles on the surface of the earth.

Faraday's Law — *See under* **laws.**

filtration — The process by which suspended matter is removed from a liquid by passing the liquid through a porous material.

First Law of Thermodynamics — *See under* **laws.**

fission — A nuclear reaction that releases energy as a result of the splitting of large nuclei into smaller ones.

fixation of nitrogen — Any process for converting atmospheric nitrogen into compounds such as ammonia and nitric acid.

flame — The glowing mass of gas and luminous particles produced by the burning of a gaseous substance.

flammable — Capable of being easily set on fire; combustible (same as *inflammable*).

fluorescence — Emission by a substance of electromagnetic radiation, usually visible, as the immediate result of (and only during) absorption of energy from another source.

fluoridation — Addition of small amounts of fluoride (usually NaF) to drinking water to help prevent tooth decay.

flux — In metallurgy: a substance that helps to melt and remove the solid impurities as slag. In soldering: a substance that cleans the surface of the metal to be soldered. In nucleonics: the concentration of nuclear particles or rays.

formula — An expression that uses the symbols for elements and subscripts to show the basic makeup of a substance.

formula mass — The sum of the atomic masses of all the atoms (or ions) contained in a formula.

fractional crystallization — The separation of the components in a mixture of dissolved solids by evaporation according to individual solubilities.

fractional distillation — The separation of the components in a mixture of liquids having different boiling points by vaporization.

free energy — *See Gibbs free energy*

freezing point — The specific temperature at which a given liquid and its solid form are in equilibrium.

fuel — Any substance used to furnish heat by combustion. *See also* **nuclear fuel.**

fuel cell — A device for converting an ordinary fuel such as hydrogen or methane directly into electricity.

functional group — A group of atoms that characterizes certain types of organic compounds, such as —OH for alcohols, and that reacts more or less independently.

fusion — A nuclear reaction that releases energy as a result of the union of smaller nuclei to form larger ones.

fusion melting — Changing a solid to the liquid state by heating.

galvanizing — Applying a coating of zinc to iron or steel to protect the latter from rusting.

gamma rays — A type of radiation consisting of high-energy waves that can pass through most materials.

gas — A phase of matter that has neither definite shape nor definite volume.

Gay-Lussac's Law — *See under* **laws.**

Gibbs free energy — Changes in Gibbs free energy, ΔG, are useful in indicating the conditions under which a chemical reaction will occur. The equation is $\Delta G = \Delta H - T \Delta S$, where ΔH is the change in enthalpy and ΔS is the change in entropy. If ΔG is negative, the reaction will proceed spontaneously to equilibrium.

Glass — An amorphous, usually translucent substance consisting of a mixture of silicates. Ordinary glass is made by fusing together silica and sodium carbonate and lime; the various forms of glass contain many other silicates.

Graham's Law — *See under* **laws.**

gram — A unit of weight in the metric system; the weight of 1 mL of water at 4°C (abbreviation: g).

gram-atomic mass — The atomic mass, in grams, of an element.

gram-molecular mass — The molecular mass, in grams, of a substance.

gram-molecular volume — The volume occupied by 1 gram-molecular mass of a gaseous substance at standard conditions; equals 22.4 L.

group — A vertical column of elements in the periodic table that generally have similar properties.

Haber process — A catalytic method for the union of atmospheric nitrogen with hydrogen to form ammonia.

half-life — The time required for half of the mass of a radioactive substance to disintegrate.

half-reaction — One of the two parts, either the reduction part or the oxidation part, of a redox reaction.

halogen — Any of the five nonmetallic elements (fluorine, chlorine, bromine, iodine, astatine) that form part of group 17 (group VII) of the periodic table.

heat — A form of molecular energy; it passes from a warmer body to a cooler one.

heat capacity (specific heat) — The quantity of heat, in calories, needed to raise the temperature of 1 g of a substance 1° on the Celsius scale.

heat of formation — The quantity of heat either given off or absorbed in the formation of 1 mol of a substance from its elements.

heat of fusion — The amount of heat, in calories, required to melt 1 g of a solid; for water, 80 cal.

heat of vaporization — The quantity of heat needed to vaporize 1 g of a liquid at constant temperature and pressure; for water at 100°C, 540 cal.

heavy water (deuterium oxide, D_2O) — Water in which the hydrogen atoms are replaced by atoms of the isotope of hydrogen, deuterium.

Henry's Law — *See under* **laws.**

homogeneous — Uniform; having every portion exactly like every other portion.

homologous — Alike in structure; referring to series of organic compounds, such as hydrocarbons, in which each member differs from the next by the addition of the same group.

humidity — The amount of moisture in the air.

hybridization — The combination of two or more orbitals to form new orbitals.

hydrate — A compound that has water molecules included in its crystalline makeup.

hydride — Any binary compound containing hydrogen, such as HCl.

hydrogenation — A process in which hydrogen is made to combine with another substance, usually organic, in the presence of a catalyst.

hydrogen bond — A weak chemical linkage between the hydrogen of one polar molecule and the oppositely charged portion of a closely adjacent molecule.

hydrolysis — Of carbohydrates: the action of water in the presence of a catalyst upon one carbohydrate to form simpler carbohydrates. Of salts: a reaction involving the splitting of water into its ions by the formation of a weak acid, a weak base, or both.

hydronium ion — A hydrated ion, $H_2O \cdot H^+$ or H_3O^+.

hydroponics — Growing plants without the use of soil, as in nutrient solution or in sand irrigated with nutrient solution.

hydroxyl — Referring to the —OH radical.

hygroscopic — Referring to the ability of a substance to draw water vapor from the atmosphere to itself and become wet.

hypothesis — A possible explanation of the nature of an action or phenomenon; a hypothesis is not as completely developed as a theory.

Ideal Gas Law — *See under* **laws**.

immiscible — Referring to the inability of two liquids to mix.

indicator — A dye that shows one color in the presence of the hydrogen ion (acid) and a different color in the presence of the hydroxyl ion (base).

inertia — The property of matter whereby it remains at rest or, if in motion, remains in motion in a straight line unless acted upon by an outside force.

ion — An atom or a group of combined atoms that carries one or more electric charges.
Examples: NH_4^+, OH^-.

ionic bonding — The bonding of ions as a result of their opposite charges.

ionic equation — An equation showing a reaction among ions.

ionization — The process in which ions are formed from neutral atoms.

ionization equation — An equation showing the ions set free from an electrolyte.

isomerization — The rearrangement of atoms in a molecule to form isomers.

isomers — Two or more compounds having the same percentage composition but different arrangements of atoms in their molecules and hence different properties.

isotopes — Two or more forms of an element that differ only in the number of neutrons in the nucleus and hence in their mass numbers.

IUPAC — International Union of Pure and Applied Chemistry, an organization that establishes standard rules for naming compounds.

joule — The SI unit of work or of energy equal to work done; 1 joule = 0.2388 calories.

Kelvin scale — A temperature scale based on water freezing at 273 and boiling at 373 Kelvin units; its origin is absolute zero. Synonymous with *absolute scale*.

kernel (atomic) — The nucleus and all the electron shells of an atom except the outer one; usually designated by the symbol for the atom.

ketone — An organic compound containing the —CO— group.

kilocalorie — A unit of heat; the amount of heat needed to raise the temperature of 1 kg of water 1° on the Celsius scale.

kindling temperature — The temperature to which a given substance must be raised before it ignites.

Kinetic Molecular Theory — The theory that all molecules are in motion; this motion is most rapid in gases, less rapid in liquids, and very slow in solids.

lanthanide series — The "rare earth" series of elements starting with lanthanum, no. 57, and ending with lutetium, no. 71.

law (in science) — A generalized statement about the uniform behavior in natural processes.

laws

Avogadro's Equal volumes of gases under identical conditions of temperature and pressure contain equal numbers of particles (atoms, molecules, ions, or electrons).

Boyle's The volume of a confined gas is inversely proportional to the pressure to which it is subjected, provided that the temperature remains the same.

Charles's The volume of a confined gas is directly proportional to the absolute temperature, provided that the pressure remains the same.

Combining Volumes *See Gay-Lussac's under* **laws**.

Conservation of Energy Energy can be neither created nor destroyed, so that the energy of the universe is constant.

Conservation of Matter Matter can be neither created nor destroyed (or weight remains constant in an ordinary chemical change).

Dalton's When a gas is made up of a mixture of different gases, the pressure of the mixture is equal to the sum of the partial pressures of the components.

Definite Composition A compound is composed of two or more elements chemically combined in a definite ratio by weight.

Faraday's During electrolysis, the weight of any element liberated is proportional (1) to the quantity of electricity passing through the cell, and (2) to the equivalent weight of the element.

First Law of Thermodynamics The total energy of the universe is constant and cannot be created or destroyed.

Gay-Lussac's The ratio between the combining volumes of gases and the product, if gaseous, can be expressed in small whole numbers.

Graham's The rate of diffusion (or effusion) of a gas is inversely proportional to the square root of its molecular mass.

Henry's The solubility of a gas (unless the gas is very soluble) is directly proportional to the pressure applied to the gas.

Hess's Law — If a series of reactions are added together, the enthalpy change for the net reaction will be the sum of the enthalpy changes for the individual steps.

Ideal Gas Any gas that obeys the gas laws perfectly. No such gas actually exists.

Multiple Proportions When any two elements, A and B, combine to form more than one compound, the different weights of B that unite with a fixed weight of A bear a small whole-number ratio to each other.

Periodic   The chemical properties of elements vary periodically with their atomic numbers.

Second Law of Thermodynamics   Heat cannot, of itself, pass from a cold body to a hot body.

Le Châtelier's Principle — If a stress is placed on a system in equilibrium, the system will react in the direction that relieves the stress.

leptons — Elementary particles that are believed to make up the electron and neutrino.

Lewis dot symbol — The chemical symbol (kernel) for an atom, surrounded by dots to represent its outer level electrons.
    Examples: K·, Sr:.

liquid — A phase of matter that has a definite volume but takes the shape of the container.

liquid air — Air that has been cooled and compressed until it liquefies.

litmus — An organic substance obtained from the lichen plant and used as an indicator; it turns red in acidic solution and blue in basic solution.

London force — The weakest of the van der Waal forces between molecules. These weak, attractive forces become apparent only when the molecules approach one another closely (usually at low temperatures and high pressure). They are due to the way the positive charges of one molecule attract the negative charges of another molecule because of the charge distribution at any one instant.

luminous — Emitting a steady, suffused light.

malleable — Capable of being hammered or pounded into thin sheets.

manometer — A U-tube (containing mercury or some other liquid) used to measure the pressure of a confined gas.

mass — The quantity of matter that a substance possesses; it can be measured by its resistance to a change in position or motion and is not related to the force of gravity.

mass number — The nearest whole number to the combined atomic mass of the individual atoms of an isotope when that mass is expressed in atomic mass units.

mass spectograph — A device for determining the masses of electrically charged particles by separating them into distinct streams by means of magnetic deflection.

matter — A substance that occupies space, has mass, and cannot be created or destroyed easily.

melting — The change in phase of a substance from solid to liquid.

melting point — The specific temperature at which a given solid changes to a liquid.

meson — Any unstable, elementary nuclear particle having a mass between that of an electron and that of a proton.

metal — (a) An element whose oxide combines with water to form a base; (b) an element that readily loses electrons and acquires a positive valence.

metallurgy — The process involved in obtaining a metal from its ores.

meter — The basic unit of length in the metric system; defined as 1,650,763.73 times the wavelength of krypton-86 when excited to give off an orange-red spectral line.

MeV — A unit for expressing the kinetic energy of subatomic particles; equals 10^6 V.

micron — One thousandth of a millimeter (abbreviation: μ).

millibar — A unit of pressure used in meteorology; equal 1000 dynes/cm^2.

millimicron — One millionth of a millimeter (abbreviation: mμ).

mineral — An inorganic substance of definite composition found in nature.

miscible — Referring to the ability of two liquids to mix with one another.

mixture — A substance composed of two or more components, each of which retains its own properties.

moderator — A substance such as graphite, paraffin, or heavy water used in a nuclear reactor to slow down neutrons.

molal solution — A solution containing 1 mol of solute in 1000 g of solvent (indicated by m).

molar mass — The mass arrived at by the addition of the atomic masses of the units that make up a molecule of an element or compound. Expressed in grams, the molar mass of a gaseous substance at STP occupies a molar volume equal to 22.4 L.

molar solution — A solution containing 1 mol of solute in 1000 mL of solution (indicated by M).

mole — A unit of quantity that consists of 6.02×10^{23} particles. Abbreviation is mol.

molecular mass — The sum of the atomic masses of all the atoms in a molecule of a substance.

molecular theory — *See* **Kinetic Molecular Theory.**

molecule — The smallest particle of a substance that retains the physical and chemical properties of that substance.

monobasic acid — An acid having only one hydrogen atom which can be replaced by a metal or a positive radical.

monosaccharide — A simple sugar, such as $C_6H_{12}O_6$.

mordant — A chemical, such as aluminum sulfate, used for fixing colors on textiles.

multiple proportions — *See under* **laws.**

nascent (atomic) — Referring to an element in the atomic form as it has just been liberated in a chemical reaction.

neutralization — The union of the hydrogen ion of an acid and the hydroxyl ion of a base to form water.

neutron — A subatomic particle found in the nucleus of the atom; it has no charge and has the same mass as the proton.

neutron capture — A nuclear reaction in which a neutron attaches itself to a nucleus; a gamma ray is usually emitted simultaneously.

nitriding — A process in which ammonia or a cyanide is used to produce case-hardened steel; a nitride is formed instead of a carbide.

nitrogen fixation — Any process by which atmospheric nitrogen is converted into a compound such as ammonia or nitric acid.

noble gas — A gaseous element that has a complete outer level of electrons; any of a group of rare gases (helium, neon, argon, krypton, etc.) that exhibit great stability and very low reaction rates.

nonelectrolyte — A substance whose solution does not conduct a current of electricity.

nonmetal — (a) An element whose oxide reacts with water to form an acid; (b) an element that takes on electrons and acquires a negative valence.

nonpolar compound — A compound in whose molecules the atoms are arranged symmetrically so that the electric charges are uniformly distributed.

normal salt — A salt in which all the hydrogen of the acid has been displaced by a metal.

normal solution — A solution that contains 1 gram of H^+ (or its equivalent: 17 g of OH^-, 23 g of Na^+, 20 g of Ca^{2+}, etc.) in 1 L of solution (indicated by N).

nuclear energy — The energy released by spontaneously or artificially produced fission, fusion, or disintegration of the nuclei of atoms.

nuclear fuel — A substance that is consumed during nuclear fission or fusion.

nuclear reaction — Any reaction involving a change in nuclear structure.

nuclear reactor — A device in which a controlled chain reaction of fissionable material can be produced.

nucleonics — The science that deals with the constituents and all the changes in the atomic nucleus.

nucleus — The center of the atom, which contains protons and neutrons.

nuclide — A species of atom characterized by the constitution of its nucleus.

octane number — A conventional rating for gasoline based on its behavior as compared with that of isooctane as 100.

orbital — A subdivision of a nuclear shell; it may contain none, one, or two electrons.

ore — A natural mineral substance from which an element, usually a metal, may be obtained with profit.

organic acid — An organic compound that contains the —COOH group.

organic chemistry — The branch of chemistry dealing with carbon compounds, usually those found in nature.

oxidation — The chemical process by which oxygen is attached to a substance; the process of losing electrons.

oxidation number (state) — A positive or negative number representing the charge that an ion has or an atom appears to have when its electrons are counted according to arbitrarily accepted rules: (1) electrons shared by two unlike atoms are counted with the more electronegative atom; (2) electrons shared by two like atoms are divided equally between the atoms.

oxidation potential — An electrode potential associated with the oxidation half-reaction.

oxidizing agent — A substance that (a) gives up its oxygen readily, (b) removes hydrogen from a compound, (c) takes electrons from an element.

ozone — An allotropic and very active form of oxygen, has the formula O_3.

paraffin series — The methane series of hydrocarbons.

pascal — The SI unit of pressure equal to 1 newton per square meter.

pasteurization — Partial sterilization of a substance, such as milk, by heating to approximately 65°C for $\frac{1}{2}$ hour.

Pauli Exclusion Principle — Each electron orbital of an atom can be filled by only two electrons, each with an opposite spin.

period — A horizontal row of elements in the periodic table.

periodic law — *See under* **laws.**

petroleum (meaning "oil from stone") — A complex mixture of gaseous, liquid, and solid hydrocarbons obtained from the earth.

pH — A numerical expression of the hydrogen or hydronium ion concentration in a solution; defined as $-\log[H^+]$, where $[H^+]$ is the concentration of hydrogen ions, in moles per liter.

phenolphthalein — An organic indicator; it is colorless in acid solution and red in the presence of OH⁻ ions.

photosynthesis — The reaction taking place in all green plants that produces glucose from carbon dioxide and water under the catalytic action of chlorophyll in the presence of light.

physical change — A change that does not involve any alteration in chemical composition.

physical property — A property of a substance arrived at through observation of its smell, taste, color, density, and so on, which does not relate to chemical activity.

pi bond — A bond between *p* orbitals.

pile — A general term for a nuclear reactor; specifically, a graphite-moderated reactor in which uranium fuel is distributed throughout a "pile" of graphite blocks.

pitchblende — A massive variety of uraninite that contains a small amount of radium.

plasma — Very hot ionized gases.

polar covalent bond — A bond in which electrons are closer to one atom than to another. *See also* **polar molecule**.

polar dot structure — Representation of the arrangements of electrons around the atoms of a molecule in which the polar characteristics are shown by placing the electrons closer to the more electronegative atom.

polar molecule — A molecule that has differently charged areas because of unequal sharing of electrons.

polyatomic ion — A group of chemically united atoms that react as a unit and have an electric charge.

polymerization — The process of combining several molecules to form one large molecule (polymer). (a) Additional polymerization: The addition of unsaturated molecules to each other. (b) Condensation polymerization: The reaction of two molecules by loss of a molecule of water.

polysaccharide — A large complex molecule with the general formula $(C_6H_{10}O_5)_n$.

positron — A positively charged particle of electricity with about the same weight as the electron.

potential energy — Energy due to the position of a body or to the configuration of its particles.

precipitate — An insoluble compound formed in the chemical reaction between two or more substances in solution.

proteins — Large, complex organic molecules, with nitrogen an essential part, found in plants and animals.

proton — A subatomic particle found in the nucleus that has a positive charge.

qualitative analysis — A term applied to the methods and procedures used to determine any or all of the constituent parts of a substance.

quantitative analysis — A term applied to the methods and procedures used to determine the definite quantity or percentage of any or all of the constituent parts of a substance.

quark — A subatomic particle with a fractional charge of one third or two thirds the charge of an electron. Six quarks have been described.

quenching — Cooling a hot piece of metal rapidly, as in water or oil.

radiation — The emission of particles and rays from a radioactive source; usually alpha and beta particles and gamma rays.

radioactive — Referring to substances that have the ability to emit radiations (alpha or beta particles or gamma rays).

radioisotopes — Isotopes that are radioactive, such as uranium-235.

reactant — A substance involved in a reaction.

reaction — A chemical transformation or change. The four basic types are combination (synthesis), decomposition (analysis), single replacement or single displacement, and double replacement or double displacement.

reaction potential — The sum of the oxidation potential and reduction potential for a particular reaction.

reagent — Any chemical taking part in a reaction.

recrystallization — A series of crystallizations, repeated for the purpose of greater purification.

redox — A shortened name for a reaction that involves reduction and oxidation.

reducing agent — From an electron standpoint, a substance that loses its valence electrons to another element; a substance that is readily oxidized.

reduction — A chemical reaction that removes oxygen from a substance; a gain of electrons.

reduction potential — An electrode potential associated with a reduction half-reaction.

refraction (of light) — The bending of light rays as they pass from one material into another.

relative humidity — The ratio, expressed in percent, between the amount of water vapor in a given volume of air and the amount the same volume can hold when saturated at the same temperature.

resonance — The phenomenon in a molecular structure that exhibits properties between those of a single bond and those of a double bond and thus possesses two or more alternate structures.

reversible reaction — Any reaction that reaches an equilibrium or that can be made to proceed from right to left as well as from left to right.

roasting — Heating an ore (usually a sulfide) in an excess of air to convert the ore to an oxide, which can then be reduced.

saccharin — $C_6H_4COSO_2NH$, a white, crystalline, organic compound derived from coal tar, and 400 times sweeter than cane sugar.

salt — A compound, such as NaCl, made up of a positive metallic ion and a negative nonmetallic ion or radical.

saponification — The reaction taking place when an alkali reacts with a fat or vegetable oil; used to make soap.

saturated solution — A solution that contains the maximum amount of solute under the existing temperature and pressure.

secondary cell — A cell (battery) that can be regenerated or "charged."

Second Law of Thermodynamics — *See under* **laws**.

shell — A level outside the nucleus of an atom that electrons can inhabit without expending energy.

sigma bond — A bond between *s* orbitals or between an *s* orbital and another kind of orbital.

significant figures — All the certain digits recorded in a measurement plus one uncertain digit.

slag — The product formed when the flux reacts with the impurities of an ore in a metallurgical process.

solid — A phase of matter that has a definite size and shape.

solubility — A measure of the amount of solute that will dissolve in a given quantity of solvent at a given temperature.

solute — The material that is dissolved to make a solution.

solution — A uniform mixture of a solute in a solvent.

solvent — The dispersing substance that allows the solute to go into solution.

specific gravity (mass) — The ratio between the mass of a certain volume of a substance and the mass of an equal volume of water (or, in the case of gases, an equal volume of air); expressed as a single number.

specific heat — The ratio between the number of calories needed to raise the temperature of a certain mass of a substance 1° on the Celsius scale and the number of calories needed to raise the temperature of the same mass of water 1° on the Celsius scale.

spectroscope — An instrument used to analyze light by separating it into its component wavelengths.

spectrum — The image formed when radiant energy is dispersed by a prism or grating into its various wavelengths.

spinthariscope — A device for viewing through a microscope the flashes of light made by particles from radioactive materials against a sensitized screen.

spontaneous combustion (ignition) — The process in which slow oxidation produces enough heat to raise the temperature of a substance to its kindling temperature.

stable — Referring to a substance not easily decomposed or dissociated.

standard conditions — An atmospheric pressure of 760 mm or torr or 1 atm (mercury pressure) and a temperature of 0°C (273 K) (abbreviation: STP).

stratosphere — The upper portion of the atmosphere, in which the temperature changes little with altitude and clouds of water never form.

strong acid (or base) — An acid (or a base) capable of a high degree of ionization in water solution. Example: sulfuric acid, sodium hydroxide.

structural (graphic) formula — A pictorial representation of the atomic arrangement of a molecule.

sublime — To vaporize directly from the solid to the gaseous state and then condense back to the solid.

substance — A single kind of matter, element, or compound.

substitution products — Products formed by the substitution of other elements or radicals for hydrogen atoms in hydrocarbons.

sulfation — An accumulation of lead sulfate on the plates and at the bottom of a (lead) storage cell.

supersaturated solution — A solution that contains a greater quantity of solute than is normally possible at a given temperature.

suspension — A mixture of finely divided solid material in a liquid, from which the solid settles on standing.

symbol — A letter or letters representing an element of the periodic table. Examples: O, Mn.

synthesis — The chemical process of forming a substance from its individual parts.

Systéme International d'Unitès — The modernized metric system of measurements universally used by scientists. There are seven base units: kilogram, meter, second, ampere, kelvin, mole, and candela. Called SI units.

temperature — The intensity or the degree of heat of a body, measured by a thermometer.

tempering — The heating and then rapid cooling of a metal to increase its hardness.

ternary — Referring to a compound composed of three different elements, such as H_2SO_4.

theory — An explanation used to interpret the "mechanics" of nature's actions; a theory is more fully developed than a hypothesis.

thermochemical equation — An equation that includes values for the calories absorbed or evolved.

thermoplastic — Capable of being softened by heat; may be remolded.

thermosetting — Capable of being permanently hardened by heat and pressure; resistant to the further effects of heat.

tincture — An alcoholic solution of a substance, such as a tincture of iodine.

torr — A unit of pressure defined as 1 mm Hg; 1 torr equals 133.32 Pa.

tracer — A minute quantity of radioactive isotope used in medicine and biology to study chemical changes within living tissues.

transmutation — Conversion of one element into another, either by bombardment or by radioactive disintegration.

tribasic acid — An acid that contains three replaceable hydrogen atoms in its molecule, such as H_3PO_4.

tritium — A very rare, unstable, "triple-weight" hydrogen isotope (H^3) that can be made synthetically.

Tyndall effect — The scattering of a beam of light as it passes through a colloidal material.

ultraviolet light — The portion of the spectrum that lies just beyond the violet; therefore of short wavelength.

USP (United States Pharmacopeia) chemicals — Chemicals certified as having a standard of purity that demonstrates their fitness for use in medicine.

valence — The combining power of an element; the number of electrons gained, lost, or borrowed in a chemical reaction.

valence electrons — The electrons in the outermost level or levels of an atom that determine its chemical properites.

van der Waals forces — Weak attractive forces existing between molecules.

vapor — The gaseous phase of a substance that normally exists as a solid or liquid at ordinary temperatures.

vapor density — The density of a gas in relation to another gas taken as a standard under identical conditions of temperature and pressure.

vapor pressure — The pressure exerted by a vapor given off by a confined liquid or solid when the vapor is in equilibrium with its liquid or solid form.

volatile — Easily changed to a gas or a vapor at relatively low pressure.

volt — A unit of electrical potential or voltage, equal to the difference of potential between two points in a conducting wire carrying a constant current of 1 ampere when the power dissipated between these two points is equal to 1 watt (abbreviation: V).

volume — The amount of three-dimensional space occupied by a substance.

VSEPR — Valence shell electron pair repulsion. This model explains the non-90° variations of bond angles for *p* orbitals in the outer energy levels of atoms in molecules because of electron repulsions.

vulcanization — The process of combining sulfur or other additives, using heat, with rubber to improve its properties, particularly to give it added hardness.

water of hydration — Water that is held in chemical combination in a hydrate and can be removed without essentially altering the composition of the substance. *See also* **hydrate.**

weak acid (or base) — An acid (or base) capable of being only slightly ionized in an aqueous solution.
  Examples: Acetic acid, ammonium hydroxide.

weak electrolyte — A substance that, when dissolved in water, ionizes only slightly and hence is a poor conductor of electricity.

weight — The measure of the force with which a body is attracted toward the earth by gravity.

work — The product of the force exerted on a body and the distance through which the force acts; expressed mathematically by the equation $W = Fs$, where W = work, F = force, and s = distance.

X rays — Penetrating radiations, of extremely short wavelength, emitted when a stream of electrons strikes a solid target in a vacuum tube.

zeolite — A natural or synthesized silicate used to soften water.

INDEX

A

Absolute zero, 12
Accuracy, 16
Acetic acid, 274
Acid:
    anhydrides, 141
    rain, 206–207
    salts, 233
Acids *See also* pH
    buffer solutions, 205
    conjugate, 199
    definitions and properties, 196–197
    formulas of common, 80
    indicators, 201
    ionization constants, 187–188
    theories of, 198–199
    volumetric analysis (titration), 201–205
Activation energy, 5, 177–178
Additivity of reaction heats, 169–171
Air pressure, 98
Alcohols, 270–272
Aldehydes, 273
Alicyclic hydrocarbons, 265
Alkali metals, 47
Alkanes, 258–262
Alkenes, 262–263
    additions to, 264–265
Alkylation, 268
Alkynes, 263–264
    additions to, 264–265
Alloys, 145, 244
Alpha particles, 286
Aluminum, 243
Amines, 277
Amino acids, 277
Ammonia, 239–240
Amphoteric substances, 206
Anhydrides, 141
Anions, 211
Anode, 211, 213
Aristotle, 135
Aromatic hydrocarbons, 266–267
Aromatization, 269
Arrhenius Theory, 197–198
Atmospheric pressure, 98
Atom, 4, 27
Atomic:
    mass unit, 31
    number, 31
    spectra, 34–38
Atomic radii, 48–49
    in groups, 50
    in periods, 50
    vs. ionic radii, 50

Atomic structure:
    atomic spectra, 34–38
    average atomic mass, 32–33
    Bohr model, 30, 34
    components of, 31–32
    electric charges, 28–30
    history, 27–28
    metals, nonmetals, and noble gases, 34
    oxidation number and valence, 33–34
    periodic table of elements, 45–53
    reactivity, 34
    sublevels and electron configuration, 41–44
    transition elements and variable oxidation numbers, 44–45
    wave-mechanical model, 38–41
Aufbau Principle, 41, 43
Average atomic mass, 32–33
Avogadro, Amedeo, 114
Avogadro's hypothesis, 114
Avogadro's Law, 114
Avogadro's number, 114

B

Balmer series, 35, 37
Barometric pressure, 98
Bases *See also* pH
    buffer solutions, 205
    conjugate, 199
    definitions and properties, 197–198
    formulas of common, 80
    indicators, 201
    theories of, 198–199
    volumetric analysis (titration), 201–205
Becquerel, Henri, 27, 285
Benzene, 266–267
Benzoic acid, 274
Beta particles, 287
Binary acids, 80
Bohr, Niels, 28, 30, 34
Bohr model, 30, 34, 38
Boiling point, 133
Bonding, 56
    covalent bonds, 58–59
    dissociation energies, 171–172
    double bond, 62–63
    electrostatic repulsion (VSEPR), 63–66
    enthalpy from bond energies, 172
    hybridization, 66–70
    intermolecular forces of attraction, 61–62
    ionic bonds, 57–58
    ionic substance properties, 71
    metallic bonds, 60
    molecular crystal and liquid properties, 71
    pi bond, 70

resonance structures, 63
sigma bond, 70
triple bond, 62–63
Boyle, Robert, 103
Boyle's Law, 103–104
Brønsted Theory, 198
Brownian movement, 130, 146
Bubble chamber, 288
Buffer solutions, 205
Butane, 267–268

C
Calorie (cal), 13
Calorimetry problems, 139–141
Carbohydrates, 278–279
Carbon, 253–255. *See also* Organic
 chemistry
 dioxide, 91, 255–257
Carboxylic acids, 273–274
Catalyst, 93, 176
Cathode, 211, 213
Cations, 211
Cavendish, Henry, 95, 135
Celsius, Anders, 12
Celsius scale, 12
Chadwick, James, 30
Change of phase, 131
Charles, Jacques, 101
Charles's Law, 101–103
Chemical:
 bond, 56
 changes, 5
 properties, 5
Chemical calculations, 113
 excess of one reactant problems, 125–128
 mass-mass problems, 121–123
 mass-volume relationships, 120
 molar mass
  and density, 117–120
  and gas volumes, 116–117
  and moles, 114–115
 mole concept, 113–114
 mole relationships, 116
 volume-volume problems, 123–125
Chemical formulas, 81–83
 Law of Definite Composition, 84
 Law of Multiple Proportions, 84
 naming compounds, 79–80
 oxidation numbers, 77–79
 phases in, 85–86
 writing, 75–77
  and balancing, 84–85
  ionic equations, 86
Chemical reactions:
 combination (synthesis), 161
  predicting, 162–163
 decomposition (analysis), 161
  predicting, 163

double replacement, 161–162
 predicting, 164–165
entropy of, 166
hydrolysis, 165–166
rates of, 175–179
reversible, and equilibrium, 181–185
single replacement, 161
 predicting, 164
Cloud chamber, 287–288
Cockcroft, Sir John, 291
Coefficients, 84
Colligative properties, 151–153
Combination (synthesis) reactions, 161
 predicting, 162–163
Combined Gas Law, 104
Common ion effect, 191
Complex organic acids, 274–275
Compound, 4
 naming, 79–80
Concentrated solution, 146–147
Concentrations (reactant), 176, 186
Conversions:
 factor-label method for, 15–16
 temperature, 12
Coordinate covalent bond, 59
Copper, 243–244
Covalent:
 bond, 58–59
 radius, 48
Cracking, 268
Critical:
 mass, 292
 pressure, 133
 temperature, 133
Crystal, 154
Crystallization, 154–155
Curie, Marie, 28, 285
Curie, Pierre, 28, 285
Cycloalkadienes, 265
Cycloalkanes, 262, 265
Cycloalkenes, 265

D
Dalton, 31
Dalton, John, 27
Dalton's Law of Partial Pressures, 105–106
De Broglie, Louis, 28, 38
Decay series, 289–290
Decomposition (analysis) reactions, 161
 predicting, 163
Dehydrogenation, 269
Deliquescent, 155
Democritus, 27
Density, 3
 and molar mass, 117–120
Deuterium, 31, 135, 138
Diffusion, 101
Dilute solution, 146

Dilution, 150–151
Dimensional analysis, 15–16
Dipole-dipole attraction, 61
Dipoles, 58, 61, 145
Disaccharides, 279
Dissociation, 211
D-orbitals, 39–40
Double bond, 62
Double replacement reactions, 161–162, 206
    predicting, 164–165
Dynamic equilibrium, 132

E
Efflorescent, 155
Effusion, 101
Einstein, Albert, 38, 290–291
Electric charges, 28–30
Electrochemistry, 211
Electrode, 213
    potentials, 213–216
Electrolysis, quantitative aspects of, 219–220
Electrolytes, 210
Electrolytic cells, 216–218
Electromotive series, 213–214
Electron, 28–30
    configuration, 41–44
    shift method, 220–222
Electronegativity, 50
Electroscope, 288
Electrostatic repulsion (VSEPR), 63–64
    and molecular geometry, 65
    summary, 66
    and unshared electron pairs, 64–65
Electrovalence bond, 57
Element, 4. *See also* Periodic table of the elements
    transition, 44–45
Empirical formula, 81, 258
Endothermic reaction, 5, 7–8, 145
Endpoint, 201
Energy:
    activation, 5, 177–178
    defined, 6
    exothermic vs. endothermic reactions, 7–8
    forms of, 7
    ionization, 50–53
    kinetic, 7
    Law of Conservation of, 8
    Law of Conservation of Mass and, 8
    potential, 7
Enthalpy, 7
    from bond energies, 172
    changes in, 167–169
    relation to entropy, 191–192
Entropy, 166
    relation to enthalpy, 191–192
Equilibrium:
    acid ionization constant, 187–188
    and changing concentrations, 186

and changing pressure, 186
and changing temperature, 186
common ion effect, 191
constant, 182–183
factors related to magnitude of K, 191–193
ionization constant of water, 188–189
Le Châtelier's Principle, 185
and reversible reactions, 181–185
solubility product constant, 189–191
systems of solids constant, 187
vapor pressure, 132
Equivalence point, 201
Esters, 277–278
Ethanol, 271
Ethers, 276
Ethylene glycol, 272
Excess reactant problems, 125–128
Exothermic reaction, 5, 7–8, 145

F
Factor-label method, 15–16, 118
Families, 47
Faraday, 219
Faraday, Michael, 210
Fermi, Enrico, 291
First ionization energy, 50
First Law of Thermodynamics, 169
Fischer-Tropsch process, 270
Fission, 291–293
F-orbitals, 39
Formula mass, 81
Fractional distillation, 262
Frasch, Herman, 231
Free energy, 192–193
Fusion, 293

G
Gamma radiation, 287
Gases, 3
    Boyle's Law of, 103–104
    Charles's Law of, 101–103
    Combined Gas Law, 104
    critical temperature and critical pressure, 133
    Dalton's Law of Partial Pressures, 105–106
    in the environment, 91–92
    Graham's Law of, 101
    hydrogen, 95–97
    Ideal Gas Law, 107–109
        deviations, 109
    Kinetic Molecular Theory, 100
    oxygen, 92–95
    pressure corrections, 106–107
    pressure of, 98–100
    pressure vs. temperature, 105
    properties of, 100–101
    volumes and molar mass, 116–117
Gay-Lussac, Joseph, 116, 135
Gay-Lussac's Law, 116

Geiger counter, 289
Gell-Mann, Murray, 294
Gibbs Free Energy Equation, 192–193
Glacial acetic, 274
Glycerine glycerol, 272
Graham's Law, 101
Gram-equivalent mass, 149
Gram-formula mass, 114, 148
Gram-molecular mass, 148
Grosse, Aristid, 135
Groups, 47

H
Half-cell, 213
Half-life, 289
Half-reactions, 214
Halogen family, 236–238
Heat, 13
    of combustion, 163, 169
    content, 7–8
    of formation, 162
    of fusion, 134, 140
    of vaporization, 140
Heavy water, 138
Heisenberg, Werner, 38
Hess's Law of Heat Summation, 169–171
Homologous series, 258
Humboldt, Alexander von, 135
Hund's Rule of Maximum
    Multiplicity, 41
Hybridization, 66–70
Hydrate, 154
Hydrated ion, 145
Hydrocarbons, 257
    alicyclic, 265
    alkanes, 258–262
    alkenes, 262–265
    alkynes, 263–265
    aromatic, 266–267
    changing, 268–270
    derivatives, 270–281
    isomers, 267–268
Hydrochloric acid, 237
Hydrofluoric acid, 237
Hydrogen, 31, 95–97
    bond, 61–62, 141–142
    peroxide, 139
    sulfide, 234–235
Hydrogenation, 269
Hydrolysis, 165–166, 279
Hydroscopic, 155
Hypothesis, 8

I
Ideal Gas Law, 107–109
Immiscible, 145
Indicators, 196, 201
Inertia, 3

Intermolecular:
    forces of attraction, 61–62
    interaction, 130
Inversion, 279
Ion-electron method, 222–226
Ionic:
    bond, 57–58
    equations, 86
    substances, properties, 71
Ionization, 210–211, 285
    constant, 187–189
    energy, 50–53
Ions, 57
Iron, 244
Isomers, 267–268
Isotopes, 31–32
IUPAC system, 260, 267, 274

J
Joule (J), 6

K
K, 187
    factors related to magnitude of,
        191–193
Kelvin, Lord William, 12
Kelvin scale, 12
K_{eq}, 182–183
Ketones, 276
Kilocalorie (kcal), 6, 13
Kilometer, 10
Kinetic energy, 7
Kinetic Molecular Theory, 100, 130

L
Laboratory setups, 298–307
Lavoisier, Antoine, 93, 135
Law of Conservation of Energy, 8
Law of Conservation of Mass and Energy, 8
Law of Conservation of Matter, 6
Law of Definite Composition, 4, 84
Law of Mass Action, 178, 181
Law of Multiple Proportions, 84
Law of Partial Pressures, 105–106
Le Châtelier's Principle, 132, 185
Leptons, 294
Leucippus, 27
Lewis, G. N., 33
Lewis dot structure, 33, 44
Lewis Theory, 199
Linear arrangement, 64
Liquids, 3
    boiling point, 133
    critical temperature and pressure of, 133
    intermolecular interaction, 130
    kinetics of, 130
    phase equilibrium, 132
    properties, 71

surface tension of, 131
viscosity of, 131
London force, 61
Lyman series, 35, 37

M
Magnesium, 243
Manometer, 99
Mass-mass problems, 121–123
Mass number, 31
Mass spectroscopy, 37–38
Mass-volume relationships, 120
Matter:
    changes in, 5–6
    composition of, 4
    defined, 3
    Law of Conservation of, 6
    properties of, 5
    states of, 3
Measurement:
    dimensional analysis, 15–16
    heat, 13
    metric system, 9–11
    precision, accuracy, and uncertainty, 16–17
    scientific method, 8–9
    scientific notation, 14–15
    significant figures, 17–20
    temperature, 12–13
Melting point, 134
Mendeleev, Dimitry, 45–46
Meniscus, 17
Mercury barometer, 98
Metallic bond, 60
Metalloids, 47, 245
Metals, 34, 47
    alloys, 244
    properties of, 241–243
    reduction methods, 243–244
Methanol, 270
Metric system, 9–11
Meyer, Lothar, 45
Millikan, Robert, 29
Miscible, 145
Mixtures, 4
    continuum of, 145–146
Model, 9
Molar heat of reaction, 167
Molar mass, 114, 148
    and density, 117–120
    and gas volumes, 116–117
Mole, 114
    concept, 113–114
    method, 118
    relationships, 116
Molecular:
    crystals, properties, 71
    mass, 81
Molecule, 4, 56

Monomers, 280
Monosaccharides, 279
Moseley, Henry, 31, 46

N
Naming compounds, 79–80
Neutralization, 197, 205
Neutron, 30
Newlands, John, 45
Nitric acid, 238–239
Nitrogen family, 238–241
Noble gases, 34, 47
Nonelectrolytes, 210
Nonmetals, 34, 47
Nonpolar covalent bond, 58
Normal:
    atmospheric pressure, 98
    salts, 233
Nuclear energy, 291–293
Nucleon, 31
Nucleonics:
    new subatomic particles, 294
    nuclear energy, 291–293
    radiation exposure, 293–294
    radioactivity, 285–291
Nucleus, 30

O
Octahedral geometry, 64
Octet, 62
Orbitals, 38
Organic acids, 273–274
Organic chemistry, 257
    hydrocarbon derivatives, 270–281
    hydrocarbons, 157–270
Oxidation, 94
Oxidation numbers, 33, 75, 77–79
    balancing equation with, 220–226
    list of, 76
    variable, 44–45
Oxidation-reduction reactions, 211–219
    balancing, 220–226
Oxygen, 92–95
Ozone, 92, 95

P
Pascal (Pa), 99
Paschen series, 35
Pauli, Wolfgang, 40
Pauli Exclusion Principle, 40
Percentage:
    composition, 81
    concentration, 147
Periodic law, 46
Periodic table of the elements, 45–47
    alkali metals, 47
    atomic radii, 48–49
        in groups, 50

in periods, 50
  vs. ionic radii, 50
electronegativity, 50
ionization energy, 50–53
metalloids, 47
metals, 47
noble gases, 47
nonmetals, 47
properties of, 48
Periods, 47
pH, 199–200
Phase:
  diagrams, 134–135
  equilibrium, 132
Phases of matter, 3
  showing in chemical equations, 85–86
Phenols, 272
Photographic plate, 287
Photons, 35
Physical:
  changes, 5
  properties, 5
Pi bond, 70
Planck, Max, 38
Polar covalent bond, 58
Polarity, 141–142
Polyatomic ion, 75
Polymerization, 269
Polymers, 279–281
P-orbitals, 39–40
Potential energy, 7
Precision, 16
Pressure, 98–100
  corrections of, 106–107
  effect on equilibrium, 186
  vs. temperature, 105
Priestley, Joseph, 93
Proportion method, 118
Protium, 31

Q

Qualitative tests, 308–310
Quantum:
  mechanics, 38
  numbers, 39–41
Quark, 294

R

Radiation exposure, 293–294
Radicals, 75
Radioactive dating, 290–291
Radioactivity, 285–291
Radon-222, 294
Reaction mechanism, 178–179
Reaction rates:
  and activation energy, 177–178
  collision theory of, 177
  factors affecting, 175–176

Law of Mass Action, 178
  measurements of, 175
  and reaction mechanism, 178–179
Reactivity, 34
Reducing:
  agent, 214
  sugars, 279
Resonance, 266
  structures, 63
Reversible reactions, 181–185
Review problems:
  acids, bases, and salts, 207–208
  atomic structure and periodic table, 53–55
  bonding, 71–72
  carbon and organic chemistry, 281–283
  chemical calculations, 129
  chemical equilibrium, 193–195
  chemical formulas, 86–88
  chemical reactions and thermochemistry, 173–174
  gases, 109–112
  introduction to chemistry, 20–23
  liquids, solids, and phase changes, 155–157
  nucleonics, 295–297
  reaction rates, 179–180
  representative groups and families, 245–248
Roentgen, 293
Röntgen, Wilhelm, 285
Rows, 47
Rutherford, Ernest, 29

S

Salts, 201–206
Saturated, 142, 146–147, 262
Schrödinger, Erwin, 38
Scientific:
  method, 8–9
  notation, 14–15
Scintillation counter, 287
Second Law of Thermodynamics, 192
Semiconductors, 245
Sigma bond, 70
Significant figures, 17–20
Silver:
  bromide, 237
  iodide, 237
Single replacement reactions, 161, 205
  predicting, 164
SI system, 9–11
Soaps, 198
Solids, 3, 133–134
  equilibrium constant for, 187
Solubility, 142–144
  product constant, 189–191
Solute, 142, 144
Solutions:
  colligative properties of, 151–153
  and crystallization, 154–155
  dilution of, 150–151

expressions of concentration, 146–147
using specific gravity in, 147–150
water, 144–145
Solvay process, 257
Solvent, 142
S-orbital, 39–40
Sp^{3}d^2 hybrid orbitals, 69
Specific gravity, 147–150
Spectroscopy, 35–37
Sp hybrid orbitals, 67
Sp^2 hybrid orbitals, 67–68
Sp^3 hybrid orbitals, 68–69
Spin quantum number, 40–41
Stable octet, 52, 56
Standard:
    enthalpy of formation, 167
    pressure, 98
    reduction potentials, 215
    state, 167
    voltage, 214
States of matter, 3
Stoichiometry, 113
STP, 101, 116
Structural formula, 258
Subatomic particles, 294
Sublevels, 41–44
Sublimation, 134
Sulfur:
    dioxide, 235
    family, 231–235
Sulfuric acid, 233–234
Supersaturated, 146
Surface:
    area, 175–176
    tension, 131

T
Temperature, 12–13, 176, 186
Ternary acids, 80
Tetrahedral shape, 64
Thermochemistry, 167
    additivity of reaction heats and Hess's Law, 169–171
    bond dissociation energy, 171–172
    changes in enthalpy, 167–169
    enthalpy from bond energies, 172
Thermoplastic polymer, 280
Thermosetting polymer, 280
Thomson, J. J., 27–28
Titration, 151, 201–205
Torricelli, Evangelista, 99
Transition elements, 44–45

Transmutation, 289
Trigonal planar, 64
Triple:
    bond, 62
    point, 135
Tritium, 31, 135, 138
True formula, 81

U
Uncertainty, 16
    principle, 38
Unified mass unit, 31
Unit cell, 154
Unsaturated, 146, 262
Urey, Harold Clayton, 135

V
Valence electrons, 33
Van der Waals forces, 61, 142
Variable oxidation numbers, 44–45
Vinegar, 274
Viscosity, 131
Visible light spectra, 36
Voltaic cells, 212–213
    applications of, 218–219
    reactions, 215
Volumetric analysis, 201–205
Volume-volume problems, 123–125

W
Walton, Ernest, 291
Water:
    calorimetry problems, 139–141
    composition of, 137–139
    heavy, 138
    history of, 135
    hydrogen peroxide, 139
    ionization constant of, 188–189
    mixtures, continuum of, 145–146
    polarity and hydrogen bonding, 141–142
    properties and uses of, 139
    purification of, 135–137
    reactions with anhydrides, 141
    solubility, 142–144
    solutions, 144–145
Wave-mechanical model, 38–41
Weight, 3
Wöhler, Friedrich, 257

Z
Zero, 18

NOTES

NOTES

NOTES

MOVE TO THE HEAD OF YOUR CLASS
THE EASY WAY!

Barron's presents THE EASY WAY SERIES—specially prepared by top educators, it maximizes effective learning while minimizing the time and effort it takes to raise your grades, brush up on the basics, and build your confidence. Comprehensive and full of clear review examples, **THE EASY WAY SERIES** is your best bet for better grades, quickly!

ISBN	Title
0-7641-1976-1	Accounting the Easy Way, 4th Ed.—$14.95, Can. $21.95
0-7641-1972-9	Algebra the Easy Way, 4th Ed.—$14.95, Can. $21.95
0-7641-1973-7	American History the Easy Way, 3rd Ed.—$14.95, Can. $21.00
0-7641-0299-0	American Sign Language the Easy Way—$14.95, Can. $21.00
0-7641-1979-6	Anatomy and Physiology the Easy Way—$14.95, Can. $21.95
0-8120-9410-7	Arithmetic the Easy Way, 3rd Ed.—$14.95, Can. $21.95
0-7641-1358-5	Biology the Easy Way, 3rd Ed.—$14.95, Can. $21.95
0-7641-1079-9	Bookkeeping the Easy Way, 3rd Ed.—$14.95, Can. $21.00
0-7641-0314-8	Business Letters the Easy Way, 3rd Ed.—$13.95, Can. $19.50
0-7641-1359-3	Business Math the Easy Way, 3rd Ed.—$14.95, Can. $21.00
0-8120-9141-8	Calculus the Easy Way, 3rd Ed.—$14.95, Can. $21.95
0-7641-1978-8	Chemistry the Easy Way, 4th Ed.—$14.95, Can. $18.95
0-7641-0659-7	Chinese the Easy Way—$14.95, Can. $21.00
0-7641-2579-6	Creative Writing the Easy Way—$12.95, Can. $18.95
0-7641-2146-4	Earth Science The Easy Way—$14.95, Can. $21.95
0-7641-1981-8	Electronics the Easy Way, 4th Ed.—$14.95, Can. $21.00
0-7641-1975-3	English the Easy Way, 4th Ed.—$13.95, Can. $19.50
0-8120-9505-7	French the Easy Way, 3rd Ed.—$14.95, Can. $21.00
0-7641-2435-8	French Grammar the Easy Way—$14.95, Can. $21.95
0-7641-0110-2	Geometry the Easy Way, 3rd Ed.—$14.95, Can. $21.00
0-8120-9145-0	German the Easy Way, 2nd Ed.—$14.95, Can. $21.00
0-7641-1989-3	Grammar the Easy Way—$14.95, Can. $21.00
0-8120-9146-9	Italian the Easy Way, 2nd Ed.—$14.95, Can. $21.95
0-8120-9627-4	Japanese the Easy Way—$14.95, Can. $21.00
0-7641-0752-6	Java™ Programming the Easy Way—$18.95, Can. $25.50
0-7641-2011-5	Math the Easy Way, 4th Ed.—$13.95, Can. $19.50
0-7641-1871-4	Math Word Problems the Easy Way—$14.95, Can. $21.00
0-8120-9601-0	Microeconomics the Easy Way—$14.95, Can. $21.00
0-7641-0236-2	Physics the Easy Way, 3rd Ed.—$14.95, Can. $21.00
0-7641-2393-9	Psychology the Easy Way—$14.95, Can. $21.95
0-7641-2263-0	Spanish Grammar—$14.95, Can. $21.00
0-7641-1974-5	Spanish the Easy Way, 3rd Ed.—$14.95, Can. $21.95
0-8120-9852-8	Speed Reading the Easy Way—$14.95, Can. $21.95
0-8120-9143-4	Spelling the Easy Way, 3rd Ed.—$13.95, Can. $19.50
0-8120-9392-5	Statistics the Easy Way, 3rd Ed.—$14.95, Can. $21.00
0-7641-1360-7	Trigonometry the Easy Way, 3rd Ed.—$14.95, Can. $21.00
0-8120-9147-7	Typing the Easy Way, 3rd Ed.—$19.95, Can. $28.95
0-8120-9765-3	World History the Easy Way, Vol. One—$16.95, Can. $24.50
0-8120-9766-1	World History the Easy Way, Vol. Two—$14.95, Can. $21.00
0-7641-1206-6	Writing the Easy Way, 3rd Ed.—$14.95, Can. $21.00

Barron's Educational Series, Inc.
250 Wireless Boulevard • Hauppauge, New York 11788
In Canada: Georgetown Book Warehouse • 34 Armstrong Avenue, Georgetown, Ontario L7G 4R9
www.barronseduc.com $ = U.S. Dollars   Can. $ = Canadian Dollars

Prices subject to change without notice. Books may be purchased at your local bookstore, or by mail from Barron's. Enclose check or money order for total amount plus sales tax where applicable and 18% for postage and handling (minimum charge $5.95 U.S. and Canada). NY State, New Jersey, Michigan, and California residents add sales tax. All books are paperback editions.

(#45) R 9/04

UNIQUE NEW STUDY GUIDES!

It's like having the best, most organized notes in class!
All the vital facts are highlighted and each Key is
compatible with every standard textbook in its subject field.

Accounting, Minars, ISBN: 0-7641-2001-8

American History to 1877, Geise, ISBN: 0-8120-4737-0

American History from 1877 to the Present, Capozzoli Ingui, ISBN: 0-7641-2005-0

American Literature, Skipp, ISBN: 0-8120-4694-3

Biology, Minkoff, ISBN: 0-8120-4569-6

Chemistry, Kostiner, ISBN: 0-7641-2006-9

College Algebra, Leff, ISBN: 0-8120-1940-7

English Literature, Griffith, ISBN: 0-8120-4600-5

Finance, Siegel & Shim, ISBN: 0-8120-4596-3

Macroeconomics, Siegel & Shim, ISBN: 0-8120-4619-6

Microeconomics, Lindeman, ISBN: 0-7641-2004-2

Physics, Gibbons, ISBN: 0-8120-4921-7

Psychology, Baucum, ISBN: 0-8120-9580-4

Statistics, Sternstein, ISBN: 0-8120-1869-9

Each Key: Paperback, approx. 144 pp., 5" 5 8", $5.95 to $7.95, Canada $7.95 to $11.50

Books may be purchased at your bookstore, or by mail from Barron's. Enclose check or money order
for the total amount plus 18% for postage and handling (minimum charge $5.95). New York State. New Jersey,
Michigan, and California residents add sales tax. Prices subject to change without notice.

Barron's Educational Series, Inc.
250 Wireless Blvd., Hauppauge, NY 11788
In Canada: Georgetown Book Warehouse
34 Armstrong Ave., Georgetown, Ont. L7G 4R9
Visit our website at: www.barronseduc.com

(#52) R 9/04

Learn Organic Chemistry and retain important details the Easy Way!

OrgoCards' Authors and Creators:
Stephen Q. Wang, M.D., Edward J. K. Lee, Babak Razani, M.D., Ph.D.,
Jennifer Wu, M.D., and William F. Berkowitz, Ph.D.

OrgoCards are an effective, high-yield way to study . . .

To master organic chemistry, you need to spend a lot of time understanding concepts and memorizing details. Using *OrgoCards* for as little as 20 or 30 minutes each day, you can preview for classroom lectures and mentally organize and retain information you'll need to get through the course. *OrgoCards* help you understand and think through the material, rather than blindly memorizing facts. By the time you've completed the course, you'll have the subject's details firmly in mind for success on the exam.

OrgoCards
176 cards, boxed, 7" × 5 ^{1}/$_2$"
$19.95, Canada $28.95
ISBN 0-7641-7503-3

OrgoCards make learning relatively easy . . .

The unique *OrgoCards* format facilitates faster comprehension of organic reactions. Each card's front presents three parts: (1.) a **sample reaction**; (2.) a **keys section** that outlines the most important information related to the reaction (for instance, catalysts, resonance structures, and the like); and (3.) a **notes section** with miscellaneous information, such as variants of the reaction. Each card's back presents a detailed **reaction mechanism** in both text and pictorial form, with space where you can make notes.

OrgoCards offer ideal preparation for mid-terms, finals, and standardized exams . . .

You'll find *OrgoCards* an indispensable study tool when you prepare for mid-terms and finals, or if you are planning to take a standardized graduate or professional school exam.

Praise for *OrgoCards* . . .

". . . a succinct review of first year organic chemistry . . . particularly helpful for the motivated student studying for MCATs and GREs . . . [and] useful for mid-semester and final exams."

—Franklin A. Davis, Professor of Chemistry
Temple University

Prices subject to change without notice. Books may be purchased at your local bookstore, or by mail from Barron's. Enclose check or money order for total amount plus sales tax where applicable and 18% for postage and handling (minimum charge $5.95). NY State, New Jersey, Michigan, and California residents add sales tax.

Barron's Educational Series, Inc.
250 Wireless Blvd.
Hauppauge, NY 11788

(#120) R 9/04

At your local bookstore or
Order toll-free in the U.S.
1-800-645-3476
In Canada
1-800-247-7160
www.barronseduc.com